MW00454202

THE NEWARK EARTHWORKS

STUDIES IN RELIGION AND CULTURE

John D. Barbour and Gary L. Ebersole, Editors

THE Newark Earthworks

ENDURING MONUMENTS, CONTESTED MEANINGS

Edited by Lindsay Jones and Richard D. Shiels

University of Virginia Press • Charlottesville and London

University of Virginia Press
© 2016 by the Rector and Visitors of the University of Virginia
All rights reserved
Printed in the United States of America on acid-free paper

First published 2016

9 8 7 6 5 4 3 2

Library of Congress Cataloging-in-Publication Data
The Newark Earthworks : enduring monuments, contested meanings / edited by Lindsay Jones and Richard D. Shiels.
pages cm.—(Studies in religion and culture)
Includes bibliographical references and index.
ISBN 978-0-8139-3777-9 (cloth : alk. paper)
ISBN 978-0-8139-3778-6 (pbk. : alk. paper)
ISBN 978-0-8139-3779-3 (e-book)
1. Mounds—Ohio—Newark. 2. Hopewell culture—Ohio. 3. Mound-builders—Ohio. 4. Indians of North America—Ohio—Antiquities. 5. Newark (Ohio)—Antiquities.
I. Jones, Lindsay, 1954- editor. II. Shiels, Richard Douglas, 1947- editor.
E99.H69N49 2016
977.1'54—dc23 2015025691

CONTENTS

ix Foreword
GLENNA J. WALLACE, *Chief, Eastern Shawnee Tribe of Oklahoma*

1 Introduction: I Had No Idea!
Competing Claims to Distinction at the Newark Earthworks
LINDSAY JONES

PART I
The Newark Earthworks in the Context of American and Ohio History

23 The Newark Earthworks Past and Present
RICHARD D. SHIELS

PART II
The Newark Earthworks in the Context of Hopewell Archaeology and Archaeoastronomy

41 The Newark Earthworks: A Monumental Engine of World Renewal
BRADLEY T. LEPPER

62 The Newark Earthworks: A Grand Unification of Earth, Sky, and Mind
RAY HIVELY AND ROBERT HORN

PART III

The Newark Earthworks in Cross-Cultural Archaeological Contexts: Nazca, Chaco, and Stonehenge

97 An Andeanist's Perspective on the Newark Earthworks
HELAINE SILVERMAN

111 Hopewell and Chaco:
The Consequences of Rituality
STEPHEN H. LEKSON

129 Beyond Newark:
Prehistoric Ceremonial Centers and Their Cosmologies
TIMOTHY DARVILL

PART IV

The Newark Earthworks in Interdisciplinary Contexts: Architectural History, Cartography, and Religious Studies

153 The Newark Earthworks as "Works" of Architecture
JOHN E. HANCOCK

164 The Newark Earthworks as a Liminal Place:
A Comparative Analysis of Hopewell-Period Burial Rituals and Mounds with a Particular Emphasis on House Symbolism
THOMAS BARRIE

180 The Cartographic Legacy of the Newark Earthworks
MARGARET WICKENS PEARCE

198 The Modern Religiosity of the Newark Earthworks
THOMAS S. BREMER

PART V

The Newark Earthworks in the Context of Indigenous Rights and Identity: American and International Frames

215 Native (Re)Investments in Ohio: Evictions, Earthworks Preservation, and Tribal Stewardship
MARTI L. CHAATSMITH

230 Whose Earthworks? Newark and Indigenous People
MARY N. MACDONALD

PART VI

The Newark Earthworks in the Context of Law and Jurisprudence: Ancient and Ongoing Possibilities

245 The Peoples Belong to the Land: Contemporary Stewards for the Newark Earthworks
DUANE CHAMPAGNE AND CAROLE GOLDBERG

262 Caring for Depressed Cultural Sites, Hawaiian Style
GREG JOHNSON

277 Imagining "Law-Stuff" at the Newark Earthworks
WINNIFRED FALLERS SULLIVAN

289 Appendix: What Is So Special about the Newark Earthworks? Fifteen Viable Replies

293 Bibliography

317 List of Contributors

319 Index

Color gallery follows page 130.

FOREWORD

GLENNA J. WALLACE

Chief, Eastern Shawnee Tribe of Oklahoma

I DISCOVERED THE Newark Earthworks just a few years ago when I traveled to Ohio to attend a lecture by British historian John Sugden. Ohio is the ancestral home of the Shawnee people, and I am the chief of the Eastern Shawnee Tribe of Oklahoma.

I have visited many impressive sites around the world, having traveled to more than seventy countries. Part of my responsibilities as a faculty member at Crowder College was to organize international travel opportunities for our students. Twice a year I led students and community residents to significant cultural and historical sites throughout the world. In that capacity I researched and visited many of the most famous and most impressive sites: the Great Wall of China, the Pyramids in Egypt, Stonehenge in the United Kingdom, the Coliseum in Rome, the Acropolis and ruins in Greece, Mount Uluru in Australia, the Vatican and the Sistine Chapel, as well as many others. I continue to travel a great deal since retiring from Crowder College and becoming chief of the Eastern Shawnee Tribe of Oklahoma.

One of my great interests is learning the history of the Shawnee people in Ohio. We were the first tribe to be forcibly removed from our homeland as part of the federal government's policy of Indian removal in the days of Andrew Jackson. Our people walked from Ohio to Missouri and Oklahoma, hundreds of miles, and large numbers of them died on the way. We had our own Trail of Tears. We had not always been in Ohio, and we were not the builders of Ohio's earthworks, but we had been there since the eighteenth century, and we had cared for the earthworks.

I want to learn about our history in Ohio, and I want my people to learn about it. For that purpose I have brought large numbers of Shawnee people to visit Ohio, not once but four times since I became the chief in 2006. And I have read the literature, of course, including biographies of Tecumseh and Blue Jacket written by John Sugden.

In the spring of 2007 I learned that John Sugden was coming to The Ohio State University to give a lecture, and I traveled to Ohio with three of my staff to attend that lecture.

Ohio State is in Columbus, of course. However, Ohio State has a regional campus in Newark, Ohio, which is also the site of incredible earthen enclosures called the Newark Earthworks. I went to Sugden's lecture in Columbus and traveled to Newark the next day simply because the Ohio State faculty who hosted the lecture were taking Sugden to Newark to tour the earthworks. I wanted to spend more time with this author, who was most knowledgeable about two of the most important leaders in my tribe's history. Sugden was my interest.

Try to imagine the shock and total disbelief I experienced when I stepped out of the car and looked out at this intricate array of earthen walls and landscapes where my people, my ancestors had lived more than three hundred years ago. It was surreal. I had spent at least thirty years researching cultures and histories of civilizations throughout the world. I had read everything I could about my Shawnee tribe: all the places they lived, wars they fought, how they dressed, how they worked, how they ate, what they built, how they believed, what they valued, and how they worshipped. I knew about Serpent Mound, but I had never heard of the Newark Earthworks. I had never even heard of Newark, Ohio. I was stunned at what I saw. I was in a state of disbelief.

There before me lay an extensive series of hills or walls—in short, earthworks built by Native Americans nearly two thousand years ago, built with earth carried one basket at a time. They appear so simplistic, yet they contain a mathematical complexity that is mind boggling. How these Native Americans who have commonly been depicted as savages could conceive and construct this massive earthen architecture is as phenomenal as Egyptian slaves constructing the Pyramids.

The Newark Earthworks are beautiful and massive. That day we visited one part of the complex, the Octagon Earthworks: earthen walls that enclose an octagon that covers fifty acres connected to a perfect circle that encloses another twenty acres. I forgot all about John Sugden.

The Newark Earthworks are owned by the Ohio Historical Society, and the Octagon is leased to a private country club and covered by a golf course. It was a warm sunny spring day, and there were many golfers. When we tried to walk to where we could view the Earthworks, we were not made to feel welcome. Quite the contrary, everyone seemed to resent our presence. "Get back," cart drivers kept shouting rather gruffly. "Get back, you are in the way."

I could not believe it. My people, my ancestors treasured these mounds. Perhaps they did not build them, but they loved them, protected them, revered them. They knew their importance, and these earthworks were sacred to them. So I experienced the beauty, the awe inspired by these beautiful ancient earthworks, but I was not permitted to walk on the sacred grounds of my people. I could not even approach, much less touch or feel these precious earthworks. Instead I as a Shawnee citizen and tribal chief could only experience the rude resentment of golfers who have somehow taken over the site.

My reaction at first was both amazement and awe. I have experienced the Pyramids, the catacombs, the Great Wall, Stonehenge. The Newark Earthworks are every bit as impressive as all of these. And they are in Ohio, where my people lived before removal. And they were built by ancestors of American Indians. How could I not have known about them? I know so much about so many of the world's great sites. How could I not have known about Newark? Are they written about? Are they talked about? Are they appreciated? The answer was no.

Then my reaction turned to anger. Who are these people telling me to get back? How can it be that they are playing golf on this sacred site? And then I thought of a biblical scripture, "Forgive them Father, for they know not what they do."

I continue to be amazed, awe stricken, disappointed, and, yes, angry.

That day I met a group of people who live in Ohio and feel much as I do. These are people who also find the current situation unacceptable, even as they appreciate the fact that the Newark Earthworks have been preserved when so many other earthworks have been destroyed. These people are associated with Ohio State's Newark Earthworks Center. These are the people who hosted a scholarly symposium that produced the essays in this book.

That day I made a commitment—to learn all I could about the Newark Earthworks, to teach others about them, and to preserve them. I have returned repeatedly to visit these people and to visit the site, and several of these people have visited me in Oklahoma. I have joined them in the effort to preserve the site and win the recognition that it deserves by winning inscription on the UNESCO World Heritage List.

I continue to be amazed, awe stricken, disappointed, and angry. But now I also feel hopeful.

THE NEWARK EARTHWORKS

LINDSAY JONES

Introduction

I Had No Idea!
Competing Claims to Distinction at the Newark Earthworks

Renowned by historians and archaeologists as one of the wonders of the ancient world,[1] the Earthworks of Newark, Ohio, nonetheless remain, for the broader public, lamentably little known. Among the largest, most geometrically precise and best-preserved earthen architecture ever constructed, these built forms have, as we'll learn in this volume, astronomical alignments no less sophisticated than those at Stonehenge and a scale no less enormous than the Peruvian geoglyphs at Nazca. And yet obscurity is also among their foremost attributes.

Indeed, incongruities abound. A two-thousand-year-old testament to another era, another civilization, and another set of socioreligious priorities, this pre-Columbian complex, irrespective of its size and precision, is, as thousands of local residents prove each day, much more easily ignored than explained. A major pilgrimage destination two millennia ago, still revered as an auspiciously sacred site by numerous Native American groups, the Newark Earthworks are, at present, host, or perhaps hostage, to a private golf club that restricts access to their principal features on all but four days per year. A contender for UNESCO World Heritage status—that is to say, a strong candidate to meet the UNESCO criterion of a site with "outstanding universal value"—the Newark Earthworks attract ever more frequent busloads of Shawnees from Oklahoma and Mormons from Utah, along with elder hostel tours, Edgar Casey aficionados, and a mix of antiquarian enthusiasts from everywhere. These earthworks are, nonetheless, never visited by the great majority of Ohioans who live just a stone's throw away.

Such are the intriguing ironies and contradictions that characterize the Newark Earthworks, alternately celebrated and snubbed. Distressingly belated appreciation of this stupendous ancient complex is an admission reechoed by nearly all of the contributors to this volume as well as by countless other Ohio natives with whom I have had essentially the same conversation over and over again. Among Ohio State University students, most of whom grew up in the state, not one in fifty can provide any informed acquaintance with the site. I too have to concede that, as a historian of religions with a special interest in sacred architecture and thus pilgrimage destinations across the globe, it was nearly twenty years before I finally made the thirty-mile jaunt from Columbus to the Earthworks of Newark. Though able to claim abundant company in my protracted indifference to Ohio's ancient architectural wonders, I should have known better. I too am among the most deeply implicated in what one contributor to this volume describes as the "I-had-no-idea" phenomenon.[2]

A Book-Making Symposium: Shared Urgency and Agreements to Disagree

Happily, however, as you will learn in Richard D. Shiels's opening essay, more dutiful and much better-informed scholars and community activists have been undertaking, in the past two decades, a whole series of spirited initiatives to bring to the Newark Earthworks a more suitably heightened respect and prominence. Notably, in 2006, The Ohio State University established the Newark Earthworks Center, which is supporting, among innumerable projects, the formulation of a proposal to have the Newark Earthworks, along with some other ancient Ohio sites, considered for UNESCO World Heritage status. It was a report on the progress of that collaborative effort that brought to light the startling realization that, irrespective of abundant site studies, there was, to this point, not even one scholarly book fully devoted to the Newark Earthworks. One has to ask, then, why, if this place is of such great consequence, has there never been a single book-length treatment of "the largest and most precise complex of geometric earthworks in the world"?[3]

Several colleagues found that lacuna to be astounding but, at the same time, a void that we could fill. And thus a small committee—composed of Richard Shiels, then the director of the Newark Earthworks Center; Marti Chaatsmith, a Comanche scholar who was then the associate director; Bradley Lepper, the curator of archaeology for the Ohio Historical Society and a longtime expert on the site; and me, then director of the Ohio State Center for the Study of Religion—was formed to organize a symposium that would have as its express purpose the assembly of a group of scholars who could, together, produce the first academic volume on the Newark Earthworks.

That symposium, titled "The Newark Earthworks and World Heritage: One Site, Many Contexts," was held at the regional campus of The Ohio State University at Newark, very near the actual site, on May 2–4, 2011. The tenor of that meeting—and, subsequently, the composition of this volume—eventually came to embrace two propositions, each of which proved somewhat contentious, even deal breaking, among various participants and onlookers.

First, while readers of this volume will easily discern that many, probably most, of the contributors are vigorously committed to winning World Heritage designation for the Newark Earthworks and related sites (Fort Ancient and the sites that make up Hopewell Culture National Historical Park, all parts of a serial nomination entitled *Hopewell Ceremonial Centers*), the volume is *not* configured as a direct advocacy for that or any other specific position on the future status and management of the site—even though some wish that it were. Alternatively, this book is an academic undertaking that admits and even encourages a wide range of opinions, including very different projections as to the most felicitous future for the Newark site. We prize rather than forestall disagreement.

Some of the contributors belong to academic traditions that allow, perhaps compel, them to move past description and interpretation to prescriptive recommendations about what *should* happen to the Newark Earthworks. These scholars do present general or specific policy recommendations. Others, by contrast, have strong personal opinions about how the site ought to be managed, but they feel compelled to withhold those opinions from their academic writing. And at least a few authors actually do abstain on policy matters insofar as they are content to engage the Newark Earthworks as a fascinating and instructive focus for a variety of larger issues. In sum, although indignation about the present circumstances wherein prime portions of the ancient earthworks are occupied by the Moundbuilders Country Club contributes considerable urgency to this project—and while this volume may well serve as a valuable resource for those committed to winning World Heritage inscription for the site—the contributors were not required to be spokespersons for that outcome.

Second, the same advocacy for multiplicity demanded the involvement of an aggressively interdisciplinary cast of characters. To be sure, archaeologists have been, and are almost certain to remain, the most prominent voices in the analysis of Ohio earthworks; and we have tried to give archaeology its due priority. But we have not granted archaeology hegemony over other disciplinary frames of reference. In this book, archaeology, not exempt from its own internal disagreements, stands as neither less nor more than the most prominent among numerous academic perspectives.

Consequently, where a more obvious strategy might have been to assemble scholars who are already well informed about these ancient earthworks—an assemblage that would have been dominated by Hopewellian archaeologists—we ventured instead to widen and to complicate the conversation by juxtaposing, on the one hand, a core of the leading experts on the Newark Earthworks with, on the other hand, an array of accomplished researchers with relevant interests but for whom the specifics of this site are much less familiar. A large portion of our symposium was, then, devoted to education about the current state of Newark Earthworks studies via lecture presentations, the sharing of bibliographies, and of course on-site visits. Building on that background, every contributor to the volume owes a large debt to archaeology. But at the same time, participants were strongly encouraged to "do their own thing," to see and to interrogate the Newark Earthworks through their own respective disciplinary lenses. As a result, the points of view range from historical and contemporary American Indian studies to art and architectural history; to archaeoastronomy, the history of religions, ritual studies, ethnohistory, and cartography; to legal studies; and to tourism and museum studies. As is apparent in the book's subtitle and its six-part structure, our incentive was to situate and then repeatedly resituate this one site in a whole series of different disciplinary, thematic, and historical contexts. Yes, we prize disagreement!

Diversity, even outright disparities, of opinions and intellectual investments is, therefore, a central and deliberate feature of this volume. Assuredly, a careful reading of the individual essays will suggest that had the authors spent more time together, and had they come to know one another's positions more fully, the intensity of disagreement would likely have been exacerbated rather than alleviated. While these widely heterogeneous scholars proved to be more or less congenial dinner companions, they are also in many cases quite robust critics of one another's interpretations of the Ohio earthworks.

For example, those of a particular scientific bent find some of the contributions quirky, ungrounded, or at times politically tendentious. Scholars with commitments to academically informed social action find other essays disturbingly detached and oriented to the site's past rather than its future. Specialists in the study of religion (including me) are uneasy with many of the allusions to ritual, "spirituality," and "sacred places" that we find in these pages. Additionally, those scholars with more finely tuned postcolonial sensibilities detect signs of lingering essentialism and ethnocentrism in their colleagues' rosy characterizations of the ancient Mound Builders' engineering talents and religious tastes. And there is also, perhaps most disputatiously, a very wide spectrum of opinions as to how we ought to evaluate the variously historical and purported connections between the pre-Columbian earthen mounds and contemporary Indian communities.

Expect, then, no shared resolution to the fundamental questions: What do the Newark Earthworks *really* mean? What sorts of activities originally went on there? And to whom, at this point, do the Newark Earthworks properly belong?

Fortunately, however, consensus was neither a goal nor an expectation. To the contrary, inquiries into the Newark Earthworks are made especially exhilarating both by the profusion of opinions that these ancient constructions engender and by the vigor with which the proponents hold these opinions. The overworked phrase "heated debate" is, in this case, fully apropos. The Earthworks of Newark are, to be sure, a "contested site" of the most intense sort, and we have aimed for a book that reflects that scrum of competing ideas.[4]

What Is So Special about the Newark Earthworks? Fifteen Viable Replies

If this collection of authors is, then, so little prone to consensus, on what can the contributors agree? What unites these essays and essayists? Two things: First, all, without exception, are by now convinced that the Newark Earthworks are an exceptional, arguably unique, place. Second, there is at this point unanimity that the Earthworks of Newark are ironically and undeservedly obscure, overlooked, and undervalued in the scholarly literature, in the public imagination, and even by the central Ohio residents who live directly on or very near them. With just one plausible exception, every contributor to this volume admits to an initial indifference to these ancient earthworks that was, only later and often very slowly, superseded by a profound realization that contemporary Ohio is built atop literally hundreds of pre-Columbian mound sites.

Both the grounds and the consequences of that belated appreciation are, however, as we shall see, as diverse as the authors. If, on the one hand, answering the UNESCO query as to the "outstanding universal value" of the Newark Earthworks demands unanimous agreement as to the most salient virtues and opportunities that the place presents, this interdisciplinary group would be hard pressed to deliver a reply. Yet, on the other hand, by directing attention to the host of different ways—or different contexts—in which Newark is special and distinctive, this set of essays also directs attention to the multitude of different ways in which the site has what might be more properly termed "audience-specific value," or perhaps "discipline-specific value." The wide array of enthusiastic endorsements demonstrates that the uniqueness and appeal of the Newark Earthworks are, in short, multidimensional. Evaluations of what is most special and most noteworthy about this place depend, in very large part, on the perspective of the evaluator.

Brief comments on each essay can, therefore, highlight both unanimous appreciation for the ancient mounds and the widely divergent opinions as to their greatest significance. In the first essay, "The Newark Earthworks Past and Present," for instance, historian and founding director of Ohio State's Newark Earthworks Center Richard Shiels places the earthworks within the context of American and Ohio history. He expresses his exuberant, still-mounting regard for this complex by noting that the configuration was, from its initial conception, "unique in at least three ways"—as the largest, the northernmost, and the most geometrically precise complex among all Hopewell earthworks. He contends that, even during the Middle Woodland era of 100–400 CE, and thus even amid considerable contemporaneous competition, this was the premier site in the region, an assessment of Newark's pre-Columbian prestige that reappears in several of the subsequent essays.

Additionally, in the context of a richly detailed discussion of the site's convoluted history of neglect and preservation during the past two hundred years, Shiels singles out the Earthworks at Newark—irrespective of rambunctious repurposings as a fairground, a military training camp, a venue for Buffalo Bill's Wild West Show, an amusement park, and a racetrack—as the most prominent exception to the wholesale destruction that befell the great majority of Ohio's earthen monuments. Fortuitously, a uniquely large share of the Newark Earthworks remains intact. Moreover, rehearsing the serpentine sequence of events that leaves the ancient complex's main features presently overlaid with the fairways of the Moundbuilders Country Club, Shiels explains how a mix of community concern and outrage paved the way for the creation of the Newark Earthworks Center, an ongoing research initiative wherein the ancient monuments become the locus of not only public, personal, and often "spiritual" interests but also more strictly academic undertakings, of which this volume is an example. In short, Shiels opens the book with a very personal and upbeat story of a once-preeminent site that is seemingly on the way to reclaiming its former prestige.

Part II, "The Newark Earthworks in the Context of Hopewell Archaeology and Archaeoastronomy," gives voice to the leading interpreters of the pre-Columbian ideas and aspirations that account for this enormous complex. Suitably enough—because no one has contributed more to our understanding of the Newark Earthworks—archaeologist Bradley Lepper begins the discussion. Convinced that "ultimately, all claims relating to the purpose and meaning of the Newark Earthworks for its ancient builders must rest upon the material evidence as revealed by archaeology,"[5] Lepper lauds the site as "unprecedented in the Hopewell world in terms of its scale and the precision of both its geometry and its embedded astronomical alignments." Like Shiels, he celebrates the opportune endurance of "the best-preserved examples of geometric earthworks in North America" and laments the sprawling destruction that so distorts present-day perceptions of Newark. Lepper's account reminds us of a theme that will reappear in many of the subsequent essays, namely, his now widely accepted assertion that Newark, at no time itself a site of dense population, nevertheless exercised remarkably wide allure as a great pilgrimage center, "like Mecca or Santiago de Compostela,"[6] which attracted religiously motivated visitors from across eastern North America.

If very largely responsible for what has become the conventional wisdom about "Newark's wonderfully preserved enclosures," Lepper here opts for a self-correcting, even iconoclastic proposal that the burial mounds were the focus of the site and key to its ultimate purpose. That is to say, having long argued that human burials were a relatively minor feature of the Newark complex, Lepper now contends that "mortuary ceremonialism was the sine qua non of the Newark Earthworks complex as it was for many of the other monumental earthwork centers." In this revised view of his own thinking about the "primary purpose" that unites the design of the entire assemblage, the Newark Earthworks were created first and foremost as a vast ritual context—a "ceremonial machine" or "monumental engine of world renewal"—in which to orchestrate a highly choreographed "Hopewellian mortuary ritual," the goal of which "may have been nothing less than the regeneration of the Earth."

The next essay, "The Newark Earthworks: A Grand Unification of Earth, Sky, and Mind," coauthored by longtime collaborators astrophysicist Ray Hively and philosopher Robert Horn, both of whom stretch their disciplinary backgrounds into the realm of archaeoastronomy, provides another instance in which the leading voices in the interpretation of the Newark Earthworks continue to extend and rethink their ideas as to what makes the site so special. Initially skeptical that careful surveys of the Newark mounds would reveal any intentional celestial alignments at all, Hively and Horn eventually became, to their great surprise, the leading proponents of the view that these massive earthen enclosures "were built to record, celebrate, and connect with the celestial actors or large-scale forces that appear to govern relations among earth, sky, and the human mind."

This essay, however, goes further. Having already provided the most thoroughgoing analyses of Newark's stunningly sophisticated referencings with respect to the movements of celestial bodies, most notably alignments to an 18.6-year lunar cycle, Hively and Horn reaffirm—but also extend and enhance—their contention that "no other site encodes with the same accuracy all of the solstice stations and all of the stations of the lunar extreme standstills." Revisiting their decades of investigatory data, they now argue that Newark is moreover unique insofar as "no other site so tightly integrates the exacting geometry of its architecture with the local terrain." In other words, while they continue to insist that the Hopewell-era builders were intensely preoccupied with the movements of the moon, this essay demonstrates also these researchers' growing appreciation of the very special allure of the Newark area's distinctive topography of local streams, valleys, and "hilltop observing stations," all of which, they argue, played determinative roles in both the site selection and orchestration of a unified earthworks design. For them, the Newark complex is a unique human construction that became plausible and meaningful only by virtue of its emplacement within a unique natural landscape.

Part III, "The Newark Earthworks in Cross-Cultural Archaeological Contexts: Nazca, Chaco, and Stonehenge," assembles the reflections of three archaeologists whose primary expertise and excavationary experience has been trained on prominent sites *other than* Newark, all UNESCO World Heritage sites that have enjoyed substantially higher public profiles than the Ohio mounds. In her essay, Andeanist archaeologist Helaine Silverman compares the Newark Earthworks and famed Nazca Lines of Peru, on the World Heritage registry since 1994. Irrespective of their stark disparities in form and appearance, Silverman is far more impressed with the similarities between the two far-spaced sets of ancient remains. In her view, both are conspicuous by the enormous scale with which they express the idea of "marking the landscape and memory"; both demonstrate "geometric precision through simple constructional techniques"; and most significantly for her, both the Nazca geoglyphs and the Newark Earthworks were, in their pre-Columbian primes, highly venerated pilgrimage destinations.

Extending to the Ohio earthworks a host of insights about pilgrimage that she has been able to test in her extensive Peruvian excavations, Silverman argues that Newark, like Nazca, was constructed as a "sacred enterprise" and visited as a "sacred obligation." That is to say, Newark, like Nazca, was conceived and then experienced as the very antithesis of the quotidian daily routine; it was a "heterotopic sacred site" with "properties of frame, scale, and perspective, which exceed or differ from that of ordinary life." Instead of being prosaic and politically expedient, the layout of Newark, like Nazca, was characterized by a striking incongruity of appearance; instead of being on a manageable human scale, Newark, like Nazca, was stupendously oversized; instead of supporting stable and permanent habitation, Newark, like Nazca, was the locus of movement, a provocateur of transient, "awe-inspiring," and therefore exceptionally memorable visitations. Accordingly, in her view, if contemporary audiences are thrilled and amazed by the "dramatic artificially created sacred landscapes" of Nazca and Newark, they are, in that respect, echoing the sentiments of pre-Columbian designers, builders, and users, all of whom similarly regarded these places as exceptional in the extreme, terrifically impressive contrasts to the broader landscapes within which they were located. From Silverman's perspective, the specialness of Newark, like Nazca, is, if humanly contrived, nonetheless deliberate, unmistakable, and permanent.

By contrast, in his essay, southwestern archaeologist Stephen Lekson juxtaposes the Ohio earthworks with another World Heritage site, Chaco Canyon, the great eleventh-century Pueblo Indian regional center, in ways that give us pause to reconsider whether or not Newark was, after all, an awe-engendering pilgrimage center wherein "spiritual" priorities prevailed over political or economic concerns. Revisiting the long history of comparing these two sites, Lekson acknowledges that unquestionably there are notable parallels insofar as both sites have large earthen constructions, precise geometric forms, extensive systems of "roads," and thus, it seems, similarly wide zones of influence. Given these apparent similarities, Lekson is not surprised that students of the Hopewell have often appealed to the New Mexico site as the preeminent model for a version of "rituality" wherein the configuration of Chaco and its network—and by extension that of Newark—are assessed as very rare even in their own time inasmuch as they were designed to support exceptionally elaborate ritual apparatuses that "cannot be reduced to . . . being the handmaiden of economic and/or political institutions."[7]

An increasingly skeptical Lekson now suspects, however, that both Chaco's uniqueness and its characterization as a religiously motivated antithesis of modern materialism have been seriously overstated. To the contrary, he presents the iconoclastic view that Chaco was simply a "garden-variety Mesoamerican *altepetl*"—that is, the northernmost example of a political formation that was exceptionally common in Classic and Postclassic Mesoamerica. According to this revisionist view, Chaco was neither uniquely mysterious nor decidedly more preoccupied with ritual than other Mesoamerican communities—and thus, in Lekson's duly cautionary view, to utilize Chaco as a kind of "poster child for rituality" that supposedly supports the presence of those sorts of ceremonial preoccupations at Newark is actually more distorting than informing. In short, here we face the disquieting possibility that neither site was the preponderantly "sacred place" that many purport it to be.

Where Lekson is most impressed by the profound differences between Chaco and Newark, British archaeologist Timothy Darvill's essay, "Beyond Newark: Prehistoric Ceremonial Centers and Their Cosmologies," the most broadly framed in the volume, directs attention to four "common themes" that he thinks are shared by virtually all ancient centers of note, whether in Eurasia, Africa, or the Americas—each of which is unmistakably exemplified at Newark. First, with respect to "sacred geography," Darvill notes that Newark's geometric ground plan is a particularly elaborate example of the use of circles, squares, and octagons as well as a strategic juxtaposition of monumental forms and natural water features, design strategies not unlike those deployed at Stonehenge, Avebury, and even the Temple of Heaven in Beijing. Second, with respect to "seasonality and communal gatherings," Darvill observes that the layout of Newark would have provided large groups of periodic visitors precisely the sort of choreographed arrivals, ceremonial performances, and departures that one observes at innumerable great pilgrimage destinations.

Third, with respect to "cosmological structuring," the intricate referencings to the movements of the moon documented by Hively and Horn provide an exceptionally sophisticated example of the "alignments to heavenly bodies" that are embedded in the design of most major ceremonial centers. And fourth, Brad Lepper's recently amplified emphasis on the significance of burials at Newark lends weight to Darvill's suggestion that this complex—again like "countless other" major sites—was configured to mark "cosmic boundaries where the mortals meet the immortals, spirits meet the living, and the quick and the dead are united." Owing to its exemplary demonstration of these four features, Darvill concludes that Newark, if still very seriously underappreciated, ought to be relocated within that class of top-tier ceremonial centers that includes the likes of Stonehenge, China's Temple of Heaven, and Çatalhöyük in Turkey. In his view, despite their modest renown, no comparative company is too fast for these Hopewell earthworks.

Part IV, "The Newark Earthworks in Interdisciplinary Contexts: Architectural History, Cartography, and Religious Studies," marks a departure from the preceding archaeologist-authored essays. While all four of the essays in this set acknowledge an explicit and essential reliance on the work of archaeologists, each is likewise explicit in its unwillingness to accept that archaeology provides the sole disciplinary frame through which we may see, interpret, and appreciate long-abandoned sites such as the Newark Earthworks, and thus each approaches the specialness of the site through a very different lens. Additionally, while not uninterested in the Hopewell-era conceptions and uses of the site, which are the primary preoccupation of the archaeological essays, all four of these essays comment both on the present status of the site and on what each of the authors imagine, albeit in quite different ways, as a lively and promising future for the Newark Earthworks. Each of these essays argues, in its own way, for the two-thousand-year-old site's continued relevance and allure in the twenty-first century.

"The Newark Earthworks as 'Works' of Architecture," is by John Hancock, architectural historian and cofounder of the Center for the Electronic Reconstruction of Historical and Archaeological Sites (CERHAS), the Cincinnati-based organization that, in collaboration with the Newark Earthworks Center, initiated the interactive web-based guide to the earthworks known as the "Ancient Ohio Trail." While CERHAS's digitally enhanced "tours" thorough Ohio's abundant earthen mounds are committed to utilizing insights from the respective "disciplinary paradigms" of archaeology, architecture, and Native studies, Hancock contends that the second alternative—wherein ancient earthworks are analyzed as "monumental works of architecture"—has at least three decided advantages. First, it provides a means of mediating the persistent tension between the prevailing "objective/scientific" investments of archaeologists and the "traditionalist" outlook of Native scholars and communities. Second, it has the practical advantage of helping to advance the World Heritage nomination process insofar as that conceptualization provides the closest match to the UNESCO requirement that, to be deserving, sites must qualify as "masterpieces of human creative genius." And, third, Hancock is persuaded that construing these ancient constructions as continuously meaningful "*works* of architecture" —rather than as, say, "archaeological *sites* that hold knowledge about distant cultures"—provides the most expeditious means of lifting the ancient constructions to their much-overdue appreciation as "objects of public engagement."

In other words, confessing his own very slow appreciation of just how marvelous these monuments are, Hancock finds in his background as an architectural historian—and more specifically, in what he terms "the primacy of experience"—both the foremost cause and the most salient antidote to the near invisibility of these fantastic built forms in the eyes of the wider public. And thus he builds a subtle and compelling case that it is, ironically enough, the contrivances of cutting-edge technologies and "augmented reality" utilized by the Ancient Ohio Trail project that provide the most promising means of helping a general audience to a fuller appreciation of these two-thousand-year-old constructions.

In his essay, another architect and architectural historian, Thomas Barrie, reopens the question of Hopewell burial mounds and rituals but then addresses it in a very different and more broadly comparative way. Informed by the interpretation of numerous better-known sites in Egypt, Europe, and elsewhere, Barrie proposes that the Newark Earthworks constitute a "liminal place" or "place of mediation" insofar as this was—and, for him, still is—an exceptional locale at which to enjoy "physical and metaphysical connections to what otherwise [would be] inaccessible." Exploring the very close connections between Hopewell funerary and domestic architectures, he suggests that the abundant "charnel houses" in which human remains were interred and later burned appropriated the symbolism of regular domiciles and thus the "ontological significance of home." Newark's charnel houses were indeed "houses of the dead," which provided Hopewell-era residents and visitors a kind of multivalent architectural expression and exploration of their comingled respect, fascination, and/or puzzlement at the conundrum of death and dying.

Be that as it may, Barrie urges students of Newark to push past the archaeological questions about Hopewellian funerary conceptions and practices in order to consider as well what this age-old architecture can teach us about our present human condition. Affirming something like John Hancock's emphasis on the continually meaningful quality of the pre-Columbian built forms, he stresses the "contemporary relevance of Newark" insofar as present-day visitors share with the ancients two fundamental human problems or "perennial conditions": first, we, like they, are challenged by a realization of "the enormity of our environmental setting (over which we have little control)"; and, second, we, like they, are no less vexed by "the mystery of death (and what may lie beyond it)." In Barrie's venturesome view, the Newark Earthworks do indeed have a kind of "outstanding universal value" required of World Heritage site designation inasmuch as they still have something relevant and profound to teach all of us about both our ongoing environmental precarities and our inescapable mortality.

Geographer Margaret Wickens Pearce's "The Cartographic Legacy of the Newark Earthworks" presents yet another disciplinary framing by locating the Newark Earthworks within the context of indigenous mapping practices. Again relying heavily on archaeology while nonetheless raising quite disparate questions, Pearce explains how the Hopewell site belongs to a tradition of "Indigenous cartography," which operates with assumptions very different from those that are taken for granted in Western cartography and thus likewise assumed in most Euro-American (mis)conceptions of this place. For instance, where Western cartographers aspire to maps that are finished products—that is, "produced for a market, intended for interpretation by a map reader, not a mapmaker"—indigenous cartography is process-oriented, dispersed and embodied so that mapping is conceived as "an ongoing series of located or situated events among people and places, and that maps themselves are . . . not intended to carry all the meanings of the process as a whole."

Given this sort of open-ended cartographic initiative, Pearce challenges us to appreciate the Earthworks at Newark not simply as an "object" of mapping but instead as an ambience that has, since the Middle Woodland era—just as it could in the future—played host to the continually regenerative mapping processes, which had been initiated by earlier generations of Hopewell surveyors. Furthermore, Pearce intimates, perhaps counterintuitively, that it is the sophisticated, forward-looking imaging of twenty-first-century technologies—such as those employed by Hancock's CERHAS project and by LiDAR (Light Detection and Ranging)—that will, in all likelihood, provide the best means of recovering those process-oriented indigenous cartographic priorities and thus the most promising antidotes to the colonialist cartographies of our nineteenth- and twentieth-century predecessors. She predicts, in other words, that, intriguingly enough, these new, high-tech mapping technologies may well enable the next "restoration" of the ancient earthworks. And thus, Pearce, like Tom Barrie, gives us reason to predict that the formerly famous site does, in that sense, have a very relevant and promising future.

"The Modern Religiosity of the Newark Earthworks," by Thomas Bremer, a specialist in the dynamics of religious travel and tourism, anticipates reservations that will reappear in Winnifred Sullivan's essay and thus demonstrates again that, at this point, it is scholars of religion who are most ill at ease with deceptively simple pronouncements that this is a "sacred place." Reiterating Lekson's concerns about a pervasive tendency—prevalent among scholars as well as wider audiences—to idealize, and thus distort, the priorities of the ancient Mound Builders as overwhelmingly "spiritual," Bremer nonetheless trains his attention on the persistence with which, even now, "modern people deem the Newark Earthworks special, sacred, and religious." But why, he asks, and in what sense, do these ancient constructions persist as religious contexts and resources?

To explore those questions, Bremer singles out two particularly instructive cases: a 2009 "pilgrimage" termed the "Walk with the Ancients," wherein some thirty walkers spent a week retracing an ancient Hopewell road that seems to have stretched about sixty miles from the earthworks near Chillicothe, Ohio, to those in Newark, and a 2007 charter-bus excursion wherein members of the now Oklahoma-based Eastern Shawnee Tribe returned to the traditional Ohio homeland from which they had been forcibly removed in 1832. Assuredly, members of both these groups—as well as many others who currently frequent the Earthworks for very heartfelt reasons—may find the designation "tourist" discomforting, perhaps even an offensive diminishment of their reverent attachments to this unique place. Tourism connotes levity. But the ironic fact that tourists invariably enrich their meaning-making journeys by imagining themselves as other than tourists actually lends support to Bremer's surmise that these Newark enthusiasts are participating in the characteristically touristic practices of "aestheticizing," "commodifying," and "ritualizing." In short, his discussion provides us both an explanation for the continued allure of the Newark Earthworks as a particularly appealing travel destination—yes, a tourist attraction!—and a more rigorously self-conscious way of assessing the Earthworks as comprising "auspicious places of modern meaningfulness" and, to that extent, a "sacred site."

While nearly every essay in this volume touches on the perhaps proprietary relationship between Ohio's pre-Columbian sites and contemporary American Indian communities, Part V, "The Newark Earthworks in the

Context of Indigenous Rights and Identity: American and International Frames," brings that issue to center stage. In "Native (Re)Investments in Ohio: Removals, Earthworks Preservation, and Tribal Stewardship," sociologist, Indian scholar, and interim director of the Newark Earthworks Center Marti Chaatsmith advances the widely held proposition that because the earthworks were built by pre-Columbian indigenous peoples, present-day Indians have both a special affinity and a special entitlement with respect to the ongoing management of the mounds. While she reminds us of the six-thousand-year history of mound building to which the Newark Earthworks belong, for Chaatsmith, the most consequential context in which to locate these geometrical remains is the colonialist history of European incursion and settlement, which included a host of broken treaties and, by 1850, the forced relocation from Ohio of all formally recognized native groups and tribes.

Another who anticipates a very bright and important future for the Newark site, Chaatsmith nevertheless describes a version of "tribal outreach" that can have doubly salutary effects both for Ohio's earthworks and for contemporary native communities. On the one hand, Indian "stakeholders," including prominent native scholars and artists as well as tribal leaders, increasingly emerge as frontline resources for the preservation and thoughtful management of the Ohio mounds. And, on the other hand, as the Eastern Shawnees' renewed interest in Newark described by Tom Bremer well demonstrates, increasing appreciation of the Ohio mounds can also serve as a vital resource for present-day Indian communities, especially those with some historical connection to this region, to revitalize a sense of their own history and cultural heritage. Chaatsmith argues, in short, that native stewardship of the mounds benefits both the ancient monuments and contemporary American Indians.

In her essay, historian of religions and specialist in the indigenous traditions of Oceania and Australia Mary MacDonald recasts the question of "Whose Earthworks?" in a more international context. Conceding that, for her, Newark is a new interest, MacDonald nevertheless immediately recognizes a number of issues pertinent from her years of reflecting on "the encounter of indigenous peoples and settler peoples," which begins in "the Age of Discovery" and era of colonialism but persists into the present. At a national level, for instance, she sees parallels in the debates, activism, and legal challenges of numerous Native American communities, in particular the Haudenosaunee in central New York; and thus she both echoes Marti

Chaatsmith and anticipates the upcoming set of essays by commenting on the qualified relevance of the federal Native American Graves Protection and Repatriation Act to the Newark site. She too believes that the Newark Earthworks "belong in a special way" to all American Indians, an affiliation that ought to be recognized and respected.

But MacDonald, moreover, urges that "we might go even further and say that [the Newark Earthworks] are indigenous constructions that the indigenous peoples of the world should celebrate in solidarity." And thus, at that international level, management of the site ought to be informed also by the United Nations Declaration on the Rights of Indigenous Peoples, a document that was adopted by the UN in 2007 in order to present "an ethical stance" with respect to "the individual and collective rights of indigenous peoples, including their rights to culture, identity, language, employment, health, and education." MacDonald, in other words, not only reaffirms that the Newark Earthworks have an "outstanding universal value" for everyone as well as a special significance for all Native Americans; she furthermore calls attention to an emergent "discourse of indigeneity" and a widening solidarity among aboriginal peoples the world over. And thus she brings to light an overlooked tier of distinctiveness that the Ohio site has for the global community of indigenous peoples.

Part VI, "The Newark Earthworks in the Context of Law and Jurisprudence: Ancient and Ongoing Possibilities," is composed of three essays that, in very different ways, continue to locate the Ohio mounds in relation to questions of law and legal contestation. Two native scholars, Duane Champagne, a professor of sociology and American Indian studies, and Carole Goldberg, a professor and practitioner of law, coauthor "The Peoples Belong to the Land: Contemporary Stewards of the Newark Earthworks," an article that reveals yet two more ways in which the Newark Earthworks stand out as particularly noteworthy. The first part, which "characterizes indigenous motivations, values, and practices about ancient holy places," not unlike Margaret Pearce, urges us to appreciate the fundamentally different presuppositions that undergird indigenous versus Western conceptions of land. Champagne and Goldberg explain that, traditionally, indigenous peoples do not "own" the land per se but claim instead that "people belong to the land, like the plants, animals, places, and even sacred bundles." They contend, moreover, that the Newark complex, as a major pilgrimage destination, was an exceptionally inclusive and "unguarded" place to which countless different groups would have felt an obligation for responsible "stewardship" but none would have claimed exclusive "ownership."

In the second, more prescriptive portion of their article, the Hopewell site emerges also as a consummate example of the difficult—but not insurmountable—legal challenges at issue in the reclamation and management of traditional "sacred sites." Bolstering optimism by recounting the successes of several recent cases, Champagne and Goldberg make a persuasive case that the traditional native notion of "belonging" to the land, and thus being compelled to share access with others, is not simply a quaint anachronism. To the contrary, that nonhegemonic ideal remains, even now, an informing principle for the effective and responsible management—in their terms, the "contemporary stewardship"—of sites such as Newark. In their upbeat forecast, like MacDonald's, were the Earthworks at Newark selected as a World Heritage site, not only American Indians of numerous tribal affiliations but also "many indigenous peoples from around the world would probably want to attend and perform ceremonies." Suitably enough, as in the Hopewell past, Newark could again be, if a permanent abode for few, a rewarding destination for many.

In "Caring for Depressed Cultural Sites, Hawaiian Style," Greg Johnson, a scholar of comparative religion and specialist in indigenous legal disputes in both North America and Hawai'i, trains his attention on the latter context in order to find clues for the management of Hopewell sites. In his considered assessment, the Newark Earthworks stand at present as "a depressed cultural site" insofar as the mounds occasion admiration, celebration, and even adulation, but most of all "concern"—that is, a sense of distress, which evokes a corrective sensibility that Native Hawaiians would call *mālama* (care). Exploring this dynamic relationship between depression and care, Johnson, another to deploy a strategic inventory of case studies, narrates the stories of three very recent Hawaiian controversies, each of which demonstrates both a different version of *mālama* "caring for sites" and a different reliance on the law, and each of which thereby sheds "comparative light on possible futures of the Newark sites."

In the first case, Hawaiian community activists—not unlike the Ohio-based Friends of the Mounds—undertake the sort of extralegal *mālama,* or "care giving" practices that, while eventuating in few if any actual legislative or policy changes, nevertheless have extremely salutary effects on people's appreciation of the "depressed" site. Johnson's second example is a kind of intermediate circumstance wherein a combination of "waves of love" for a revered site and legal action result in partial victories or "uneasy compromises" that forestall development at one of Hawai'i's most prized stretches of beach. His third case study directs attention to more fully legalist caregiving strategies such as grant writing, petitioning state and federal agencies for action, or working with UNESCO for heritage site designation. Johnson then concludes that, encouragingly enough, counterparts to all three versions of Hawaiian *mālama,* or "caring for sites," are presently at work in central Ohio. Thus, as in Champagne and Goldberg's contribution, his own concerns about the neglect of this distressed site are superseded by a qualified optimism with respect to the mounting attention—and thus diverse versions of nurturance and care—that Newark is attracting.

In the final essay, "Imagining 'Law-Stuff' at the Newark Earthworks," another seasoned expert on religion and law who comes fresh to the consideration of Hopewell sites, Winnifred Fallers Sullivan, shifts the attention from contemporary legal wrangling over the site to consideration of the role that law and legal jurisdiction, broadly conceived, may have played in the pre-Columbian conception and use of the place. Another who is wary of seemingly laudatory designations of Newark as a "ceremonial center" or place of pilgrimage, her expansive and duly tentative comments thereby reopen the debate as to whether Newark was primarily a place of "rituality" or whether the geometric mounds might have been designed instead as a forum in which to undertake a distinctive sort of "law-stuff," that is, "a space for regularizing human relations, resolving disputes, and performing justice." Redoubling the theme of inclusiveness from Champagne and Goldberg's article—but then transferring that notion of pluralism from the realm of multiple worship styles to that of "legal multiplicity"—Sullivan proposes that ancient Newark may have been "a context of overlapping jurisdiction," that is to say, a site at which, instead of one state-sponsored authority enjoying absolute control, many sorts of indigenous "law-stuff" were "all jostling up against each other."

In sum, then, Sullivan exemplifies the spirit of the "One Site, Many Contexts" subtitle of the symposium and the exploratory aims of this volume by pushing against the grain, widening the range of conceptual options, and encouraging us to consider that the ancient Hopewell site may have been a place at which to formulate, disseminate, debate, and adjudicate topics and policies that are not less suitably imagined as "legal matters" than as "religious matters." In that sense, she provides a befitting ending to this interdisciplinary collection—wherein shared consensus was never an aspiration—insofar as she brings to the table yet one more provocatively plausible reply to that driving question *What is so special about the Earthworks at Newark?* To be sure, only cursory readers will find agreement among the fifteen responses that follow.[8]

Notes

1. See, for example, Scarre, *Seventy Wonders of the Ancient World.*
2. See John E. Hancock's contribution to this volume.
3. See Richard D. Shiels's contribution to this volume.
4. I borrow the term "contested site" from the editors' introduction to *American Sacred Space,* edited by Chidester and Linenthal, 16ff.
5. This and all subsequent phrases that are quoted in this introduction are, unless otherwise noted, drawn from the essays in this book.
6. Lepper, "The Great Hopewell Road and the Role of Pilgrimage in the Hopewell Interaction Sphere," 128.
7. Yoffee, Fish, and Milner, "Communidades, Ritualities, Chiefdoms," 265–66, quoted by Lekson in this volume.
8. For a concise enumeration of fifteen replies to the question "What is so special about the Newark Earthworks?" see the appendix to this volume.

PART I The Newark Earthworks in the Context of American and Ohio History

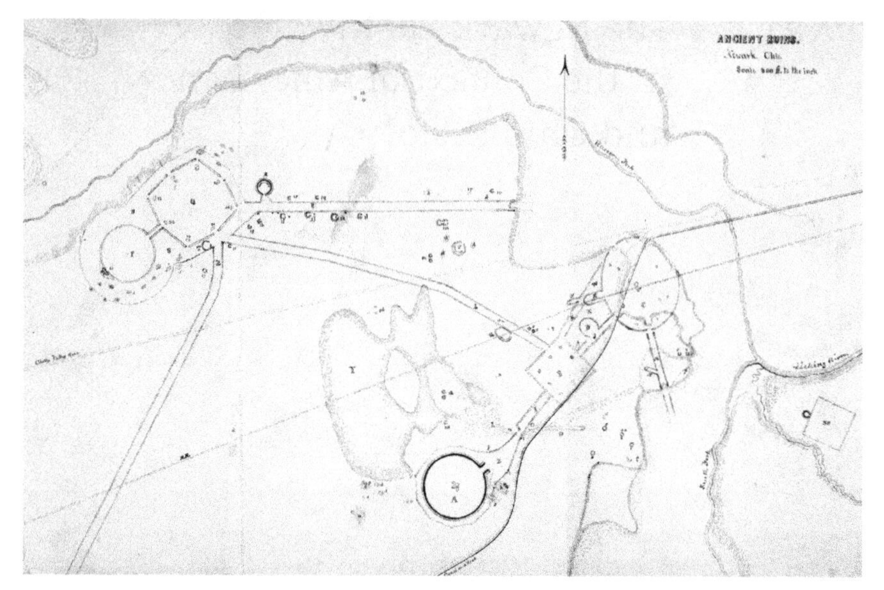

FIGURE 1. Map of the Newark Earthworks drawn by James and Charles Salisbury in 1862, the most complete and comprehensive map of the Newark Earthworks ever produced. (Courtesy American Antiquarian Society)

RICHARD D. SHIELS

The Newark Earthworks Past and Present

THE LARGEST and most precise complex of geometric earthworks in the world was built in what are today the adjoining cities of Newark and Heath, Ohio, roughly two thousand years ago. Two huge features remain. The Newark Earthworks are today among a small number of sites that the US Department of the Interior is considering nominating for inscription on the UNESCO World Heritage List. Built by ancestors of today's American Indians, the site has been preserved by several generations of Newark residents (fig. 1).

The original complex covered four and a half square miles and included four huge geometric enclosures defined by earthen walls:

- a Great Circle nearly twelve hundred feet in diameter with walls fourteen feet high at the gateway and an interior ditch eight to fifteen feet deep
- an Octagon consisting of six walls 550 feet long and two somewhat shorter, encompassing 50 acres and connected by two parallel walls to a circle (Observatory Circle) encompassing twenty more
- a square 931 feet on each side
- an ellipse surrounding eleven burial mounds, a platform mound, and a small circular enclosure[1]

The height of the walls varies from one feature to another and from the outside of the enclosure to the inside. For anyone walking on the outside of the Octagon and its circle, the height of the walls seems to vary from four to eight feet because the ground rises and falls. From within the Octagon the height of the walls is consistently just about eye level. At the base of this circle what may have once been a gateway is blocked shut with a mound fourteen feet high, which has long been called "Observatory Mound" and seems to have been built for sighting. The Great Circle was built on a flat piece of land that slopes down toward the east. The height of the wall varies from fourteen feet on the eastern side to five feet at the opposite end. The effect from within the Great Circle is a level horizon. A ditch immediately inside the wall is fifteen feet deep at the gateway, eight feet deep at the opposite end.

Passageways defined by parallel walls two hundred feet apart led from one feature to another: from the ellipse to the square and then to either the Great Circle or the Octagon. Smaller circles, one hundred to two hundred feet in diameter, were situated along these passageways. Two of these were immediately outside the Octagon, and several others were located along the passageways. An outer wall surrounded the complex. The builders left no way to enter the circles or the Octagon, square, or ellipse without climbing over the walls other than following these passageways from the waterways.

The entire site is surrounded by water on three sides: Raccoon Creek to the north, Ramp Creek two miles to the south, and the Licking River to the east. Parallel walls similar to those within the complex defined passageways into it: from the Licking River into the ellipse, from close to Raccoon Creek into the Octagon, and from Ramp Creek into the Octagon. The passageway from the Octagon to Ramp Creek is two miles long. It crossed the creek and continued for at least another six miles and may have extended much further. Calling it "The Great Hopewell Road," Brad Lepper argues that it connected these earthworks to others near today's city of Chillicothe.[2] Among the numerous geometric enclosures built in the same period in Chillicothe is a great circle identical in size to the one in Newark and the only other octagon connected to a circle ever built.

Nineteenth-century maps portray the Newark Earthworks complex as though viewed from the air.[3] Of course ancient people could not have viewed it from above, but it is impossible to envision the entire complex from any other perspective. Such maps present the viewer with a seemingly random set of shapes. However, twentieth-century surveys and twenty-first-century LiDAR images reveal something else entirely. It now seems certain that the Newark Earthworks must have been built according to a plan that was highly informed by knowledge of both geometry and astronomy.[4]

The architects of this ancient site knew geometry and thought it important to encode geometric knowledge within these earthen walls. Consider the location of each feature in the complex. The circle attached to the Octagon is 1,054 feet in diameter. The distance between the centers of the two large circles is six times 1,054 feet. The distance between the center of the Octagon and the center of the square is the same: six times 1,054 feet. LiDAR studies of other geometric earthworks built across Ohio reveal the same distance (1,054 feet) repeatedly, as well as half that (526 feet) and half that (263.5 feet). Consider also the size of three key features: the circumference of the Great Circle is equal to the perimeter of the square, while the area of Observatory Circle is equal to the area of the square, suggesting that its builders were capable of squaring a circle, a difficult geometric feat.[5]

Similarly, Newark's ancient architects knew astronomy and considered the movements of the moon important to the architecture of the Octagon. The moon moves according to a complex 18.6-year cycle. The spot on the horizon where the moon first rises changes every night, but its pattern is anything but random. Rather, that spot moves toward the south for fourteen nights and then returns over the same length of time. The distance between its rising point on the first night and on the fourteenth increases each month for 9.3 years, then decreases each month for the same length of time. Careful observers can identify four standstill points in the cycle: a northernmost rising and a southernmost, a northern minimum and a southern minimum. The pattern is repeated, with four more standstill points, when one tracks the spot on the horizon where the moon sets. As noted in their contribution to this volume, Ray Hively and Robert Horn demonstrate that the central axis of the Octagon aligns to the northernmost rising of the moon and that alignments for four of the other standstill points can also be identified.[6] The Octagon was built to align with the movements of the moon. How many generations observed the moon's movements before they understood this cycle? Why did they consider this lunar cycle sufficiently important to build massive earthen walls that align to it? No written records have been discovered for these people archaeologists have called the Hopewell culture, but surely the earthworks themselves constitute a kind of writing—Native knowledge written on the land.

The Newark Earthworks are unique in at least three ways: it is the largest, the northernmost, and the most geometrically precise of the earthworks built in the Hopewell era. Today they include the only observable Octagon, and it is the only observable Hopewell earthwork built to align with the 18.6-year cycle of the moon.

On the other hand, the Newark Earthworks is representative of a great many earthen enclosures built in and immediately surrounding the current boundaries of Ohio in the Middle Woodland period. The 1914 *Archaeological Atlas of Ohio* provides a map showing 587 sites within the state that included one or more earthen enclosures from the Hopewell era. In recent years archaeologist Jarrod Burks has located many of these sites and verified their existence, shape, and size using geophysical technology. Along the way, Burks has discovered earthworks that the *Atlas* omits: four more entire complexes and numerous enclosures at the sites that the *Atlas* includes. Burks considers it highly likely that there were *more than six hundred* ancient sites within the current boundaries of Ohio. The Newark Earthworks complex was the largest, the northernmost, and the most precise, but it was one among many.[7]

Very few geometric earthworks enclosures remain. Literally hundreds have been destroyed. The massive pair of circles for which the town of Circleville, Ohio, was named is gone, replaced by a town square much like so many other town squares built within Ohio in the nineteenth century. Yet another massive circle that once stood a short distance from what archaeologists called the "Mound City Group" in Chillicothe has been leveled; today it is covered by a prison. Hundreds of lesser enclosures disappeared beneath the farmer's plow or gave way to city streets as European settlers changed Ohio's landscape over the past two hundred years. One, the Mound City Group enclosure, was rebuilt in the 1920s by the federal government and is preserved in Hopewell Culture National Historical Park.

Two huge features of the Newark Earthworks were never destroyed: the Great Circle and the Octagon (connected to Observatory Circle). The Ohio Historical Connection (formerly called the Ohio Historical Society and originally named the Ohio Archaeological and Historical Society) owns them both.

The Great Circle is today a park, accessible to the public nearly every day, with a small museum that is kept open to the public by the Licking County Convention and Visitors Bureau. The Octagon is not so accessible. For more than a century the Octagon has been leased to a private country club. Since 1963 both the Great Circle and the Octagon have been National Historic Landmarks. Since 2008 these two monuments have been included on the "tentative list" kept by the US Department of the Interior from which the United States will make nominations to UNESCO for inscription on the World Heritage List.[8]

Preserving the Earthworks: 1853, 1893

How did these two geometric enclosures survive when virtually all of the others were destroyed? How did one of them become a public park and the other a private country club?

Numerous individuals and groups in the Newark community deserve credit for preserving them. No one person or organization was responsible.

Rather, groups of Newark residents, representatives of several interests and generations, acted to save the Great Circle and the Octagon—first in the 1850s, then in the 1890s, then in the 1930s, and again in more recent times. It must be noted that the people of Newark also destroyed more than half of the site: the ellipse and the burial mounds within it, almost all of the square (named "Wright Square"), and most of the passageways. Given what happened elsewhere, what is remarkable, however, is how much they preserved.[9]

Euro-American settlement of the Newark area began at the very end of the eighteenth century. Newark Township was first surveyed in 1797. Isaac Stadden, one of the town's first residents, came upon the Great Circle in October 1800. His wife, Catherine, left a written record of what happened the next day: the two of them explored it on horseback, riding together on top of the circular embankment. Hers is the first written record of the Newark Earthworks.[10]

Over the next half century two publications brought the site to the attention of the educated public. The American Antiquarian Society published *A Description of the Antiquities Discovered in Ohio and Other Western States* by Caleb Atwater in 1820, including a map of the Newark Earthworks on the very first plate. Thomas Jefferson read Atwater's account and wrote a congratulatory letter to the society. The Smithsonian Institution published *The Ancient Monuments of the Mississippi Valley* by Ephraim Squier and Edwin Davis in its very first publication, *Contributions to Knowledge, No.1,* in 1848. The map of the Newark Earthworks included in that work showed that already the Ohio and Erie Canal and a road that would become Newark's Main Street cut through two sets of parallel walls, the square, and the ellipse. Two more maps were produced in the 1860s, one by David Wyrick (1860) and the other by James and Charles Salisbury (1862). By that time only the Great Circle and the Octagon remained largely intact.[11]

However, some of Newark's civic-minded citizens began the first effort to preserve the earthworks as early as 1853. They formed an organization, the Licking County Agricultural Society, for this purpose. Raising money by public subscription, they purchased plots of land that included most of the Great Circle from two local couples and one man from New Jersey for a total of $3,833.90. Over the next decade they made three other much smaller purchases and included all of the Great Circle. Their strategy was to preserve the Great Circle by finding new uses for it. They made it a fairground. Both the county fair and Ohio's fifth state fair were held in the Great Circle in 1854, and many other fairs followed at the site, the last of which was in 1934. The Great Circle became a community gathering place, serving the public at large (fig. 2).[12]

FIGURE 2. The Great Circle Earthworks as the Licking County Fairgrounds. (Licking County Historical Society Archives)

Fairs were the most common function hosted at the site, but there were others as well. During the Civil War the site was called "Fort Sherman" and served as a military training ground. In the decades following the Civil War it was the venue for a wide variety of very large events. The Grand Army of the Republic gathered for a reunion on July 22, 1878, an event that featured President Rutherford B. Hayes and former president Ulysses S. Grant. Fifteen to twenty thousand people are said to have attended. The scaffolding that held chairs for prestigious guests collapsed that day, forcing the presidents and the others to leap for safety. Buffalo Bill's Wild West Show brought cowboys and Indians, guns and horses to the Great Circle in 1884. By that time the fairgrounds, as the site was commonly called, included a cinder horse-racing track, cinder paths, public toilets, display buildings, and a hotel. Weekend dancing occurred in a special dance pavilion.[13] In 1898, a local banker named James Lingafelter leased the grounds from the Agricultural Society and opened a summer resort that he called Idlewilde Park. Over time the establishment became an amusement park with a Ferris wheel, roller coaster, casino, theater, bowling alley, ponds for boating and swimming, and a European-style hotel and restaurant. Bicycle, motorcycle, and even automobile racing were added to harness racing after World War I. Idlewilde Park

FIGURE 3. The Great Circle Earthworks as Idlewilde Park. (Licking County Historical Society Archives)

was declining by that time, however. Lingafelter was facing legal difficulties for forgery, embezzlement, and theft, and other sites were being developed to satisfy the public demand for recreation. Idlewilde Park disappeared from the Newark city directory after 1924. The public began referring to the site as Moundbuilders Park instead. For seven decades the Great Circle had been preserved by local citizens and served as a community resource (fig. 3).[14]

The Octagon has also been preserved thanks to the efforts of community leaders. In 1892, nearly forty years after the first efforts to preserve the Great Circle, local people acted to preserve the Octagon. Once again they found a new use for an ancient site. Once again they set out to benefit the community at large.

In 1892, the state was seeking a site that might be used as a summer training ground for the National Guard. Newark leaders—members of the city council and county commissioners—proposed to purchase the Octagon (the land was owned by several parties, as the Great Circle had been) and to give it to the state for this purpose. Nearly seven thousand men would come to Newark every summer, they argued, creating a market for food and many other items. Equally important, the ancient earthen enclosure would be preserved. Voters approved a tax levy, choosing to raise their own taxes for this purpose.

Voters within the city of Newark supported the measure by a margin of three to one. County residents approved the levy by a smaller margin, and the site was purchased by the joint efforts of the city and county governments. The deed was initially made out to the Newark Board of Trade, a private organization that functioned much like a chamber of commerce. However, the state accepted the deed, and the National Guard came to Newark each summer from 1893 to 1907.[15]

How did this site, purchased with taxpayer dollars, come to be leased to a private country club?

The State of Ohio moved the National Guard out of Newark in 1907 and then returned the deed to the Octagon to the Newark Board of Trade. For the next few years the site was unoccupied, but a small group of local men had begun playing golf on it in 1901. The decision to lease the site to a newly formed country club was made in a Board of Trade meeting in April 1910.

A lively public debate about the best use for the Octagon took place over three months preceding that meeting. A series of letters on the topic was published in the *Newark Advocate* beginning in January 1910. Eli Hull, a successful businessman who later purchased land that included an earthen enclosure and built his home within it, wrote in favor of making the site a public park. Several others did the same. During these same months, however, Hull and other local businessmen who were members of the Board of Trade formed two new organizations: first a businessmen's club intended to provide recreational opportunities for men such as themselves and then a country club. Historian Neal Hitch writes that the Board of Trade meeting in April was called "in order to propose a lease agreement between the Board of Trade and the country club."[16] Seventeen men attended the meeting and voted. All seventeen were members of the Board of Trade; most were also members of the new businessmen's organization; fourteen were charter members of the country club. Still, the *Newark Advocate* reported their decision as what Hitch describes as a "benevolent act."[17] The city could not afford to maintain a park, according to the *Advocate*. The golf course would not require public money, but the site would be "open to the public at all times." The Board of Trade wished "to encourage the public to regard the property as belonging to them and to feel that they have the right to use them."[18] The final vote in favor of the lease included a provision that the lease could be canceled and the site reclaimed after one year if dissatisfaction arose (fig. 4 and fig. 5).[19]

The Licking County Country Club opened for business on the site in the summer of 1911. Over time it became Moundbuilders Country Club, and

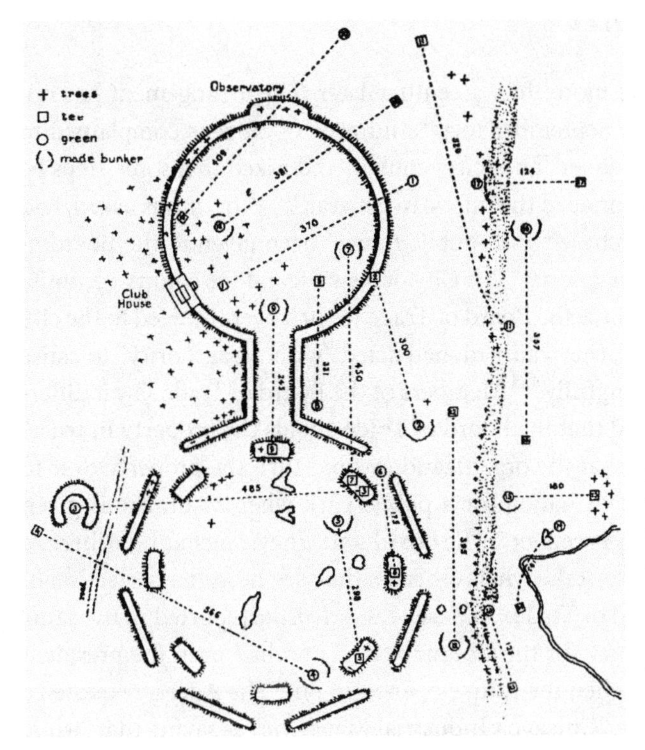

FIGURE 4. Map showing the location of the eighteen holes for golfing at the Octagon Earthworks, in *Moundbuilders Country Club 75th Anniversary* (1985), 11. (Licking County Historical Society Archives)

FIGURE 5. Scorecard for golfing, in *Moundbuilders Country Club 75th Anniversary* (1985), 10. (Licking County Historical Society Archives)

it remains on the site more than a century later. The arrangement has not been uncontested. By September 1913, "a number of citizens complained to the [county] commissioners and [city] council and asked that some steps be taken to open the grounds to the public in general."[20] Why, it was asked, had the deed to a site purchased with public money been given to the Board of Trade, a private organization?[21] The City of Newark and the county commissioners filed a suit to force the Board of Trade to turn over the deed to the city and county. Someone, they charged, had acted "without authority" to cause the deed to "read wrongfully."[22] Members of the Board of Trade saw it differently. They considered that the Board of Trade "holds the property in trust" and "repeatedly said that the organization would turn the property over to the city, county, state or nation for a public park when assurance is given that it will be properly cared for." The board's attorneys, including Albert A. Stasel, were directed to "take whatever steps they see fit in the protection of the rights of the Board of Trade."[23] Stasel, the *Advocate* reported in the same story, was also "attorney for the Country club" and had been the president of the Country Club when the course opened in 1911. The *Advocate* quoted a story about Newark in "Chicago's Industrial Magazine" as saying that "while the Country Club is not strictly a Board of Trade institution, it is made up largely of Board of Trade Members."[24]

The case was settled in 1917. Judge Park B. Blair found that "the Board of Trade holds the same [i.e., the deed] in trust for the City of Newark and the County."[25] However, the judge continued by noting that, by 1917,

> the Board of Trade having ceased to transact the business for which it was incorporated, and being now dormant if not defunct, it would seem advisable that this trust be terminated, or that such Board of Trade be removed as such trustee and that a successor be appointed by this court. . . . A. A. Stasel, the member of the Board of Trade who has had actual and active control of these lands, will be appointed as trustee to succeed the Board of Trade.[26]

The lease with the country club was found to be valid although the deed was now held in trust for the City of Newark and Licking County, the two entities that had bought the site originally.

Preserving the Earthworks: 1933

The next critical period for the Newark Earthworks began during the decade following World War I and culminated in the midst of the Great Depression. Idlewilde Park closed after 1924. Visitation to the Great Circle declined over the next several years, according to the *Newark Advocate*. In 1927, the county commissioners bought the site from the Licking County Agricultural Society, which had fallen $25,000 in debt; the sale price was $25,339. The Licking County Conservation League was founded the same year to assume responsibility for preservation and beautification of what was now being called Moundbuilders Park. The league hosted a meeting between the county commissioners and Mr. Henry Shetrone, director of the Ohio State Archaeological and Historical Society, the next year and began working to acquire state funding to maintain the site in 1929.[27]

Even as the nation sank into the Great Depression, new organizations arose to preserve the earthworks. The Licking County Historical Society was organized in 1931. Its president, Frank A. Woolson, was the editor of the *Newark Advocate*. Its stated purpose was "to preserve the mounds" and to "foster a movement to have the group of mound enclosures incorporated into a State Park under the State [Archaeological and Historical] Society." Clarence Jones, chairman of the historical committee of the Newark Chamber of Commerce, emerged as a tireless promoter of the site.[28] Jones published an article in the *Columbus Journal Dispatch* on August 7, 1932, in which he argued that "while the Newark earthworks are in local public ownership, the fact that they are the only mounds of the large geometric enclosure type left in the United States demands that they be properly maintained and preserved for all time. This can only be done by the State or federal government."[29]

However, neither the state nor the federal government took the lead. Rather, it was citizens like Woolson and Jones who acted to "maintain and preserve" the Newark Earthworks. County commissioners, city councilmen, and leaders in the Licking County Conservation League, the Licking County Historical Society, and the chamber of commerce each played a role. The key decision—which Hitch has called "a community decision"—was made in December 1932, while the nation was experiencing the Great Depression at its worst.[30]

It was decided that the deeds to both the Octagon and the Great Circle (which were now held jointly by the City of Newark and Licking County) would be given to the Ohio State Archaeological and Historical Society. Both the Octagon and the Great Circle would henceforth be considered state parks. By the end of 1933, the Ohio Historical Society was operating thirty-three sites, all of which were called state parks. Henry R. McPherson, curator of parks for the society, summarized what had happened as follows: "Culminating an active campaign of three years or more on the part of citizens of Newark and Licking County the final Chapter has recently been written in the transfer of Moundbuilders State Park and Octagon State Park to the Custody of this Society."[31] Presumably, Octagon State Park would become a public site when the country club's lease expired in 1940.

Local government officials gave the deeds to the statewide historical society, which in turn enlisted the aid of the federal government. Franklin Roosevelt's New Deal included federal jobs programs that were looking for projects such as this one. Moundbuilders Park was physically transformed by federal government workers over the next several years. The Civilian Works Administration (CWA) employed local men who repaired damage to the walls of the Great Circle and to Eagle Mound in its center in 1934. The Civilian Conservation Corps (CCC) succeeded the CWA the next year. Two hundred twenty CCC employees camped out within the Great Circle in the summer of 1935. Working under the direction of the National Park Service and the Ohio State Archaeological and Historical Society, they removed buildings and facilities that had been part of the fairgrounds and the amusement park and made the site a public park. They also built a house at the Great Circle intended to house a park superintendent.

The Octagon was supposed to be transformed as well, but federal money ran out. Congress reduced the CCC budget in March 1936. However, the CCC proceeded to commission Newark resident Richard Fatig to draw the master plan for the Octagon. It presents a park in which the country club has been removed, a train leads visitors from the parking lot to Observatory Mound, and a segment of the parallel walls along Thirtieth Street are restored. Dated February 19, 1937, the plan is signed by Richard McPherson, who is identified as the "State Park Authority." The CCC left Newark for Fairlawn, Ohio, later that year. By then CCC employees had built a parking lot at the Octagon and a fence around the entire property and had begun a superintendent's house but left before the house had been completed or a train had been installed.[32]

In summary, much was accomplished at the Great Circle earthworks in the 1930s. Picnic facilities and open space enhanced the ancient mounds that had been hidden too long among the trappings of an amusement park. The earthen walls had been repaired and now appeared even more impressive. Much less changed at the Octagon, where Moundbuilders Country Club members continued to play golf.

In 1938 the club asked for another extension of the lease, and the Ohio State Archaeological and Historical Society granted the request. Only five years earlier the Octagon as well as the Great Circle had been deeded to the society with the expectation that both sites would become state parks. Over those years, New Deal government programs had come and gone. Money to create Newark's state park had become difficult to secure. Public visitation had soared but had never been as great at the Octagon as the Great Circle. The Ohio State Archaeological and Historical Society reported 68,000 visitors to Moundbuilders Park in the first nine months of 1938 and 32,050 visitors to the Octagon. Hence the lease was renewed to run through 1945 and has been renewed repeatedly to the present day. In 1985 the club celebrated seventy-five years on the site; in 2010 it celebrated a century (plates 9–10). The current lease will expire in 2078.[33]

Ongoing Efforts

The Newark Earthworks continue to inspire local people and others to take up the task of preservation and to increase public access. A group calling itself the Licking County Archaeological and Landmarks Society took up the cause in the 1990s, while another group calling itself the Friends of the Mounds followed in the summer of 2000. The Ohio State University's Newark Earthworks Center, officially recognized by the university's board of trustees in 2006, includes several people who had been part of both of these groups and many more who have joined the effort within the last decade. We work to win inscription on the UNESCO World Heritage List for the same reasons that previous generations purchased these sites, created fairgrounds, and strove to create a state park: we are committed to preserving Newark's Great Circle and Octagon in perpetuity and making both of these enclosures fully public sites.

Notes

1. See Lepper, "The Newark Earthworks: Monumental Geometry and Astronomy at a Hopewellian Pilgrimage Center."

2. See Lepper, "Tracking Ohio's Great Hopewell Road"; and Lepper, "The Great Hopewell Road and the Role of Pilgrimage in the Hopewell Interaction Sphere."

3. See, for instance, Atwater, "Description of the Antiquities Discovered in the State of Ohio and other Western States" (1820); Squier and Davis, *Ancient Monuments of the Mississippi Valley* (1848); Salisbury and Salisbury, "Accurate Surveys and Descriptions of the Ancient Earthworks at Newark, Ohio" (1862); and Wyrick, "Ancient Works near Newark, Licking County, Ohio" (1866).

4. See Romain and Burks, "LiDAR Imaging of the Great Hopewell Road"; and Romain and Burks, "LiDAR Imaging of the Newark Earthworks."

5. See Romain, *The Mysteries of the Hopewell*, chapters 2–3.

6. Hively and Horn, "Geometry and Astronomy in Prehistoric Ohio"; and Hively and Horn, ""Hopewell Cosmography at Newark and Chillicothe, Ohio."

7. See W. Mills, *The Archaeological Atlas of Ohio;* Burks and Cook, "Beyond Squier and Davis"; Burks, "Geophysical Survey at Ohio Earthworks"; and Burks, "Recording Earthworks in Ohio, Historic Aerial Photography, Old Maps and Magnetic Survey."

8. "The World Heritage Tentative List" is online as of June 10, 2014, at http://www.nps.gov/oia/TentativeList/Tentative_map.htm.

9. Historian Neal Hitch compiled much of the material that follows, and the *Newark Advocate* provides documentation of preservation efforts by local people over the past two centuries. The history that Hitch (in collaboration with others) compiled is included in *Historic Site Management Plan for Newark Earthworks State Memorial,* appendix I, "A Brief History of the Newark Earthworks" (Columbus: Ohio Historical Society, 2003), A.41–A.66, online at http://ohsweb.ohiohistory.org/places/c08/hsmp2.shtml.

10. "A Relation of Incidents and Events in the Licking valley, during the Year 1800, Made at the Meeting of the Pioneer Society, May 18, 1870," Licking County Pioneer, Historical, and Antiquarian Society, *Pioneer Paper* no. 69 (1870).

11. See note 3 above.

12. See "A Brief History of the Newark Earthworks," A.52–A.54; "State Fair Decided," *Zanesville (Ohio) Courier,* Jan. 19, 1854; "Ancient Works near Newark, Ohio," *Ohio Cultivator,* Apr. 1, 1854; "Ohio State Fair," *Cleveland Daily Herald,* Sept. 12, 1854; "The Ohio State Fair," *New York Times,* Oct. 18, 1854; and "Fair Board Owned Grounds Free of Debt in Tenth Year," *Newark Advocate and American Tribune,* Sept. 24, 1932.

13. See "Buffalo Bill," *Newark Daily Advocate,* July 15, 1896. Warren Moorehead attended Buffalo Bill's show in the Great Circle as a child and later published a description of it in Moorehead, *The American Indian in the United States,* 303.

14. See "Two More Days of Work at the Old Fort," *Newark Daily Advocate,* Feb. 1, 1898; and "A Brief History of the Newark Earthworks," A.53.

15. The local press urged voters to pass a levy raising $23,000 to purchase the Octagon and give it to the State of Ohio to serve as an encampment grounds. See "Observations on the Grounds Chosen for Encampment," *Newark Daily Advocate,* Nov. 11, 1891; and "The Encampment Grounds," *Newark Weekly Advocate,* Feb. 18, 1896. Support was strong within the city council; see "Action of the City Council," *Newark Weekly Advocate,* Feb. 18, 1892. Residents of the city of Newark and the county voted

on separate levies to buy the Octagon; see "Separate Propositions," *Newark Weekly Advocate,* Feb. 25, 1892. Both levies passed; see "OVERWHELMING! The Majority in Favor of Encampment. ALMOST UNANIMOUS," *Newark Weekly Advocate,* Mar. 10, 1892. After the levies passed and the land was purchased, the state authorized $12,000 to "improve the grounds"; see *Newark Daily Advocate,* Apr. 7 and 26, 1893. Within months National Guard units from Hebron, Columbus, Cincinnati, and Springfield were scheduled for encampments in the Octagon; see *Newark Weekly Advocate,* Aug. 5, 1893.

16. "A Brief History of the Newark Earthworks," A.56.

17. Ibid.

18. Ibid.

19. Among newspaper articles that engage the issue, see "Would Beautify Old Encampment Grounds," *Newark Advocate,* Dec. 10, 1909; Carey Montgomery, "Camp Ground," *Newark Advocate,* Dec. 17, 1909; E. M. P. Brister, "Judge Brister Favors Idea of a Public Park," *Newark Advocate,* Dec. 17, 1909; M. R. Scott, "Grounds Should Be Converted into Fine Public Park," *Newark Advocate,* Dec. 23, 1909; "Should Convert Encampment Ground into Public Park," *Newark Advocate,* Jan. 20, 1910; "Country Club Leases Encampment Grounds," *Newark Advocate,* Apr. 7, 1910; "Country Club Will Erect Fine $10,000 Club House," *Newark Advocate,* May 26, 1910; and "Handsome New Home of Country Club Will Be Opened June 15," *Newark Advocate,* June 5, 1911.

20. See "Want Lease of Grounds Returned," *Newark Weekly Advocate,* Sept. 4, 1913.

21. See "No Authority to Make Deed," *Newark Weekly Advocate,* Sept. 4, 1913.

22. See "Petition in Country Club Case Is Filed," *Newark Daily Advocate,* Sept. 15, 1913.

23. Ibid.

24. Ibid.

25. "Old Board of Trade Vindicated by Judge P. B. Blair's Decision in Country Club Case Recently Tried," *Newark Daily Advocate,* Nov. 17, 1917.

26. Ibid.

27. See "A Brief History of the Newark Earthworks," A.54–A.56.

28. "Preservation of Mounds in County Urged by Ohio Press," *Newark Advocate and American Tribune,* June 1, 1932; "Reports Many Mounds Visitors," *Newark Advocate and American Tribune,* Aug. 12, 1932; "Mounds Publicity Plan Wins Support of Many Groups," *Newark Advocate and American Tribune,* Aug. 17, 1932; "Show Interest in the Mounds Here," *Newark Advocate and American Tribune,* Oct. 10, 1932; "Newark's Mounds," *Newark Advocate and American Tribune,* Oct. 17, 1932; and Ben Hoover, "At Last—Mound Conscious," *Newark Advocate and American Tribune,* Nov. 4, 1932.

29. See Jones, *Columbus Journal Dispatch,* Aug. 7, 1932.

30. See "A Brief History of the Newark Earthworks," A.58.

31. Ibid.

32. See "Begin Work at Barn at Park," *Newark Advocate and American Tribune,* Oct. 30, 1935; "Start Work on Park Cottage," *Newark Advocate and American Tribune,* Nov. 9, 1935; and "CCC Campers Interested in Mound History," *Newark Advocate and*

American Tribune, Sept. 25, 1935. Richard Fatig's plan for the Octagon hangs on the wall at the Newark Earthworks Center, Ohio State University at Newark.

33. Records of the Licking County Archaeological and Landmarks Society, the Friends of the Mounds, and the Newark Earthworks Center are available at the Ohio Native American Archive at The Ohio State University at Newark. All of the leases and other public documents are available as of June 10, 2014, online at http://ohsweb.ohiohistory.org/places/c08/hsmp2.shtml.

PART II The Newark Earthworks in the Context of Hopewell Archaeology and Archaeoastronomy

BRADLEY T. LEPPER

The Newark Earthworks

A Monumental Engine of World Renewal

THE NEWARK EARTHWORKS represent the pièce de résistance of the florescence of monumental architecture that is an important part of what archaeologists have referred to as the Hopewell culture. I make this claim not simply because it is the largest complex of geometric earthworks in the Hopewellian world but because it is an integrated combination of functionally discrete architectural elements incorporating astronomical and geometrical knowledge with uncanny precision and on a scale that is overwhelming to on-the-ground observers. Elsewhere I have compared the Newark Earthworks to "a North American Kaaba, Sistine Chapel, and *Principia* all rolled into one,"[1] but in some ways the site may be more like a pre-Columbian Large Hadron Collider—a vast machine, or device, designed and built to unleash primordial forces.

The principal surviving elements of the Newark Earthworks are the best-preserved examples of geometric earthworks in North America. Until rather recently, however, the site was conspicuous by its absence in most overviews of the Hopewell culture. I think the most important reason for this was that archaeologists had defined the Hopewell culture largely in terms of its extravagant burial practices and especially the glittering array of mortuary and other offerings crafted from exotic materials obtained from across much of the continent. Since Newark's burial mounds had been leveled without systematic study, it appeared to have nothing substantive left to contribute to the corpus of Hopewellian iconography. As an example of the apparent disciplinary disappointment with Newark, when Emerson Greenman excavated the so-called Eagle Mound within the Great Circle in 1928, the absence of burials and sumptuous mortuary offerings consigned the important results of that excavation to a revealingly terse two sentences in Henry Shetrone's 1930 book *The Mound Builders:* "The so-called Eagle Mound, situated at the center of the fair-ground circle, was explored by the Ohio State Museum in

1928 and found to be without burials. It apparently was erected as a strictly religious or ceremonial structure."[2]

Beginning in the 1980s, a renaissance of research undertaken at Newark's wonderfully preserved enclosures has led to extraordinary insights about the Hopewell culture generally and the structure and function of the Newark Earthworks in particular, which have brought Newark to greater prominence in Hopewell studies.[3] The Newark Earthworks are now seen to incorporate a sophisticated understanding of geometry and astronomy into their architecture, a fact that would not have been demonstrable had it not been for the remarkable degree of preservation of the earthworks.

In contrast, the assumption that there is a nearly complete absence of data from the lost burial mounds at Newark has led to a practical inability to determine directly what role this component of the earthworks may have served in the functioning of Newark's monumental ceremonial machine.[4] The purpose of this essay is to review what is known about the Newark Earthworks from an archaeological perspective with an emphasis on the Cherry Valley Ellipse, the earthwork that encompassed the most important burial mounds at this site. In so doing, I will argue that there are sufficient data to shed some light on the nature of Newark's burial mounds and, indeed, to allow us to recognize the centrality of mortuary ceremonialism in the design and operation of the Newark Earthworks.

An Archaeological Perspective on the Newark Earthworks

Archaeology provides the bedrock foundation upon which all interpretations of the Newark Earthworks fundamentally must be based. Without knowledge of the structure of the earthworks, the activities associated with them, and the cultural context in which they were created and operated, all inferences about their use and meaning (for the aboriginal builders in particular) would be reduced to speculation. The only reliable clue to these aspects of the site is the material evidence preserved in the archaeological record.

American Indian oral traditions have been sieved for insights into the ancient earthworks by numerous authors, including me, but with only indifferent success.[5] The principal reasons for the absence of such traditions are the relative antiquity of the architecture and the degradation of authentic historic content in oral traditions over that much time.[6] And, of course, the normal process of loss of reliable content over time was catastrophically accelerated by the impact of European contact on indigenous American societies, including the loss of the oral archives of traditional knowledge, through such factors as the premature deaths of elders and the forced assimilation of subsequent generations of American Indians.

As a result, the only reliable source of information about the historical context in which this florescence of extraordinary architecture, art, and ceremony emerged, the structure of the earthworks, and what transpired within and outside the confines of the monumental walls, is the archaeological study of the material traces of those aspects of ancient Newark. Many additional disciplines can contribute vitally important insights, such as astronomy, ethnography, geology, geometry, and others, but ultimately, all claims relating to the original purpose and meaning of the Newark Earthworks for its ancient builders must rest upon the material evidence as revealed by archaeology.

THE HOPEWELL CULTURE

It long has been accepted that the Newark Earthworks were built by the Hopewell culture,[7] sometime between 100 BCE and 400 CE. The few documented radiocarbon dates associated with the Newark Earthworks suggest this site belongs to the later portion of this period, between approximately 100 and 400 CE.

The people of the Hopewell culture lived in small, scattered villages, or hamlets, with little or no evidence of a political hierarchy.[8] Their corporate ceremonial lives were focused upon the great earthen enclosures, which must have required the collective effort of scores of such dispersed communities to construct, and whose vast interior spaces could have accommodated many thousands of celebrants.

The great earthworks are concentrated in the principal river valleys of southern Ohio—the Great and Little Miamis, Scioto, and Muskingum—but these people were participants in a much broader "interaction sphere" that extended from the Atlantic Coast to the Rocky Mountains and from the Great Lakes to the Gulf of Mexico.

THE NEWARK EARTHWORKS

The extraordinary significance of the Newark Earthworks has been recognized widely. It is a National Historic Landmark and the State of Ohio's official prehistoric monument, and it is on the US Department of the Interior's Tentative List for nomination to the UNESCO World Heritage List. According to Samuel Haven, Daniel Webster was so impressed with the Newark Earthworks that he "desired to have [them] preserved in perpetuity at the

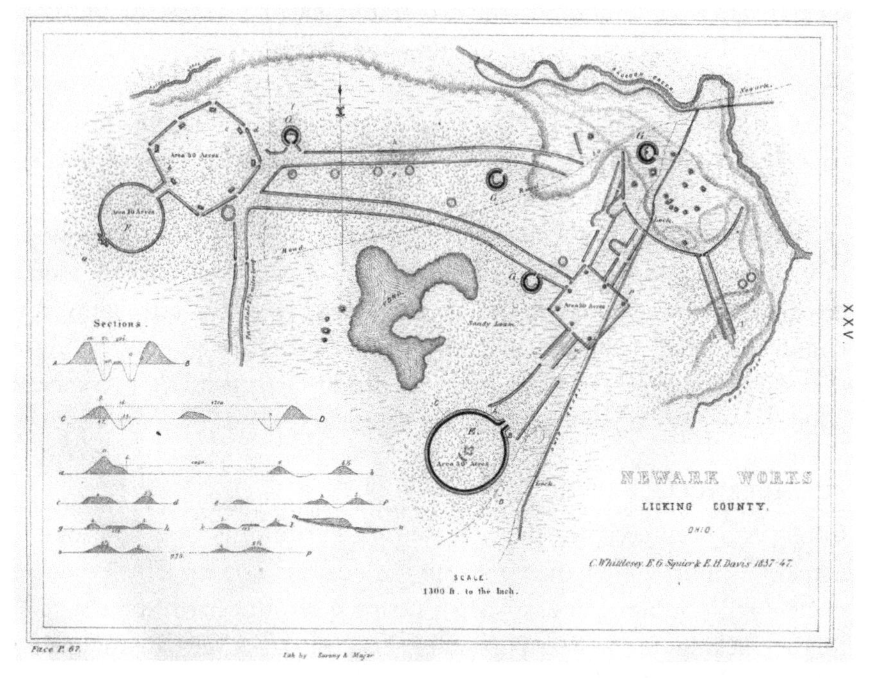

FIGURE 1. Squier and Davis map of the Newark Earthworks. Based on an original survey by Charles Whittlesey, the Squier and Davis map is not the most comprehensive map of the Newark Earthworks, but it is the most historically significant, as it appeared in 1848 in the Smithsonian Institution's first scientific publication.

national charge."[9] If he had achieved his desire, the Newark Earthworks would have become the first of America's National Parks and would have been preserved nearly in their original grandeur.

Because of the massive scale and complexity of the Newark Earthworks, the site has often been featured as an illustration of the apogee of Hopewell earthwork construction.[10] Nevertheless, from the glory days of Warren Moorehead, William Mills, and Henry Shetrone to the relatively recent Chillicothe Conference,[11] Newark appears to have been largely ignored. Its location on the northern periphery of the classic Hopewell world usually was explained as having something to do with the proximity of the Flint Ridge flint quarries.[12]

The Newark Earthworks originally consisted of a series of monumental geometric enclosures connected by a network of walled roads (fig. 1). The primary enclosures included a circle, the so-called Observatory Circle (fig. 2), connected to an octagon (fig. 3); a circle with an interior ditch—the Great

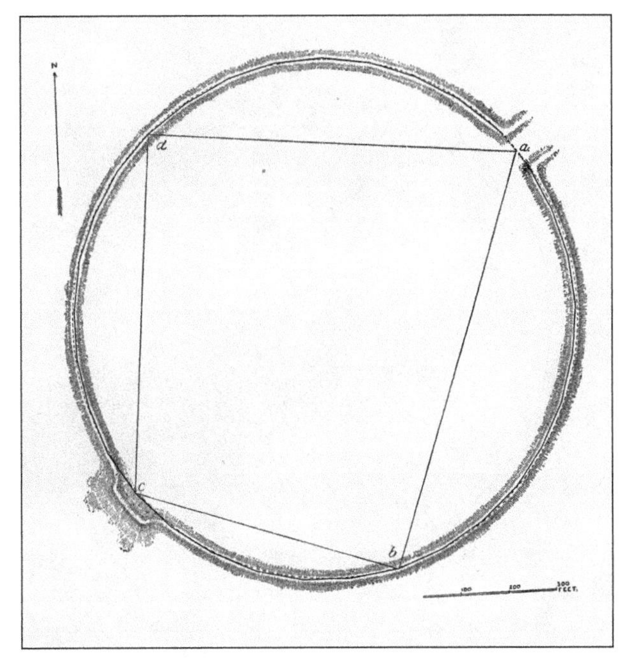

FIGURE 2. Middleton map of the Observatory Circle. James Middleton's surveys represent the best data available for the individual components of the Newark Earthworks. Unfortunately, he never integrated these data into a comprehensive map of the Newark Earthworks as a whole incorporating the other surviving elements. (Thomas, *Report on the Mound Explorations of the Bureau of Ethnology,* facing page 462)

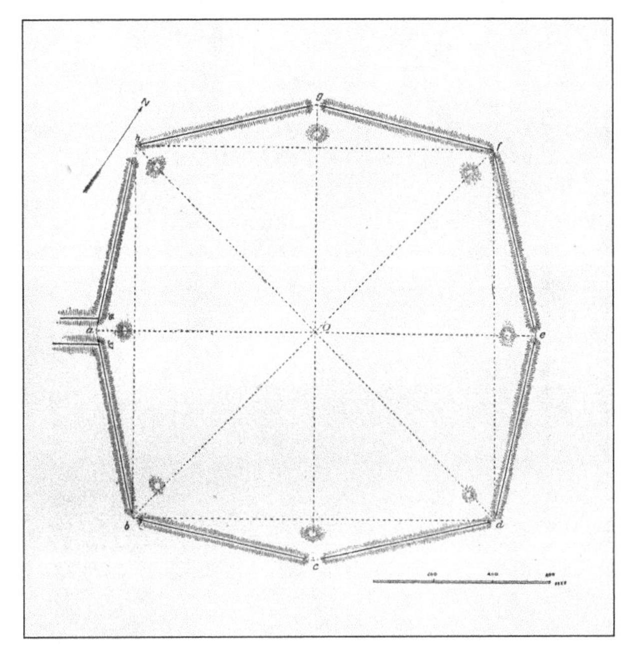

FIGURE 3. Middleton map of the Octagon Earthworks. (Thomas, *Report on the Mound Explorations of the Bureau of Ethnology,* facing page 464)

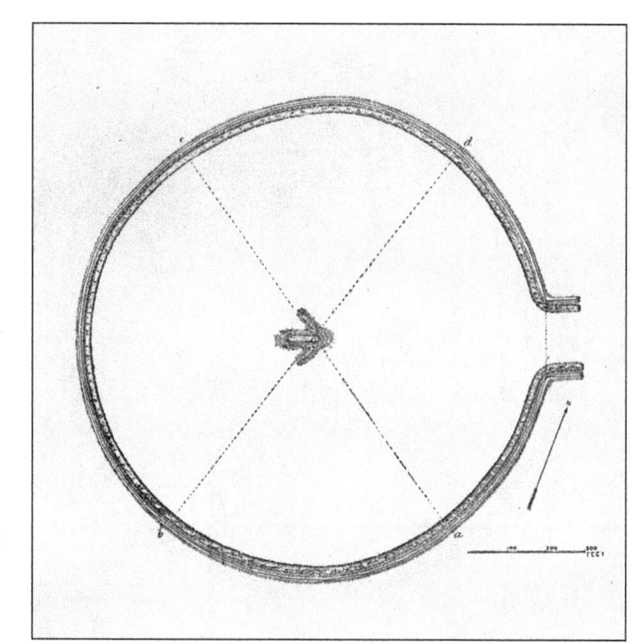

FIGURE 4. Middleton map of the Great Circle Earthworks. (Thomas, *Report on the Mound Explorations of the Bureau of Ethnology*, facing page 460)

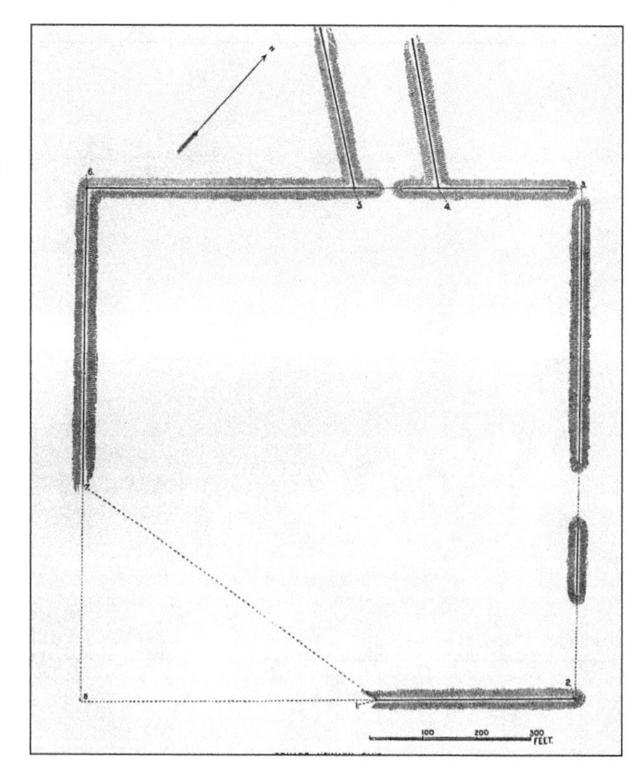

FIGURE 5. Middleton map of the Wright Square. (Thomas, *Report on the Mound Explorations of the Bureau of Ethnology*, facing page 466)

Circle (fig. 4); a square, often referred to as the Wright Square (fig. 5); and an ellipse surrounding a number of large and small, conical and loaf-shaped mounds—the Cherry Valley Ellipse.[13]

GREAT CIRCLE

The Great Circle (fig. 4), sometimes referred to as the "Fairground Circle," since it was the site of the Licking County Fairgrounds between 1854 and 1933, is a large circular enclosure ranging in diameter from between 1,163 and 1,189 feet (355–62 m). The walls are between 5 and 14 feet (1.5–4.3 m) in height and between 35 and 55 feet (11–17 m) in width.[14]

In 1992, excavations through the enclosure wall revealed a three-stage construction sequence that utilized different colors and textures of soil for the inside and outside portions of the embankment.[15] An intact, buried A-horizon at the base of the embankment yielded a radiocarbon date of 2,110 ± 80 years BP (before present) on soil humates (Beta-58449). This date, however, does not relate directly to the construction of the earthwork and only permits the inference that the embankment cannot be older than about 2,100 years. Pollen and phytoliths recovered from the paleosol indicate the environment at the time the earthworks were constructed was dominated, at least locally, by a prairie.[16]

At the center of the Great Circle is a three-lobed mound that has been named "Eagle Mound" for its supposed resemblance to a bird, although its shape is amenable to numerous alternative interpretations and is not likely to have been built as an effigy mound. It was excavated in 1928 by Emerson Greenman for the Ohio Historical Society.[17] At the base of the mound, Greenman uncovered rows of postmolds from a large rectangular wooden structure that occupied the site prior to the erection of the mound. This building was almost 100 feet (30 m) long by about 23 feet (7 m) wide with walls like wings extending outward on each side at a forty-degree angle from the main axis. In the center of this structure there was a large, rectangular prepared-clay basin, similar to crematory basins excavated at Mound City and other Hopewell sites. The Eagle Mound basin, however, contained no cremated human remains, although it exhibited clear evidence of repeated episodes of intense burning.

The Great Circle has an interior ditch that is between 28 and 41 feet (8.5–12.5 m) wide and, in 1888, ranged from 8 to 13 feet (2.4–4.0 m) in depth. Caleb Atwater, author of the earliest published description of the Newark Earthworks, observed that when he saw the Great Circle, "the ditch was half filled with water."[18] Since ditches of any sort were not used with any of the other major earthworks at Newark, the presumably water-filled ditch at the Great Circle must then reflect the unique requirements of this structure.

WRIGHT SQUARE

The Wright Square (fig. 5) was a nearly perfectly square enclosure with walls averaging 931 feet (284 m) in length. Little of the square has survived, but a remnant is preserved at the Wright Earthworks, which includes a short segment of one of the sides of the square and a segment of one of the parallel walls that framed a passage leading from the square to the Cherry Valley Ellipse. Another set of parallel walls led from the Newark Square to the Great Circle, and yet another led to the Octagon. The Wright Square, therefore, appears to be a key nexus for the site, as it is the only enclosure with direct links to each of the other components with the exception of the Observatory Circle, which only connects directly to the Octagon.

It is interesting to note that the perimeter of the square earthwork is precisely equal to the circumference of the Great Circle, and its area is equal to the area of the Observatory Circle. These are indications of the remarkable sophistication of the geometry incorporated into the architecture of the Newark Earthworks.

OCTAGON EARTHWORKS

The Observatory Circle (fig. 2) is a nearly perfectly circular earthwork with a diameter of 1,054 feet (321 m). It is connected by a short set of parallel walls to the Octagon, which encloses an area of 50 acres (20 hectares). The walls of the octagonal enclosure were each about 550 feet (168 m) long and from 5 to 6 feet (1.5–2 m) in height. There were gateways or openings at each corner of the Octagon varying from about 50 to 90 feet (15–27 m) in width. Each opening of the Octagon is partially blocked by a rectangular or oblong platform mound about 100 feet (30 m) long by 80 feet (24 m) wide at the base and between 5 and 6 feet (1.5–2 m) high.

The circle is named for the Observatory Mound, a large, loaf-shaped platform mound located on the outside of the circle opposite the gateway leading into the Octagon.[19] It appears to have been built across another opening into the circle consisting of a short segment of parallel walls.

The Octagon Earthworks are a remarkable testament to the architectural and engineering genius of the Hopewell culture, but scholars recently have come to realize that there is an even more remarkable aspect to the architecture. Ray Hively and Robert Horn have shown that the Hopewell builders aligned these earthworks to the cyclical movements of the moon across the sky.[20]

The Hopewell builders appear to have encoded these astronomical landmarks into the architecture of the Octagon, not to create some kind of protoscientific lunar observatory, but more likely to represent and reproduce the larger cosmos—"to provide a context in which ceremonies could occur" allowing participants to participate directly in cosmic rhythms.[21]

A RITUAL ENGINE FOR RENEWING THE WORLD

In every culture, the sky and the religious impulse are intertwined. I lie back in an open field and the sky surrounds me. . . . And when I concentrate on the stars, the planets, and their moons, I have an irresistible sense of machinery, clockwork, elegant precision working on a scale that, however lofty our aspirations, dwarfs and humbles us.—Carl Sagan, *Pale Blue Dot: A Vision of the Human Future in Space*, 1994

The remarkable degree of precision in the geometry and astronomy incorporated into the architecture of the Newark Earthworks, unequaled in the Hopewell world, together with the high degree of functional and ritual integration of the various components of the complex suggest to me that the entire complex was conceived, designed, and used as a ceremonial machine with a particular primary purpose.[22] Given the monumental scale and cosmological implications of the architecture, that purpose must have addressed a need of profound significance. Martin Byers makes a compelling case that this purpose was nothing less than world renewal.[23]

Byers posits that "the primary medium of these world renewal rituals was the mortuary sphere" and that a complicated *chaîne opératoire* of mortuary ceremonies was involved.[24] I agree with this general formulation and propose that the various components of the Newark Earthworks, such as the Great Circle with its dual circle of earth and water and the Octagon with its lunar alignments, all provided the necessary contexts for fulfilling some or all of these various intentions in an intricate series of sequential steps from the arrival of deceased persons at the site to their eventual interment in one of the Cherry Valley mortuary facilities or their transfer to other locations.

Given the centrality of the Cherry Valley Ellipse in this conception of the purpose of the Newark Earthworks, it is most unfortunate that so much of the archaeological record of this component of the ceremonial machine has been lost. Yet perhaps enough information can be salvaged or reconstructed from archival sources to test the implications of this model.

The Cherry Valley Ellipse: The Engine Powering the Newark Earthworks Machine

The Cherry Valley Ellipse takes its name from the fact that the Raccoon Creek valley originally was known as Cherry Valley because of the profusion of cherry trees encountered in the area by the earliest European American settlers. The ellipse was approximately 1,800 feet (550 m) long by 1,500 feet (450 m) wide and enclosed approximately 50 acres (20 ha).

Within the Cherry Valley Ellipse there were at least eleven discrete mounds of varying sizes and shapes, many or perhaps all of which contained interments of human remains. In addition, there was a small circular enclosure with a diameter of about 200 feet (60 m), and a subrectangular platform mound similar to the Hopewellian platform mounds at Marietta,[25] but unlike the Marietta mounds, the Newark platform apparently had no ramps. These features do not all appear on any single map of the earthworks, so it is necessary to consult several versions to gain this more comprehensive picture.[26]

The earliest description of the Cherry Valley Mound Group appeared as a footnote to Ephraim Squier and Edwin Davis's 1848 description of the Newark Earthworks. It consists of excerpts from a letter written by Israel Dille, a former mayor of Newark and an avid antiquarian. Dille refers to one of these mounds that was destroyed in 1827 by excavations for a canal lock pit. It was said to contain fourteen human skeletons, buried about 3 feet (1 m) beneath the surface. Associated with these remains were a large quantity of mica sheets, some of which measured "eight to ten inches long [20–25 cm] by four and five wide [10–13 cm], and all from half an inch to an inch thick [1–2.5 cm]."[27] Dille reported an estimate that "*fifteen or twenty bushels of this material* were thrown out to form the walls or supports of the lock."[28] Dille also mentions the discovery of a large marine shell from a separate mound located a short distance to the south.[29]

The Cherry Valley Mound was the central and dominant mound of the group. Most of the maps of this mound show it as a cluster of four or five separate mounds. According to John Wilson, a local physician and antiquarian, it was composed of four separate mounds "all joined together at their base."[30]

Wilson stated that three mounds "stood in a line north and south; the fourth one was a little east and between the two northern ones."[31]

The size of the Cherry Valley Mound was impressive. It was about 140 feet (43 m) long and 52 feet (16 m) wide at its widest point.[32] The original height of the largest part of the mound was about 20 feet (6 m), but by the time of the earliest recorded measurements it had been reduced to 16 feet (5 m).[33] These dimensions would have made it comparable to the Harness Mound (160 ft. long) and the Tremper Mound (200 ft. long).[34] The spectacular objects that were excavated from these other large Hopewell mounds provide tantalizing hints at what might have been found if the Cherry Valley Mound had been properly studied before it was obliterated. Byers suggests that Newark's mounds were relatively impoverished in mortuary offerings in comparison to other great Hopewell mounds, but the meager available evidence does not warrant a confident assessment of the extent of the wonders Newark's mounds did or did not hold.[35]

According to Wilson, the southernmost of what he perceived as four conjoined mounds "was included in the [railroad] embankment,"[36] whereas "the other three were greatly injured" by the removal of earth to add to the embankment.[37] In May of 1868, the northern and largest section of the Cherry Valley Mound "was leveled to form a site for a rolling mill."[38] Wilson evidently was present to observe much of the destruction of this part of the mound. He made some notes regarding the internal structure of the mound, collected at least a few artifacts, some of which subsequently were observed and described by Charles Whittlesey.[39]

Based on Wilson's sketchy observations, presented in a paper read at the July 4, 1868, meeting of the Licking County Pioneer, Historical and Antiquarian Society, I have identified a minimum of five strata in the mound representing apparently distinct episodes of construction, which I have designated as strata 1 through 5.

Wilson stated that approximately 4 feet (1.2 m) of material had been removed from the top of the mound prior to its destruction by the railroad. Nothing is known about the nature of this missing material.

Wilson noted a marked difference between the upper 8 feet (2.5 m) of the mound and the lower 8 feet (2.5 m). The upper portion, which I have designated stratum 5, was "composed almost entirely of black loam, which appeared in layers." The layers were discontinuous, lenticular "seams" and they "often overlapped." Usually, the layers were separated by "marks of fire." In one case, the band of charred material was 4–6 inches (10–15 cm) thick and extended "across the mound." Wilson stated that "no human or other bones" were noted in the black loam, but he observed "several sheets of mica" in these layers,[40] and Whittlesey recorded that "8 copper fluted ornaments" of very thin "rolled copper" were found at a depth of 4 feet (1.2 m). The crudely sketched cross-section of one of these copper artifacts in his notebook suggests that these may have been panpipes. They were 7 to 8 inches (18–20 cm) in length.[41] The lenticular bands of charcoal associated with a relatively high frequency of artifacts crafted from exotic materials may represent cremation burials with funerary offerings. Considering the manner in which the mound was being excavated it would not be surprising if small fragments of burned bone went unobserved or unrecognized.

Stratum 4 was a poorly described zone consisting of "layers of blue clay, then sand," which extended to a depth of about 11 feet (3.4 m) below the surface.[42] Whittlesey's notebook includes a crude illustration of a bear canine ornament from this stratum.[43]

Stratum 3 was a layer of cobblestones laid over an intensely burned surface that extended continuously across the mound at a depth of 11 feet (3.4 m) below the surface and 5 feet (1.5 m) above the base of the mound.[44]

Stratum 2 was an apparently undistinguished layer of fill between the cobble layer and the base of the mound. Finally, stratum 1 consisted of deposits extending from the original ground surface to a depth of about 4 feet (1.2 m). Wilson stated that "the whole base of this mound was of disturbed earth four or more feet below the surrounding surface." He referred specifically to several "human buryings," including "a part of the lower jaw of a human being with one tooth in it" recovered at a depth of 3 feet below the base of the mound in a pit dug for the rolling mill's flywheel.[45] This mandible fragment may have been part of the poorly preserved remains of a burial or an ornament crafted from a detached human mandible.

In addition to the submound tombs, there were numerous "post holes" at the base of the mound, most of which only penetrated to a depth of "a few feet." One, however, "on the east side was filled with fine charcoal and ashes, and extended fully four feet below the surrounding surface." Some postmolds may have originated at higher levels of the mound. One in particular appears to have originated in stratum 5. It was located "near the center" of the mound and "was observed to continue down very near to the bottom of the mound." Wilson noted that "in some places" this postmold was "filled with sand differing from the earth around it."[46]

In subsequent years, discoveries relating to the submound tombs continued to be made. The most celebrated involved the recovery of the remarkable stone figurine known variously as the Wray Figurine or the Shaman of Newark (plate 8).

The earliest report of this discovery appeared in the *Newark American* on August 12, 1881. The rolling mill had been demolished, and workmen employed in excavating foundations for new buildings on the site encountered "portions of a human skeleton and a stone image supposed to be of ancient manufacture."[47]

A few additional details concerning the discovery of the Newark figurine are included in some brief notes made by James C. Wright, who purchased the sculpture from its original discoverer, Jacob Holler: "There were perhaps other trinkets buried with him. A large conk [*sic*] shell was discovered but in the excitement was covered up and lost."[48]

The *Newark Daily Advocate* for November 22, 1899, referred to additional material that had been recovered from the remnants of the Cherry Valley Mound without mentioning any particular provenience within it. The account mentions "remains of an ancient altar," "charred bones and embers," and "some flint arrow heads, etc."[49]

In summary, a large Hopewellian big house, or renewal lodge, likely comparable in shape and size to the Harness Mound Big House originally stood on the site of the Cherry Valley Mound.[50] Numerous burials were interred in submound tombs. At least one of these burials appears to have included an offering of a marine shell, a stone figurine, and other unspecified artifacts.

At some point, the ceremonies having been concluded, the big house was dismantled, and the site was buried beneath the approximately 5 feet (1.5 m) of fill composing stratum 2. Apparently, some artifacts and possibly some burials were interred in this mound fill. The end of this stage of construction was marked by a major burning event over the entire surface of the mound, followed by the emplacement of a layer of cobbles directly on the still-hot surface. The cobblestone layer makes up stratum 3.

After an undetermined period of time, approximately 3 feet (0.9 m) of blue clay and sand was added to the mound to form what I have designated as stratum 4. Then, at least 8 feet (2.4 m) of black loam, possibly incorporating many apparent cremation burials was added to the mound over an indefinite but probably relatively long period of time. This series of accretive deposits defines stratum 5.

An unknown amount of material was removed from the mound prior to Wilson's observations. This missing material may have been a continuation of stratum 5 but may also have included other distinctive episodes of construction.

Conclusions

The Newark Earthworks complex is unprecedented in the Hopewell world in terms of its scale and the precision of both its geometry and its embedded astronomical alignments. The diversity of discrete earthwork components suggests that each enclosure fulfilled a particular function, and the integration of all these components into a unified design expressed through various geometrical, astronomical, and architectural elements suggests those separate functions all were necessary to achieve a more comprehensive ultimate purpose.

Byers argues that the culminating outcome of the *chaîne opératoire* of Hopewellian mortuary ritual may have been nothing less than the regeneration of the Earth.[51] Such a grand goal certainly would be commensurate with the magnitude of the architectural achievement embodied in the Newark Earthworks. The operation of this ceremonial machine involved synchronizing the architecture and the ceremonies to be performed therein with the deep rhythms of the cosmos.

Hopewell priests may have used the calendrical capabilities of the architecture to determine the appropriate times for gatherings. The corps of priests and shamans would have assembled the necessary regalia, while other functionaries collected the food to provide for the needs of the participants. Groups from varying distances came to the earthworks bearing offerings along with the remains of their honored dead. In what likely was a highly choreographed sequence, the bodies were carried through the ceremonial spaces undergoing a series of sequential operations, including ceremonies of mourning, spirit release, spirit adoption, and burial, each with their particular ritual and architectural requirements.

If this argument is valid, then the key elements of Newark's architecture, even if not directly associated with the Cherry Valley Ellipse, were dictated to some extent by the requirements of mortuary ceremonialism. The Great Circle's water barrier, for example, may have contained the spirits of the dead within the circle for the duration of the particular ceremonies performed therein.[52]

Finally, the creation of such a wonder of the world would have endowed this valley with an immense spiritual magnetism. According to Preston, spiritual magnetism is a consequence of "historical, geographical, social and other forces that coalesce in a sacred center."[53] Quite apart from its intended purpose, by its massive presence alone it would have become a place to experience the ineffable magic and majesty of what must have been perceived as a marvel of creative expression.

Elsewhere, I have described the Newark Earthworks as a pilgrimage center, and I still argue that this is a vital aspect of the way the site was perceived and used by its builders and the thousands of people who traveled sometimes great distances to touch the mystery, leave offerings of thanksgiving or supplication, and participate in the various ceremonies. Byers has criticized this interpretation as being somehow ethnocentric in its reliance upon analogy with Christian and Islamic religious traditions,[54] but pilgrimage is a nearly universal cultural practice.[55] And thus it is not at all implausible to suggest that it served the spiritual needs of ancient American Indians.[56]

One of the lines of evidence that has led me to the interpretation of Newark as a pilgrimage center is what I have called the Great Hopewell Road.[57] The Great Hopewell Road is the set of parallel walls that extended from the southeastern opening of the Octagon earthwork an unknown distance to the southwest.[58] The remarkably straight trajectory of this road and its length, which has been determined by multiple independent observers to have exceeded six miles (10 km), is formally similar in many respects to Mayan roads that were explicitly identified by Mayan people as sacred routes of pilgrimage.[59]

Even after the monumental Hopewellian earthworks ceased to function as ceremonial machines, they likely continued to be pilgrimage destinations. The so-called Intrusive Mound culture of the Late Woodland period possibly reflects a revitalization movement predicated on the hope that the ancient ceremonial machinery still could be activated to send deceased loved ones safely to the land of the dead. Historic Creek oral traditions refer to pilgrimages northward in the spring and autumn to "special mounds."[60] Whether the mounds in question were as far north as Ohio is an open question, but Warren K. Moorehead recorded a story heard in his youth that the pioneer Simon Kenton observed Shawnee Indians, in spite of having no traditions concerning the builders of the Fort Ancient Earthworks, nonetheless visiting "the place en route to the Ohio and [doing] homage to the spirits of its makers."[61]

Today, heritage tourism brings many people, including American Indians from diverse tribal affiliations, to Newark and the other surviving Hopewell centers. In this way, they continue to function as pilgrimage centers.

The Great Circle and Octagon Earthworks are justly celebrated as the best-preserved remnants of Hopewellian geometric earthworks. Nevertheless, the vagaries of history have given us a truncated view of the Newark complex. The aspect of Hopewell architecture that we know more about than any other at the classic sites clustered around Chillicothe—that is, the burial mounds—were among the first parts of the Newark Earthworks to be obliterated by the growth of the city of Newark and so are the least understood aspect of Newark. The only sources of information on this component of the site were scattered in forgotten newspaper articles and in other even more obscure records. This essay represents an attempt to recover as much as possible of what was lost and to develop a more complete understanding of the purpose and meaning of the entire Newark Earthworks.

What could we have learned if the various Cherry Valley mounds had been systematically studied—even if only by archaeologists in the era of Moorehead and Mills? For one thing, we could hope to have access to the remains of the people that were buried at Newark with all the rich bioarchaeological evidence of the lives of these individuals.[62]

Human remains excavated from the Newark mounds and curated at museums could have revealed biological relationships among the Newark Hopewell, the Chillicothe Hopewell, other contemporary populations, and modern American Indian tribes through a study of epigenetic traits[63] as well as DNA analysis.[64] Isotopes of strontium and oxygen sealed in their teeth could have revealed whether the people buried at Newark were pilgrims, missionaries, or migrants from other regions or were local people who had perhaps traveled to Chillicothe to obtain the necessary knowledge or franchise rights[65] from the southern groups and returned to the Raccoon Creek valley to synthesize and realize an even grander vision.[66]

Another source of invaluable clues to the extraordinary events that transpired at Newark two thousand years ago would be the panoply of material culture that would have been associated with the human remains as well as other nonmortuary offerings deposited in the Cherry Valley mounds. Funerary objects and ceremonial regalia accompanying the burials would have provided important clues to individual social status and clan affiliations and would have revealed the extent of Newark's involvement in the broader Hopewell interaction sphere. Were there mortuary deposits of obsidian and other exotica to rival what was found at the Hopewell Mound Group and Mound City? Byers assumes not,[67] but the strong similarities between the form and scale of the Harness Mound and the Cherry Valley Mound as well as the recovery of between fifteen and twenty bushels of mica plates from one of the more nondescript mounds in the Cherry Valley Ellipse[68] suggest that Byers's assumption is unwarranted.

Throughout my career of interpreting the Newark Earthworks, I have emphasized to visitors that, contrary to their preconceptions, there were no burial mounds preserved at Newark and the near total absence of burials at the surviving components indicated that mortuary ceremonialism was only one among many activities engaged in by the Hopewell people at this site. I was not actually wrong in these assertions, but I now think the implication that mortuary ceremonialism was simply one among many activities that took place at Newark is misleading. Certainly, many other social and religious activities likely were conducted within Newark's varied sacred spaces. Ultimately, however, mortuary ceremonialism was the sine qua non of the Newark Earthworks complex as it was for many of the other monumental earthwork centers.[69] The Newark Earthworks represent a kind of monumental ceremonial machine, and the facilities and rituals associated with the treatment of the honored dead, which culminated in the burial of their remains beneath the Cherry Valley mounds, constituted the engine of the machine.

Notes

I extend my sincere thanks to Dick Shiels for his efforts to organize the symposium of which this book represents a culmination and for his broader efforts on behalf of the Newark Earthworks. I thank Robert Horn, Ray Hively, N'omi Greber, Mary Borgia, Jeff Gill, and my wife, Karen Lepper, for their support and encouragement through the occasionally tumultuous process of getting from the presentation I gave at the symposium to this rather different contribution. The views expressed in this essay are those of the author and do not necessarily reflect those of the Ohio History Connection.

1. Lepper, "The Ceremonial Landscape of the Newark Earthworks and the Raccoon Creek Valley," 124.

2. Shetrone, *The Mound Builders,* 265.

3. E.g., Byers, "Is the Newark Circle-Octagon the Ohio Hopewell 'Rosetta Stone'?"; Hively and Horn, "Geometry and Astronomy in Prehistoric Ohio"; Hively and Horn, "Hopewell Cosmography at Newark and Chillicothe, Ohio"; Lepper, "An Historical Review of Archaeological Research at the Newark Earthworks"; Lepper, "The Newark Earthworks and the Geometric Enclosures of the Scioto Valley"; Lepper, "The

Archaeology of the Newark Earthworks"; Lepper, "The Great Hopewell Road and the Role of Pilgrimage in the Hopewell Interaction Sphere"; and Lepper, "The Ceremonial Landscape of the Newark Earthworks and the Raccoon Creek Valley."

4. Lepper, "The Ceremonial Landscape of the Newark Earthworks and the Raccoon Creek Valley."

5. E.g., Hall, "Ghosts, Water Barriers, Corn, and Sacred Enclosures in the Eastern Woodlands"; Lepper, "Tracking Ohio's Great Hopewell Road"; Lepper, "The Great Hopewell Road and the Role of Pilgrimage in the Hopewell Interaction Sphere"; Lepper, "The Ceremonial Landscape of the Newark Earthworks and the Raccoon Creek Valley"; Lepper, "Commentary on DeeAnne Wymer's 'Where Do (Hopewell) Research Answers Come From?'"; and Wymer, "Where Do (Hopewell) Research Answers Come From?"

6. Mason, *Inconstant Companions.*

7. The "Hopewell" designation has become a problematic label for some people. As used by archaeologists, it is merely a handy name for those indigenous groups that shared elements of material culture, architecture, and burial practices in various portions of eastern North America, especially in southern Ohio, between about 100 BCE and 400 CE. Given the area encompassed, it is likely that many more or less diverse "tribes," whose names now are unknowable, are subsumed under the label. The name comes from the "Hopewell Mound Group" in Ross County, Ohio, which is the "type site" for the Hopewell culture. The site was named, by long archaeological convention, for the owner of the land on which the site was located.

8. Lepper, *Ohio Archaeology,* 109–69.

9. Haven, "Report of the Librarian," 41.

10. E.g., Fagan, *Ancient North America,* 431; Jennings, *Prehistory of North America,* 231; Prufer, "The Hopewell Cult," 224; D. Thomas, *Exploring Native North America,* 91; and Whittlesey, "Historical and Archaeological Map of Ohio."

11. Brose and Greber, *Hopewell Archaeology.*

12. E.g., Bernhardt, "A Preliminary Survey of Middle Woodland Prehistory in Licking County, Ohio."

13. Lepper, "An Historical Review of Archaeological Research at the Newark Earthworks"; Lepper, "The Newark Earthworks and the Geometric Enclosures of the Scioto Valley"; Lepper, "The Archaeology of the Newark Earthworks"; Lepper, "The Newark Earthworks"; Lepper, "The Great Hopewell Road and the Role of Pilgrimage in the Hopewell Interaction Sphere"; Lepper, "The Ceremonial Landscape of the Newark Earthworks and the Raccoon Creek Valley"; and Lepper and Yerkes, "Hopewellian Occupations at the Northern Periphery of the Newark Earthworks."

14. C. Thomas, *Report on the Mound Explorations of the Bureau of Ethnology,* 462.

15. Lepper, "The Newark Earthworks and the Geometric Enclosures of the Scioto Valley," 233.

16. Lepper, "The Archaeology of the Newark Earthworks," 126.

17. Lepper, "An Historical Review of Archaeological Research at the Newark Earthworks."

18. Atwater, "Description of the Antiquities Discovered in the State of Ohio and Other Western States," 127.

19. Squier and Davis, *Ancient Monuments of the Mississippi Valley,* 70.

20. Hively and Horn, "Geometry and Astronomy in Prehistoric Ohio"; Hively and Horn, "Hopewell Cosmography at Newark and Chillicothe, Ohio"; and Hively and Horn in this volume.

21. Deloria, "Power and Place Equal Personality," 26.

22. Lepper, "The Ceremonial Landscape of the Newark Earthworks and the Raccoon Creek Valley."

23. Byers, *Sacred Games, Death, and Renewal in the Ancient Eastern Woodlands.* See also Romain, "Hopewell Geometric Enclosures"; and Carr, "Social and Ritual Organization," 632–34.

24. Byers, *Sacred Games, Death, and Renewal in the Ancient Eastern Woodlands,* 284–85.

25. Squier and Davis, *Ancient Monuments of the Mississippi Valley,* 70 and plate 25, facing p. 67.

26. Salisbury and Salisbury, "Accurate Surveys and Descriptions of the Ancient Earthworks at Newark, Ohio"; Squier and Davis, *Ancient Monuments of the Mississippi Valley;* and Wyrick, "Ancient Works near Newark, Licking County, Ohio."

27. Squier and Davis, *Ancient Monuments of the Mississippi Valley,* 72.

28. Ibid., emphasis in original.

29. Ibid.

30. Wilson, "Mounds near Newark," 69.

31. Ibid.

32. Salisbury and Salisbury, "Accurate Surveys and Descriptions of the Ancient Earthworks at Newark, Ohio."

33. Wilson, "Mounds near Newark," 69.

34. W. Mills, "Explorations of the Seip Mound"; and W. Mills, "Exploration of the Tremper Mound."

35. Byers, *Sacred Games, Death, and Renewal in the Ancient Eastern Woodlands,* 458.

36. This note leaves open the intriguing possibility that a significant portion of the Cherry Valley Mound may be preserved intact beneath the railroad embankment. The irony of such a possibility is striking, but the potential for addressing our profound lack of data for this component of the Newark Earthworks should be kept in mind as opportunities for investigation present themselves in the future.

37. Wilson, "Mounds near Newark," 69.

38. Ibid.

39. Whittlesey, "Field book, July 1, 1868."

40. Wilson, "Mounds near Newark," 69.

41. Whittlesey, "Field book, July 1, 1868," 41.

42. Wilson, "Mounds near Newark," 69.

43. Whittlesey, "Field book, July 1, 1868," 42.

44. Wilson, "Mounds near Newark," 69.

45. Ibid.

46. Ibid.

47. Dragoo and Wray, "Hopewell Figurine Rediscovered," 197.

48. Ibid., 196–97.

49. "Bones of an Ancient People Found by Henry Barrett Short Distance below Earth's Surface near His Home in West Newark on Tuesday," *Newark Daily Advocate*, Nov. 22, 1899, 4.

50. Greber, "Recent Excavations at the Edwin Harness Mound, Liberty Works, Ross County, Ohio."

51. Byers, *Sacred Games, Death, and Renewal in the Ancient Eastern Woodlands.*

52. Hall, "Ghosts, Water Barriers, Corn, and Sacred Enclosures in the Eastern Woodlands."

53. Preston, "Spiritual Magnetism," 33.

54. Byers, *Sacred Games, Death, and Renewal in the Ancient Eastern Woodlands*, 413.

55. Morinis, *Sacred Journeys;* and E. Turner, "Pilgrimage."

56. E.g., Malville and Malville, "Pilgrimage and Periodic Festivals as Processes of Social Integration in Chaco Canyon"; Joy McCorriston, *Pilgrimage and Household in the Ancient Near East* (Cambridge: Cambridge University Press, 2011); Silverman, "The Early Nasca Pilgrimage Center of Cahuachi and the Nazca Lines"; and Silverman in this volume.

57. Lepper, "Tracking Ohio's Great Hopewell Road."

58. Lepper, "The Great Hopewell Road and the Role of Pilgrimage in the Hopewell Interaction Sphere."

59. David A. Freidel and Jeremy A. Sabloff, *Cozumel, Late Maya Settlement Patterns* (Orlando, FL: Academic, 1984), 82; and Stanzione, "Walking Is Knowing."

60. Chaudhuri and Chaudhuri, *A Sacred Path*, 9.

61. Moorehead, "Fort Ancient," 31.

62. E.g., Agarwal and Glencross, *Social Bioarchaeology;* and Buikstra and Beck, *Bioarchaeology.*

63. Pennefather-O'Brien, "Biological Affinities among Middle Woodland Populations Associated with the Hopewell Horizon."

64. Bolnick, "The Genetic Prehistory of Eastern North America"; Shook, Schultz, and Smith, "Using Ancient mtDNA to Reconstruct the Population History of Northeastern North America"; L. Mills, "Mitochondrial DNA Analysis of the Ohio Hopewell of the Hopewell Mound Group"; and L. Mills, "Mitochondrial DNA Analysis of the Ohio Hopewell of the Hopewell Mound Group."

65. Byers, *Sacred Games, Death, and Renewal in the Ancient Eastern Woodlands.*

66. Ironically, the 2010 Native American Graves Protection and Repatriation Act "final rule" for the disposition of culturally unidentifiable human remains, which mandates the disposition of these remains to Native American claimants with no demonstrable biological or cultural affiliation to them, may result in the loss of the rich biohistorical archives preserved in the bones of people excavated from the classic

Hopewell sites, which are, for now, curated at museums around the country. And this erasure of history will be due, not just to the callous vandalism of European Americans who did not value the American Indian heritage they wantonly destroyed at Newark, but to the political maneuverings of a few activist bureaucrats in the Department of the Interior who believe that scientific efforts to listen to and share these faint voices made of bone, to use the evocative phrase of the Choctaw archaeologist Dorothy Lippert, "In Front of the Mirror," are a combination of colonialist exploitation and ghoulish perversity. See Seidemann, "Altered Meanings," for an analysis of these deeply flawed regulations.

67. Byers, *Sacred Games, Death, and Renewal in the Ancient Eastern Woodlands,* 458.

68. Squier and Davis, *Ancient Monuments of the Mississippi Valley,* 72.

69. Byers, *Sacred Games, Death, and Renewal in the Ancient Eastern Woodlands,* 284–85.

RAY HIVELY & ROBERT HORN

The Newark Earthworks

A Grand Unification of Earth, Sky, and Mind

O, swear not by the moon, the inconstant moon
That monthly changes in her circled orb,
Lest that thy love prove likewise variable.
—*Romeo and Juliet*

ROMEO INVOKES "yonder blessed moon" to seal his pledge of love for Juliet, and Juliet reminds him that the blessed moon is also fickle, a poor sponsor for a constant love. Today it is difficult to fathom either Romeo's awe or Juliet's doubt.

The famous NASA image of the Apollo 11 moon landing that depicts a half-Earth visible in the background as the lunar module lifts off from the moon to rendezvous with the Apollo command module on July 21, 1969, is an appropriate place to begin thinking about the blessed but inconstant moon and its place in the story of the Newark Earthworks. Even today the sight of the moon can inspire awe. But this image of the Apollo 11 lander lifting off from the moon as the blue earth rises on the horizon has another fascination. It calls attention to the remarkable human achievement of space exploration. These two—awe still inspired by the sight of the moon and fascination with the strange match between human wit and the perplexing cosmos—may bring us closer to Newark than we might imagine.

We cannot juxtapose Newark and NASA without reviving the valid criticism that there are vast differences between modern celestial mechanics and prehistoric observation of the heavens, however practiced and disciplined it may have been. Brad Lepper has suggested Newark is something like "a pre-Columbian Large Hadron Collider—a vast machine, or device, designed and built to unleash primordial forces."[1] The gap between CERN and Newark remains great. Yet they may share the grand assumption that there is a fundamental affinity between our human aspiration to comprehend the world and the world we try to comprehend. The Large Hadron Collider plays a fundamental role in the continuing effort at a Grand Unified Theory of the forces at work in the cosmos. It may not be folly to guess that the planners and builders of Newark supposed they had in one vast geometric design mirrored the fundamental forces of their world. They had mapped the annual travel of the sun and found constancy in the odyssey of the inconstant moon, a grand unification of earth, sky, and mind. They appear to have had an answer for Juliet. Romeo may, after all, "swear by the moon."

Hopewell Background

Over the past two centuries there has been a gradual accumulation of evidence showing that during the centuries between 100 BCE and 500 CE the American Indian peoples of the eastern American Woodlands shaped a remarkable culture that archaeologists call Hopewell. Many believe that the core of this cultural explosion was in Central and South Central Ohio. Remnants are still visible on the ground in the form of massive geometrical earthworks and in museum displays of sophisticated textiles, ingenious artwork, pottery, and exotic raw materials. But analysis of the evidence, which traces the origins of Hopewell or Middle Woodland culture to Adena, or Early Woodland, culture, has not resolved the essential mystery of the geometric earthworks. Archaeologists have suggested variously that they were ceremonial centers; corporate centers encompassing periodic social, civic, and trade exchanges, burial rituals, sacred games; goals for pilgrimage. Still there has been no widespread consensus about the ultimate motivation for the spectacular geometric accuracy and scale of the earthworks. Nowhere is this puzzle more provocative, or answers more within reach, than at Newark.

Our work at Newark began in 1975 as a field exercise in data collection and analysis for an undergraduate interdisciplinary course at Earlham College. The scope of the course included the cosmology and the astronomical knowledge of prehistoric and ancient cultures. Our aim at that point was to teach students the rudiments of surveying by mapping the remnants of the Newark Earthworks. We did not expect to find any particular geometrical or astronomical pattern. Indeed, given the difficulty of showing that any such pattern was intentional rather than fortuitous, we doubted any persuasive hypothesis regarding design of the earthworks could be formed.

Much to our surprise, our continued analysis of the Newark Earthworks over the past thirty years has revealed repetitive patterns of earthwork and topographical features oriented or aligned to the extreme rise and set points of both the sun and the moon on the horizon. These alignments combined with the massive scale, geometrical symmetry, and regularity of the earthen enclosures suggest that the Newark Earthworks were built to record, celebrate, and connect with the celestial actors or large-scale forces that appear to govern relations among earth, sky, and the human mind.

Geometry and Scale of the Site

The major geometric enclosures associated with the Newark Earthworks are shown in figure 1. These figures include a Circle-Octagon joined by an avenue, a second larger circle known as the Great Circle, two squares (the Wright Square and the Salisbury Square), and an oval earthwork enclosing some forty-nine acres that we refer to as the Cherry Valley Ellipse. The first notable feature of the Circle-Octagon combination is the geometrical precision and plan involved in their design and construction. The scale of both Observatory Circle and the Octagon is based on a common unit of length (which we call the Observatory Circle diameter, or OCD) of 1,054 feet. The shape of the Octagon conforms to a simple geometrical construction involving circles (centered on the corners of a square) with radius equal to the diagonal of the length of the associated square. The square has a side equal (to within the errors of measurement) to the OCD. This plan is illustrated in figure 2. The intentional nature of this design is supported by the fact that a similar design was employed in the only other circle-octagon earthwork constructed by the Hopewell, the so-called High Bank Works located at Chillicothe, Ohio.[2]

The importance of the OCD as a geometrical length in Hopewell earthwork design is revealed by the fact that the High Bank Circle has (within the errors of measurement) the same diameter as the Newark Observatory Circle. The same multiple of this distance (6 OCDs) separates the centers of the Observatory Circle and Great Circle and also the centers of the Octagon and Wright Square. Other geometrical regularities can be found as well. The Observatory Circle encloses an area equal to that of the Wright Square within ~0.6 percent. The perimeters of the Wright Square and the Great Circle correspond within less than .02 percent. These together with repetitive and accurate, though often not equally precise, geometrical figures in the valleys of Paint Creek and the Scioto River (near Chillicothe) show that the Hopewell were experimenting on a monumental scale with geometrically regular shapes and dimensions.

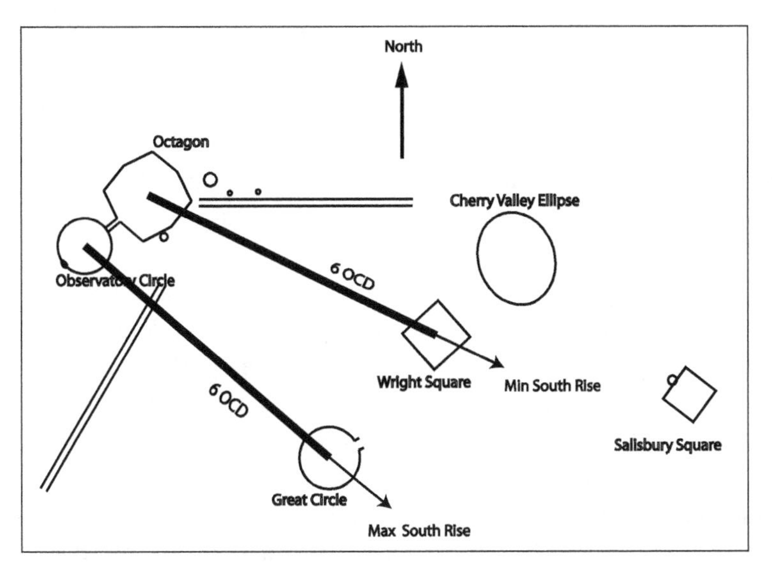

FIGURE 1. Schematic map of the Newark Earthworks. The only surviving major components are the Observatory Circle, the Octagon, the Great Circle, and a small part of the corner of the Wright Square. The scale of the map is established by noting that the Observatory Circle diameter (OCD) is 1,054 ft. It should be noted that the distances between the two sets of major figures is the same (6 OCDs). Lines between the figure centers align with the southern extreme moonsets at major and minor standstills.

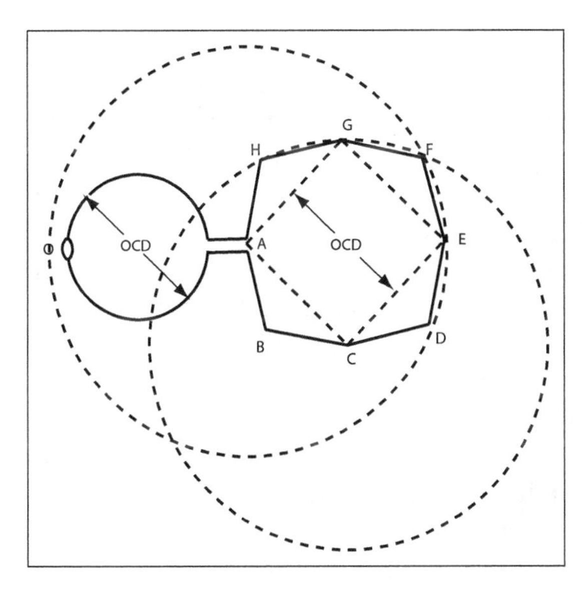

FIGURE 2. The geometrical plan that relates the Octagon to the Observatory Circle. The plan suggests that the Octagon was laid out and constructed using a simple but elegant construction. The Octagon begins with a square ACEG with a side equal to the Observatory Circle diameter. Then circles (or arcs) are drawn around each square vertex with a radius of the diagonal of the square. The intersection of these vertex-centered circles then determines the Octagon vertices B, D, F, and H.

Any assessment of the intentionality of a geometrical or astronomical design for the earthworks must consider objective quantitative measures of the accuracy and precision with which the earthworks embody such a plan. Accuracy here refers to a measure of how closely the earthwork structure conforms to the proposed plan. Precision refers to a measure of how well the earthwork structure can actually be determined from measurements at the site.

The accuracy and precision of the Newark Earthworks and other comparable geometrical sites were first demonstrated by meticulous surveys undertaken by the Smithsonian Institution. As reported in the *Twelfth Annual Report of the Smithsonian Bureau of Ethnology*, "In the summer of 1887 a resurvey of some of the more important ancient works described and figured by Squier and Davis was made in order to determine the accuracy of the measurements and figures of these authors."[3] The accurate and precise geometry of the Newark Earthworks came as a major surprise to James Middleton and Gerard Fowke, the surveyors who did the Bureau of Ethnology resurvey. Fowke described the care with which the measurements were made and their astonishment at the result: "Greater care was taken in getting bearings and distances than is usually employed in railway or canal surveys. Middleton and I, who did the work, stand by our figures, and with all the more reason too, that in some cases they completely upset our antecedent ideas and opinions."[4]

For example, the Observatory Circle wall is constructed so accurately that its surveyed circumference of 3,309 feet is within 2 feet of the ideal circumference of a perfect circle with the measured diameter of 1,054 feet (the OCD). The angles formed by the diagonals through the Octagon vertices BF and DH (shown in fig. 2) differ by only 10 minutes of arc from a right angle; similarly, the Octagon diagonals AD and EG show a comparable accuracy.

The monumental scale of the Newark Earthworks is certainly one of its most remarkable features and suggests a purpose larger than that of an enclosure whose symmetry and structure could be utilized or even appreciated on the ground. The main features of the site covered four square miles, and their construction required the placement of seven million cubic feet of earth. The amount of labor required to construct this complex would have been comparable to the several hundred thousand person-days estimated for structures such as Stonehenge and Avebury.[5]

The precise symmetry and construction of the Newark Earthworks become even more remarkable when one considers that neither of these features can be seen or appreciated by casual observation from the ground. The scale of the figures is so grand that no observer from ground level or even on the surrounding hills can see the geometric enclosures in a manner that makes their accuracy and precision readily apparent. Only an observer situated high above in the sky can see this, a notable fact in formulating a probable interpretation of the site.[6]

Archaeoastronomy at Newark

A possible and plausible answer for the design, location, and scale of the Newark Earthworks comes from the field of archaeoastronomy, the study of the artifacts of prehistoric and more recent societies for evidence of astronomical knowledge. Typically such an analysis involves a search for the systematic alignment of linear architectural features to important periodic celestial rise and set events on the local horizon. In the absence of ethnographic evidence indicating a society's interest in and knowledge of specific astronomical events, the primary challenge in such studies is to demonstrate that astronomical alignments found in the architecture were deliberate and not simply fortuitous.

At present the only methodology for establishing intentionality in the astronomical alignments of Hopewell earthworks involves an attempt to find repetitive patterns of alignments to the same astronomical phenomena. If such consistent patterns can be found, statistical methods (such as Monte Carlo simulations of earthwork shapes) can be utilized to determine whether such patterns are likely by chance alone. Unfortunately, there is no established consensus about the procedures to be used in establishing repetitive patterns or determining an unambiguous way of computing the probability that such patterns are the result of chance. However, the Newark Earthworks present an almost unique opportunity among prehistoric sites for developing a methodology to lead to confident conclusions. Specifically, meaningful statistical analyses require a significant number of intentional linear structures that define azimuthal alignments to specific points on the horizon to within a fraction of a degree. No other prehistoric site meets these criteria more impressively than the Newark Earthworks.

It has been evident since the Smithsonian resurveys of Newark in 1887 that the geometrical features at Newark were designed with careful attention to their accuracy. Until our surveys, beginning in 1975, there had been no comparable attention to their orientation.[7] In fact, it was the remarkable geometric accuracy of the Circle-Octagon that forced us to think more seriously about orientation. The level of accuracy of the Octagon design is so high that it is possible both to discover departures from the achievable accuracy and to suspect they are not accidental. Might these departures from exact symmetry be explained by an attempt to orient the walls in question? What structures or properties of the natural world could have been used to provide such reference directions? There would seem to be only two possibilities: (1) directions defined by topographical features in the local terrain and (2) directions defined by astronomical phenomena, most likely the locations of the rise and set points of the sun and moon on the local horizon. Studies of the monumental architecture of ancient and prehistoric societies around the globe indicate that both earth and sky were frequently used to orient and locate major constructions. Our study of the geometry of the earthwork figures and their relationship to the surrounding terrain has revealed substantial evidence that the Newark site was designed and located to achieve an integrated harmony with the features of the local terrain and directions established by extreme rise and set points of the moon.

Analysis of the relation between the internal geometry of the site and the surrounding terrain suggests that for many generations astronomical observations were made from specific identifiable hilltops surrounding Cherry Valley. The hilltops would have been chosen in part because they were connected by sightlines that marked the sunsets and sunrises that occurred at the winter and summer solstices. The earthen geometric figures constructed in the valley below these prominent observation points were then apparently located so that lines between the designated hilltops and the centers of major figures marked the extreme north and south moonsets and moonrises. If the case for deliberate design and planning can be firmly established, the result will offer insight into the mentality and worldview of the Hopewell not accessible in any other fashion. An advance in understanding how the builders sought to bring together their experiments in geometry, their grasp of their terrain, and periodic events at the margin of earth and sky would undoubtedly help us infer more reliably some of the social, political, and ceremonial practices that structured and gave meaning to daily life.[8]

Astronomical Events Marked at the Newark Earthworks

Among the wide variety of celestial phenomena that prehistoric observers recorded, the most vivid were the cycles of the sun and moon. Few if any societies until recently have lacked a sense of reverence, awe, and curiosity about the periodic movements of the sun, the moon, the stars, and the "wandering stars" we call planets. Observers at the latitude of Newark would have noticed that the regular annual movements of the sun were related roughly to the cyclic passage of the seasons. The most widely recorded aspect of the sun's motion is the oscillation of the rise and set points of the sun between a northern extreme in summer and a southern extreme in winter. At Newark the northern extreme rise and set points occur about 30° north of east and west, respectively, at the summer solstice. The southern extremes occur about 30° south of east and west at the winter solstice.

Even casual observers would note that the rise and set points of the moon undergo a similar periodic movement between northern and southern extremes. The primary difference with the moon, however, is that the period of the movement between northern and southern extremes is much more rapid than that of the sun. The moon completes its north-to-south-and-back excursion in only 27.3 days. A more careful and persistent observer would note over time that the precise location of the lunar extreme rise and set points oscillates much more slowly between maximum extremes and minimum extremes, spanning a period of 18.6 years. The directions to the extreme rise and set points of the sun and moon are illustrated in figure 3.

The rate of movement of the moon's rise and set points varies dramatically near the northern and southern extremes. When near a monthly extreme point, the moon's rise point will vary by no more than 0.5° over three days. Near the midpoint of the monthly cycle the moon's rise point can change by as much as 7.0° per day. In a similar fashion the location of the extreme north and south rise and set points changes much more rapidly during the middle of the 18.6-year cycle than it does when the cycle is at the maximum and minimum extremes. During the middle of the cycle, when the lunar extremes are near the solstice rise and set points, the position of the extreme lunar rise and set points changes by about 3.0° per year. In contrast, when the lunar extreme rise and set points are near a maximum or minimum value (as shown in fig. 4), the extreme rise and set points remain fixed in position within 0.5° for a period of some three years. During this three-year period the rise and set points of the moon appear to linger at the extreme points with little change. Hence when the moon is at the maximum of this cycle, it is said to be at a major standstill. When the moon is at a minimum of the cycle, it is said to be at a minor standstill.

There is some evidence for the alignment of prehistoric structures to the

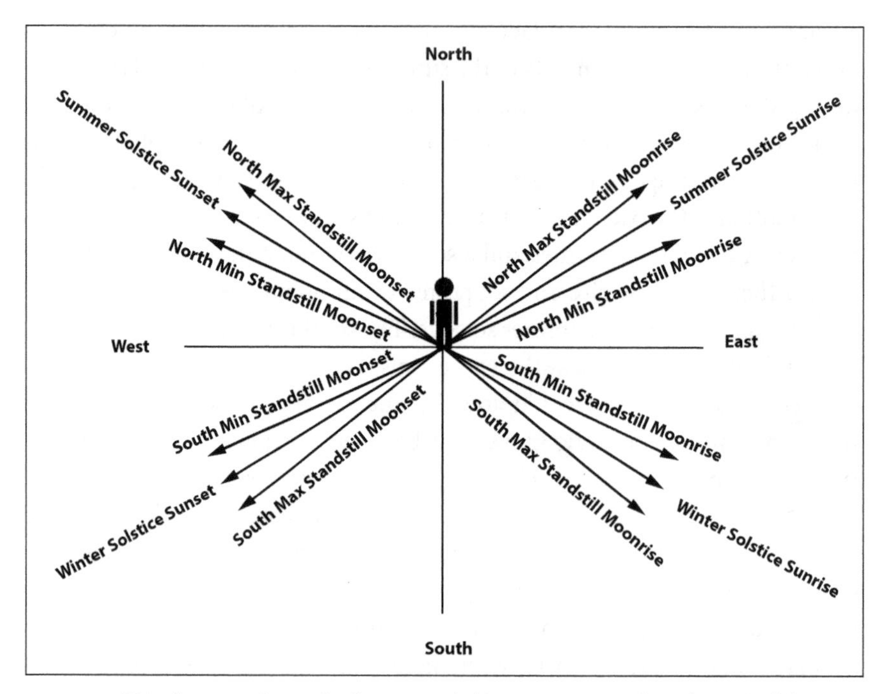

FIGURE 3. This diagram shows the directions (relative to true north at the top) of the northern and southern extreme rise and set points for both the sun and the moon. The solar extreme rise and set points are attained on the days of the summer and winter solstices. The lunar extremes vary between a maximum value at major standstill and minimum value at the minor standstill. The major standstill is achieved every 18.6 years.

lunar standstills in ancient Britain and Ireland, Mesoamerica, and the American Southwest.[9] However, no clear ethnographic evidence indicating that these cultures observed the 18.6-year cycle has survived. As a result the issue of the existence and importance of an ethnographic precedent for such knowledge remains a topic of continuing conjecture, debate, and controversy. Nevertheless, it should be noted that the 18.6-year motion of lunar extreme rise and set points is not a trivial or subtle effect if one is concerned with following and understanding the moon. The swing between maximum and minimum extremes at Newark is 14°, some twenty-eight lunar diameters. This shift would be highlighted in dramatic fashion viewed against a horizon background punctuated with hills and valleys. Such a punctuated horizon would invite and facilitate both observing and recording this motion of extreme lunar rise and set points. This is the case at Newark, and even more

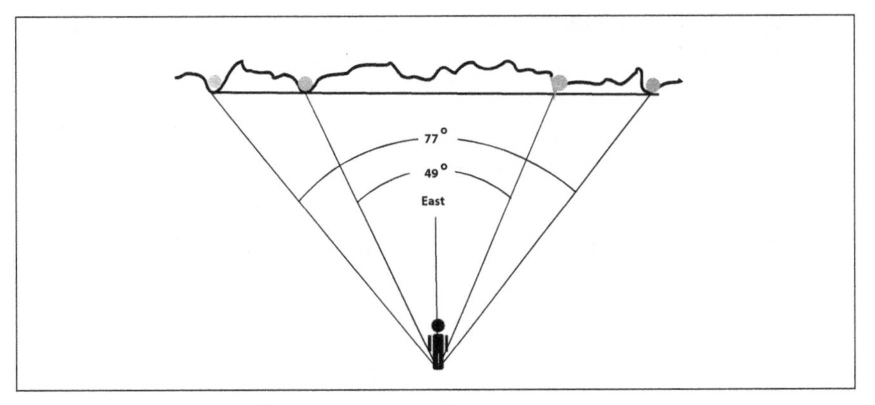

FIGURE 4. This diagram shows a hypothetical observer looking toward the east at the latitude of Newark to observe the northern and southern extreme moonrises at the major and minor standstills. The full moon is seen rising relative to prominent hills and valley on the local horizon. Notice the large difference in the monthly angular motion of the moonset: a swing of 77° at major standstill compared to only 49° at the minor standstill.

remarkably so among the many related Hopewell geometric sites in the valleys of the Scioto River and Paint Creek, as we have shown in recent work.[10]

Observing and recording the 18.6-year lunar cycle would perhaps be much more difficult if the motions occurred against a background of distant, relatively featureless horizons. The situation at Newark is quite different. Many local hills, streams, valleys, and distinct topographical features that make the motion of the moon from major to minor standstill quite conspicuous would have been well known to the inhabitants. Indeed we believe that evidence supports the notion that the location and orientation of the Newark Earthworks were chosen in part to record and celebrate the correspondences to be seen between the local topography and the motions of the moon. Skeptics concerning prehistoric knowledge of lunar standstills generally make two points: (1) observation of the lunar standstills has no pragmatic or survival value, since it is not correlated with agricultural seasons, the prediction of tides, or other phenomena of immediate importance; and (2) establishing and recording the lunar extreme rise and set points require a long period of multigenerational observation and persistence. Many lunar events would not be visible because they occur in daylight, when the moon is near new phase, or when the weather is unfavorable.

With respect to practical reasons for observing the lunar standstills, the most practical "reason" we can imagine is a response rather than a reason: reverence, awe, recognition of mystery, and fear of the cosmic power represented by the moon. Something of this survives, if only as a literary trope, in Romeo's appeal to the "blessed moon" and in Juliet's wariness of the moon's caprice.

If we need a reminder of why prescientific societies might have had a passionate concern for observing the moon, consider the *Natural History* of the Roman polymath Pliny the Elder (29–79 CE), an Old World contemporary of the Hopewell:

> But the moon, which is the last of the stars, and the one most connected with the earth, the remedy provided by nature for darkness, excels all the others [celestial bodies] in its admirable qualities. By the variety of appearances which it assumes, it puzzles the observers, mortified that they should be the most ignorant concerning that star which is the nearest to them. She is always either waxing or waning; sometimes her disc is curved into horns, sometimes it is divided into two equal portions, and at other times it is swelled out into a full orb; sometimes she appears spotted and suddenly becomes very bright; she appears very large with her full orb and suddenly becomes invisible; now continuing during all the night, now rising late, and now aiding the light of the sun during a part of the day; becoming eclipsed and yet being visible while she is eclipsed; concealing herself at the end of the month and yet not supposed to be eclipsed.
>
> Sometimes she is low down, sometimes she is high up, and that not according to one uniform course, being at one time raised up to the heavens, at other times almost contiguous to the mountains; now elevated in the north now depressed in the south; all which circumstances having been noticed by Endymion, a report was spread about that he was in love with the moon.
>
> We are not indeed sufficiently grateful to those, who, with so much labor and care, have enlightened us with this light; while, so diseased is the human mind, that we take pleasure in writing the annals of blood and slaughter, in order that the crimes of men may be made known to those who are ignorant to the constitution of the world itself.[11]

Here we find ample "reasons" for attending to the moon. Pliny does more. He alludes to the ancestors, like the legendary Endymion, who brought these lunar phenomena to light. And he shames his contemporaries for ignoring them. This is not to say that Pliny's text or any other text we know unambiguously records the long lunar cycle. What Pliny does report are lunar phenomena that in the right circumstances could engender awe, wonder, and the attentive observation that could lead to discovery of the long lunar cycle. Indeed one would have to establish the nature of the lunar cycles in order to determine whether such knowledge had practical predictive power.

With respect to the difficulty of measuring the standstill cycle, it is undoubtedly the case that marking the extreme points with a typical accuracy of 0.5° would require strong motivation. It would demand the persistence to make regular and disciplined observations and to average the results of observations extending over many human generations. It would also require some way of transmitting the knowledge from one generation to the next. This would require a certain minimum of social, political, and ritual stability. However, these requirements of curiosity, intelligence, persistence, and the ability to transmit information from one generation to the next are precisely the traits suggested by the construction of the earthworks themselves. That was not a feat likely to be confined to a single generation.

Analysis of Internal Geometry for Astronomical Alignment

An investigation into the possibility of astronomical alignments within the Newark Earthworks has two major prerequisites: (1) an accurate survey of the azimuths (directions relative to true north) of major linear features in the earthworks before the structure was significantly altered by agriculture and by undocumented or poorly documented construction at the site and (2) an accurate representation of the physical relation of each of these elements of the site to the others. Our previous archival work and surveys of the site have established that only the aforementioned survey by James Middleton and Gerard Fowke for the Smithsonian Institution's Bureau of Ethnology in 1887 meets these prerequisites. The 1887 survey, published in the *Twelfth Annual Report of the Bureau of Ethnology*, includes detailed drawings and data tables for surveys of the Circle-Octagon, the Great Circle, and the Wright Square.[12] It also maps the ensemble formed by these figures, including the parallels that connect the Octagon and the Great Circle. Other surveys and maps, including the celebrated map of Charles Whittlesey, published in Squier and Davis's pioneering *Ancient Monuments of the Mississippi Valley*, give valuable information about the Newark site but do not satisfy these criteria. Consequently our study of the orientation and internal geometry of the Newark

Earthworks relies solely on the Middleton survey, on our confirmation of that survey through archival study, and our own surveys and work at the site.[13]

The first and most important fact emerging from a survey of the Circle-Octagon combination is that the 2,839-foot symmetry axis through the two figures and the connecting avenue points with subdegree accuracy (less than 1.0°) toward the northern extreme rise point of the moon at the major standstill. This alignment is shown in figures 14 and 15 and in plate 1. Taken by itself, there is no reason to believe that this alignment is anything but a random accident. It would take a pattern of repeated alignments to lunar standstills to give the intentionality of this alignment any credibility. Remarkably, we do find as many repetitive alignments to lunar standstills within the Octagon geometry as would be allowed by the constraints of a symmetrical equilateral structure.

If we are to believe that the Hopewell designers went to great effort to align the axis of the structure to the northern extreme moonrise at the major standstill, then we must think it probable that they would give some attention to its southern counterpart. Our expectations in this regard are met when we find that a sightline along the Octagon wall AB (see fig. 5), some 570 feet in length and 5.5 feet high, points directly (again with subdegree accuracy) to the southern extreme moonrise at the major standstill. It is also notable that this structure achieves another standstill alignment of comparable accuracy along the Octagon wall CB to the southern lunar extreme set point at the minor standstill.

If the Octagon had been constructed perfectly in conformity with the geometrical plan shown in figure 2, an alignment along Octagon wall FE would duplicate the alignment along wall AB, and an alignment along wall FG would parallel the alignment along wall CB. Here we encounter a most significant property of the Octagon: its largest deviation from its nearly perfect symmetry. The most notable "error" in the construction of the Octagon is in the placement of vertex F. This vertex alone has been built in a location some 20 feet closer to the center of the figure than its ideal position. The result of this "distortion" of vertex F is to create two additional standstill alignments to rise and set events when the walls are used as sightlines in the reverse directions EF and GF. By pushing vertex F inward by a significant amount, sightlines have been produced to the northern extreme moonset at major standstill along wall EF and to the northern extreme moonrise at minor standstill along wall GF.

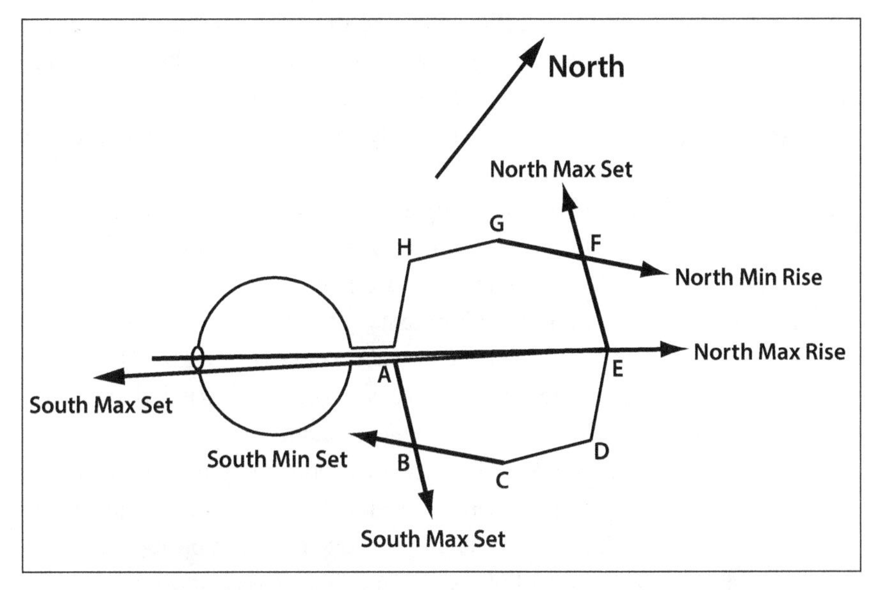

FIGURE 5. The five major lunar standstill alignments found in the internal geometry of the Newark Circle-Octagon. The six alignments have an impressive average accuracy of about 0.5°. No possible octagon of significantly different orientation or shape could align with more standstills than we find here.

The immediate question that presents itself, then, is this: does the position of vertex F constitute a random construction error, or is it evidence of a deliberate distortion of the geometry to achieve intentional alignments to additional standstill events? We can evaluate the likely standard of accuracy for the construction of parallel walls by looking at all the pairs of walls that would be parallel in ideal geometrical figures at the site. There are six such pairs of parallel walls, four pairs in the Octagon and two pairs in the Wright Square. When we look at the four wall pairs that are not astronomically aligned, their average deviation from true parallelism is 0.4°. In contrast, the pairs of Octagon walls aligned to the lunar standstills (BC-FG and EF-AB) differ from true parallelism by 1.4° and 1.8°, respectively. Thus the four nonastronomical pairs have an angular divergence that is three to four times smaller than the astronomically aligned pairs. This difference amounts to three lunar diameters, and we believe that, given the entire context, the hypothesis of deliberate distortion is more probable than that of random error.

Altogether, then, the Octagon incorporates five accurate alignments to five of the eight lunar standstill events. The pressing question that cannot be answered reliably through intuition alone is, What is the likelihood that a randomly constructed octagon would produce five or more alignments with comparable or greater accuracy? A randomly constructed octagon is one with a randomly chosen orientation and a single randomly chosen vertex angle between 90° and 180°. With one vertex angle chosen, octagonal symmetry determines the remaining angles. A randomly chosen octagon has only these two degrees of freedom: orientation and vertex angle. This makes it much more difficult to encode more than two alignments without distortion of symmetry.

The only means of determining the probability of five alignments (to either sun or moon) of comparable accuracy in a randomly constructed octagon is to perform a Monte Carlo analysis, in which a computer algorithm actually counts the number of accurate alignments on sides or symmetry axes produced by large numbers of randomly constructed octagons. We have conducted such a Monte Carlo analysis of some one hundred billion randomly chosen octagons.[14] A variety of plausible models and assumptions were considered. The analysis firmly yielded the conclusion that the probability of accidental alignments producing these data was on the order of one in a million. The study also showed that no symmetrical, equilateral octagon with a significantly different shape or orientation could possibly have captured as many standstill alignments as the Newark Octagon. The Newark Octagon appears to have been optimally designed for that purpose.

We have argued earlier that the general shape of the Octagon was determined by geometrical experimentation that resulted in the elegant plan shown in figure 2. Given that geometrical plan, some distortion was necessary to achieve the five standstill alignments. We believe the Hopewell planners were energized and fascinated by the discovery that a single structure of their own design could simultaneously encode important geometrical and astronomical regularities. This result could only have been achieved after a long period of careful experimentation with geometrical figures and observations of astronomical events.

This is not the place to discuss the sociopolitical aspects of this disciplined observation and experimentation. Martin Byers has argued recently that this was the work of dispersed nonkinship groups who actively sought, used, and maintained this kind of esoteric knowledge.[15] At the same time, it seems to us reasonable to suppose that such an elite effort could only have succeeded if a significant portion of the population shared enchantment with the moon and fascination with the power of those who could reliably anticipate it. To encode that schedule in an earthen symbol with unprecedented scale and accuracy would have been a stunning achievement.

Evidence of Lunar Astronomy beyond the Octagon

If the Hopewell deliberately conceived the Octagon structure as a simultaneous encoding of geometrical and astronomical regularity, we should expect the priority of this interest to appear as well in the additional degrees of freedom afforded by other enclosures in the earthwork complex at Newark. These added alignments are shown in figure 6 and figure 7.[16]

Evidence of Hilltop Observing Stations

Observing and recording the location of the lunar standstill directions with subdegree accuracy would require accumulating astronomical information extending over many human generations. How could such observations be most effectively carried out? A consideration of that question is involved in a resolution of a problematic aspect of our account of the Octagon "alignments." One of the most puzzling and troubling features of our original interpretation was the relative inaccuracy of the alignment along wall EF.[17] This alignment to the northern extreme moonset at the major standstill would have an impressive accuracy of 0.2° if viewed against a zero-altitude or plane horizon. However, the actual moonset at the standstill as seen from vertex E is displaced by 1.7° from the wall line. The accuracy of the other four standstill alignments within the Octagon geometry averages 0.25° when viewed against the local (and small) horizon altitudes. Why should the builders have tolerated this "error" in the alignment of wall EF, especially when the "error" is created by a local hill located only 5,905 feet (1.8 km) to the northwest of wall EF? This local hill creates a horizon altitude of 1.7°, which shifts the direction of moonset by 1.7° away from the wall alignment. The anomalously large "error" is some six times greater than the average error of the remaining four alignments.

This "error" is troublesome for our initial interpretation because it might have been avoided by moving the Octagon. The whole array of the enclosures as we have described it thus far could be preserved unchanged by moving the entire ensemble about a mile to the southwest along the northern

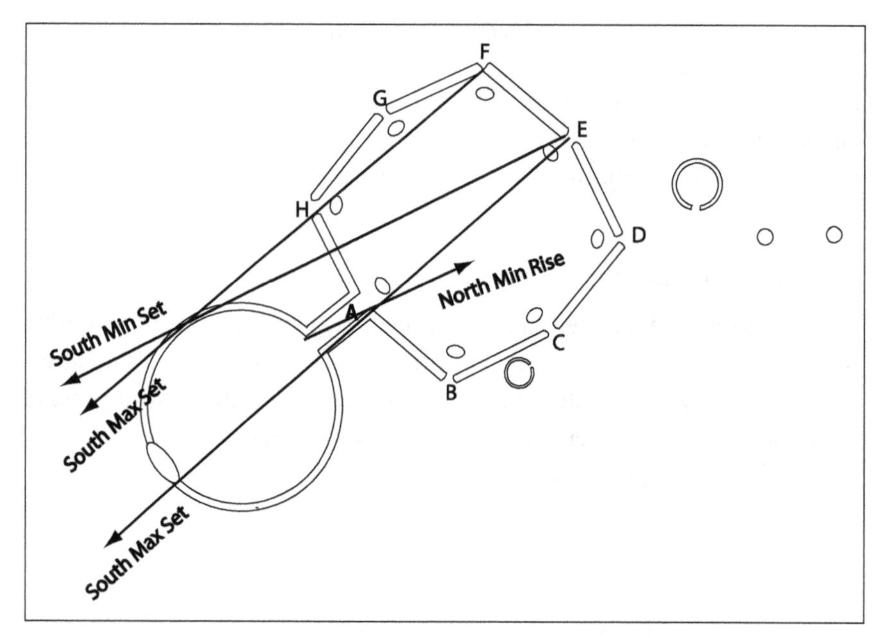

FIGURE 6. This diagram shows additional accurate alignments to the lunar standstills (major and minor) in the Circle-Octagon. If intended as part of the design, these alignments "explain" the length and width of the Circle-Octagon avenue, the shortening of wall HA, and the width of the Observatory Mound on the Observatory Circle. Otherwise these features remain unexplained.

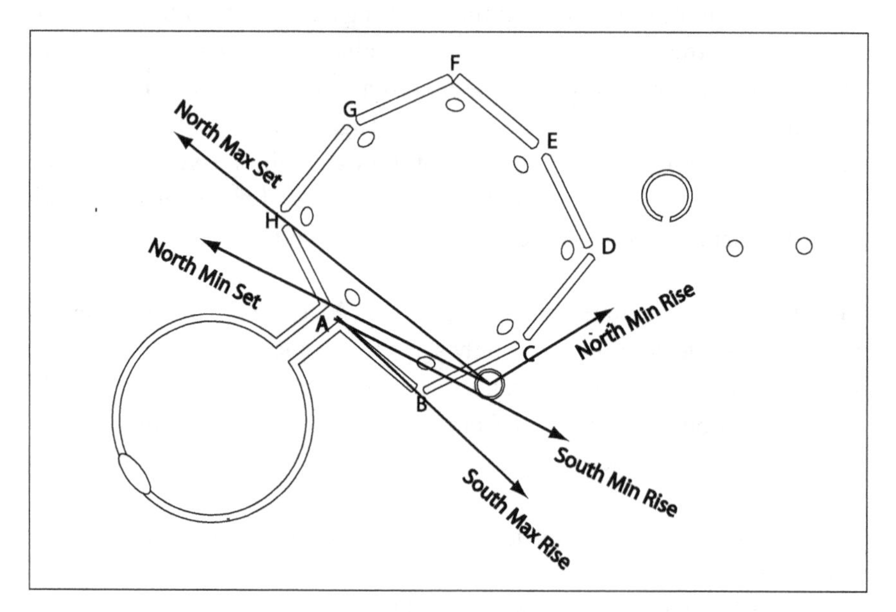

FIGURE 7. This diagram shows the lunar standstill alignments (major and minor) that can be achieved from the placement, size, and orientation of the circle BC. The hypothesis of deliberate alignment to the standstills provides a consistent explanation for these features.

maximum rise line. That would reduce the EF error to about 0.4°. As we were to discover later, probable constraints on the overall design made this option unavailable.[18]

It finally occurred to us that there would be no anomalous error in the bearing of wall EF if the enclosure walls were built to align with directions established by horizon rise and set observations made from backsights on elevated hilltops toward foresights on hilltops across Cherry Valley. In that case the altitude of distant horizons would be negligible. On further reflection we believe this is by far the best interpretation of the alignments.

The repetitive and long-term observations required for establishing and marking the lunar 18.6-year cycle would have been most easily made in places with distant clear horizons in all directions and where there would be little interference from other human activity or seasonal changes in horizon visibility—specifically, high elevations such as hilltops or ridges.[19] Thus we came to believe that the most probable scenario for establishing Octagon alignments to the lunar standstills was a two-step procedure involving (1) long-term observations from elevated points in the local topography that had small and distant horizons and (2) the projection of these alignments into Cherry Valley to establish the standstill directions, which would then be encoded by the Octagon.

This scenario requires that the Hopewell builders had the capacity to project parallel lines over distances of several miles. We know in fact from the archaeological record at Newark that they had this capability. One of the best-known features associated with the Newark Earthworks is a set of parallel walls (separated by about 190 ft.), which are known to extend at least six miles from the site along a very straight course through terrain of varying slope and altitude at an azimuth of ~211° (fig. 1).[20] There is some evidence that these walls extended in an undeviating course for at least 12 miles from the Octagon. Brad Lepper has suggested that this road may have extended some 55 miles all the way to the "core" region of Ohio Hopewell earthworks, on the Scioto River and Paint Creek near Chillicothe, Ohio.[21] This hypothetical passage has become known as the Great Hopewell Road. Whether the road extends all the way to Chillicothe is currently a matter for continued research. What is certain is that the Hopewell had the ability to extend straight lines (for the planning and construction of earthen walls) for distances of several miles.

Tests of the Hilltop Hypothesis

We will refer to our hypothesis that the Hopewell observed and recorded lunar standstills from elevated positions with negligible horizon altitude as the zero-altitude hypothesis and call it H_Z. One clear testable consequence of H_Z is this: if any Octagon standstill alignment is extended in the reverse direction, the line will pass over a prominent elevation where the critical astronomical observations were made.[22] At first glance this prediction would appear to have little significance. One would expect that any line drawn from the earthworks in the valley back toward the surrounding hills would pass over some high point that could be claimed as the sought observing position.[23] Consequently our initial enthusiasm for even testing the hypothesis was very small. When we did test the hypothesis, we were astonished at the result.

The most obvious test of H_Z is to extrapolate the standstill alignment on the symmetry axis of the Circle-Octagon back along a line to the southwest and determine whether this line passes over a prominent overlook that could have been used for establishing the alignment. As we have said, we expected any such line drawn from Cherry Valley to the surrounding hills to pass over some hill or ridge. But we were impressed to find that the line through the Octagon and the Observatory Circle passes over one of the most prominent overlooks on the valley to the southwest, an elevated plateau some five miles from the center of the Octagon and more than 200 feet above the valley floor. This position is designated H1 in figure 8.

After locating H1 as the optimum site for establishing the standstill alignment through the Circle-Octagon, we were surprised to discover that a line from H1 through the center of the Great Circle passes through the gateway opening of that circle and points accurately to the extreme northern moonrise at the minor standstill. An observer at H1 would see the northern extreme moonrise move from a line through the Circle-Octagon at major standstill to a line through the center of the Great Circle at minor standstill. H1 thus serves as common backsight for complementary extreme northern moonrises occurring over the centers of the two major circular earthworks at major and minor standstills. Still a skeptical reader might reasonably assume this is accidental.

The strongest evidence that this was intentional would be finding that the six remaining standstill alignments are marked in a similar fashion. We did not expect such a confirmation. Indeed we were quite surprised to find that this is the case. Consider the alignment shown in figure 1 (to the extreme

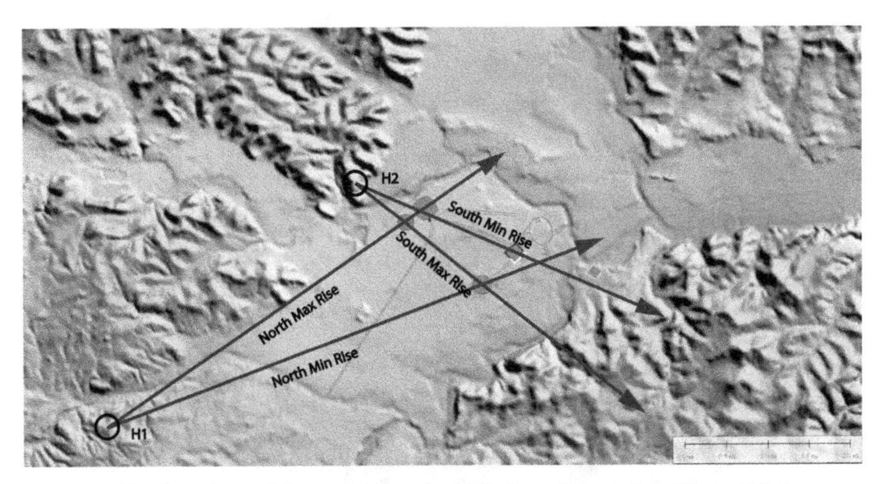

FIGURE 8. The locations of the prominent elevated observing stations H1 and H2 (some 200 ft above the valley floor) are shown in this map. Both H1 and H2 serve as common backsights for accurate (subdegree) alignments to major and minor standstill extreme moonrises passing through the centers of the major geometrical figures in the earthworks. This suggests that these figures were intentionally located with respect to H1 and H2 to achieve those alignments.

southern moonset at major standstill) that passes through the centers of the Observatory and Great Circles. Extending this alignment in the reverse direction to the northwest shows that it passes directly over the top of a very well-defined hill (designated H2), offering a splendid view of both circles in the valley below.

H2 is located about 0.8 miles from the center of the Observatory Circle and is about 200 feet above the valley below. The most remarkable fact about H2 is that the line from H2 through the center of the Wright Square marks the southern extreme moonrise and the minor standstill. Thus H2 plays the same role as H1: it provides the backsight for subdegree alignments (through the centers of major earthen figures) to complementary extreme southern moonrises at both standstills. If the major geometric figures were placed deliberately to mark all four of the standstill moonrises as seen from H1 and H2, we would expect that the Hopewell designers would have situated other earthwork features to mark the corresponding standstill set points from other prominent observation points.

Next we look at the northern moonset alignment along wall EF. If we extend this alignment backward toward the southeast, it passes over an elevated ridge top that is well positioned to view the valley below. This point is

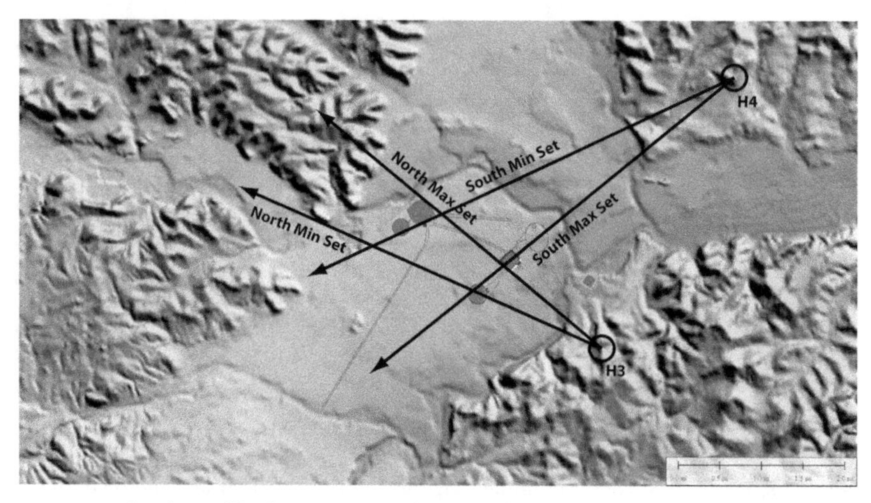

FIGURE 9. The elevated high points H3 and H4 were located by backward extrapolation of standstill alignments along the Octagon walls EF and CB. These overlooks (some 150 ft above the valley floor) provide common backsights for both major and minor standstill extreme moonsets as seen through the centers of major geometric figures in the earthworks. They play a role similar but complementary to H1 and H2.

designated H3 in figure 9. It is located about 2.9 miles east of vertex E and is elevated some 150 feet above valley below. As predicted by our hypothesis of standstill alignment, H3 also serves as a backsight for a subdegree alignment to the complementary extreme northern moonset at the minor standstill that passes through the center of the Great Circle. Thus H3 plays the same role in the alignment scheme as H1 and H2.

The final test for the standstill alignment hypothesis would be a convincing location for an elevation H4 that would provide alignments to the remaining two southern moonsets (fig. 9). The prediction of H4 can be verified by taking the Octagon alignment to the southern extreme at minor standstill along wall CB and extending it back to the northeast until it passes over a well-defined high point on a 150-foot ridge some 4.5 miles from vertex C. Remarkably, the location of H4 conforms to the pattern established by H1–H3. A line from H4 through the center of the Wright Square marks the complementary southern extreme moonset at the major standstill with subdegree accuracy. Thus, from H4 as the 18.6-year cycle unfolds, the southern extreme moonset moves from Octagon wall CB at minor standstill to a line through the center of the Wright Square at major standstill.

The four overlooks H1–H4 were all located in the same fashion: finding the first prominent overlook along the reversed directions of the standstill alignments previously established in the Octagon or between the centers of two major figures (the Observatory Circle and Great Circle). The overlooks were located without any reference to one another or to any other geometrical earthworks at the site. In each case, however, we find that each overlook serves as a backsight for observing extreme lunar events at both standstills over the centers of major earthwork figures. Moreover, taken together the alignments mark all eight standstill events with no duplication.

Readers who imagine that it is possible to read into the earthworks any comparable pattern can test this hypothesis very simply. Try the exercise for constructing a similar set of alignments to the solstice rise and set points and cardinal directions. We have tried, and we know it cannot be done. Our own remaining doubts about the use of H1–H4 as key observations points were further reduced when we discovered an unexpected relation among the four overlooks.

Solar Astronomy at the Newark Earthworks

From the beginning of our work at Newark, one of the puzzling aspects of our analysis of the site was the complete absence of any credible alignment to solar rise and set events at the solstices. This raises the question of why the Hopewell would invest such effort and give such priority to marking lunar standstills and ignore the comparable solar solstices completely. Historical and archaeological evidence from the study of ancient and prehistoric cultures across the world shows that most societies with an interest in observing the moon also closely follow the sun. The absence of solar alignments within the earthworks was hard to explain. We looked for credible solar alignments, and they were not there. Why not? We had no explanation.

Now the mere act of inquisitively drawing lines between H2 and H3 and between H1 and H4 revealed the long-missing solstices. Lines connecting pairs of these high points chosen by the Hopewell builders picked out the solstice events quite accurately. The geometric earthworks in Cherry Valley were drawn on a template formed by solstice stations on the surrounding hills. We had found the sun. Or better, the American Indian planners already had found it.

If a line is extended from H2 through H3 (a distance of 3.7 miles), the line marks with subdegree accuracy the winter solstice sunrise (when viewed

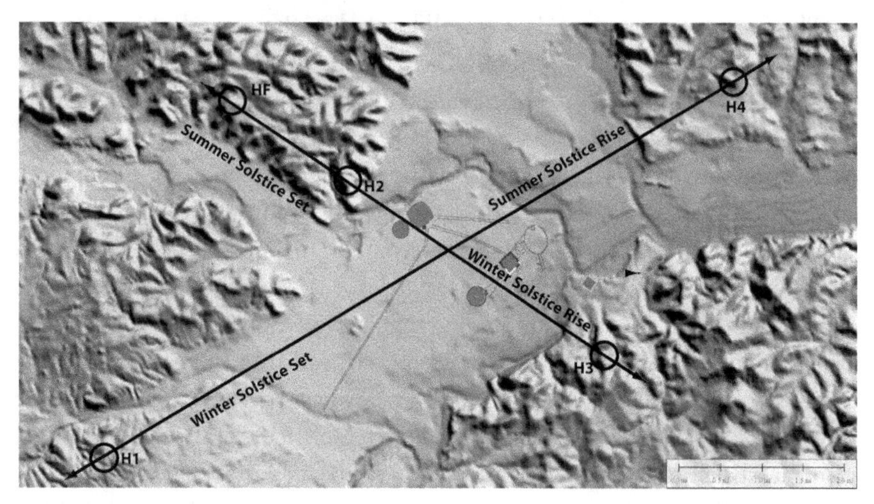

FIGURE 10. This map shows the surprising solstice alignments connecting H1 with H4 and H2 with H3. This set of alignments suggests the reason that H1–H4 were chosen as observation sites for establishing lunar alignments. The circle labeled HF shows the hilltop fort on the summer solstice sunset alignment.

from H2 across H3) and the summer solstice sunset (when viewed from H3 across H2). The obvious test of the likelihood of the intentional use of high points related in this way is to check H1 and H4 for a similar relation. Indeed, we find that the elevations H1 and H4 have a precisely similar but complementary relation, as one would expect if solstice alignments were deliberately sought out for high points in the local topography. A line from H1 to H4 (a distance of 8.6 miles) provides subdegree accuracy in marking alignments to the winter solstice sunset and summer solstice sunrise. These solstice alignments are shown in figure 10.

Further supporting evidence for the significance of the H3–H2 alignment to the summer solstice sunset comes from the presence of a hilltop earthwork along that line. This hilltop earthwork (shown in fig. 10) described by Squier and Davis forms the natural horizon as viewed from H3.[24] The major earthwork enclosure on this hill was known to have a smaller earthen circle of some 100 feet in diameter at the very highest point, near the center of the enclosing earthwork. According to Squier and Davis, this smaller circle contained two earthen mounds that upon excavation contain what appeared to be altars, which showed evidence of fire. Fire or smoke from this location would have made it easily visible as an astronomical foresight as viewed from H3.

These relations and alignments suggest two reasons why the four points H1–H4 would have been singled out as astronomical observing stations: (1) they all provide a commanding view of the valley in which the earthworks were constructed, and (2) lines between the stations accurately record the positions of the sunrises and sunsets associated with the solstices. This scenario, then, makes it plausible that the Hopewell did indeed make observations of solar events as well as lunar events.

Integration of Topography with Standstill Alignments

Our analysis thus far suggests that the Hopewell attempted to integrate individual geometric enclosures with standstill alignments passing through significant topographical features. This idea offers a plausible explanation for the general location of the earthworks in Cherry Valley. In fact it now appeared that the entire earthwork ensemble was a coherent structure that could not be moved from its present position. Our earlier guess that the anomaly at Octagon wall EF could be resolved by moving the whole ensemble to the southwest now appeared untenable in the light of tight azimuthal relations we had found between the array of earthworks in the valley and the hilltop solstice stations.

Once it seemed plausible that local topographical features had played a role in observing and marking astronomical events, we examined what could be seen from the highest point near the Newark Earthworks, Coffman Knob, some 3.2 miles southeast of the center of the Great Circle and 320 feet above the valley (fig. 11). As viewed from Coffman Knob, the northern extreme moonset at the standstills has a vivid and impressive alignment with two major topographical features across the valley to the northwest. A line from the top of Coffman Knob along the west edge of the linear Sharon Valley aligns with the northern extreme moonset at major standstill. Similarly, a line from the peak of Coffman Knob along the eastern edge of the valley of Raccoon Creek aligns with the northern extreme moonset at minor standstill.

As viewed from the highest elevation in the locality, the extreme northern moonset moves (during the standstill cycle) back and forth between two prominent valleys in quite dramatic fashion. This correspondence between major topographical features and lunar standstill events would certainly draw the attention of those interested in understanding or connecting terrestrial and celestial phenomena. So we conjecture that exactly this location

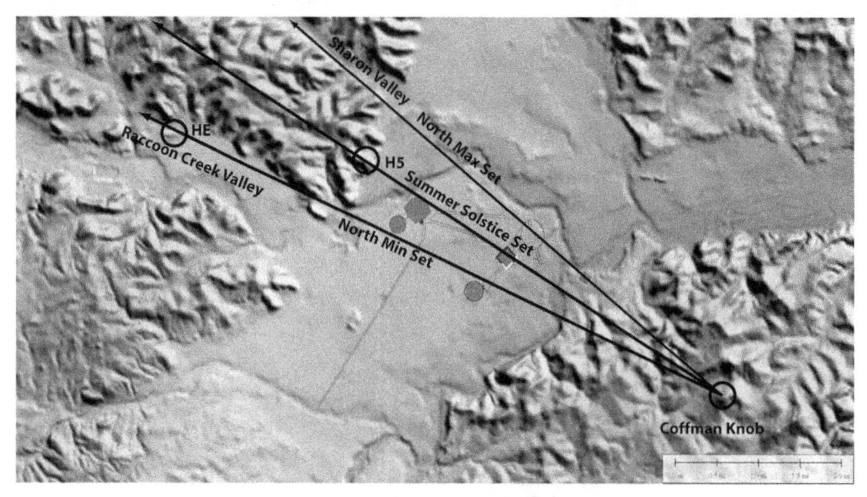

FIGURE 11. This map shows the fortuitous but striking topographical alignment of the northern extreme moonsets at major and minor standstills along major valleys as seen from the highest point in the area (the top of Coffman Knob some 320 ft above the valley). The north maximum moonset alignment passes through the centers of the Salisbury Square and the Cherry Valley Ellipse. The northern minimum moonset passes through the Hill Earthwork shown as circle HE.

of the major geometric earthworks in the valley was chosen to highlight the correspondence. There is independent evidence in support of this suggestion. The line from Coffman Knob along the northern extreme moonset at major standstill also passes through the centers of two significant earthwork figures, the Salisbury Square and the Cherry Valley Ellipse, known to contain the "main focus" of burials found at the Newark Earthworks.[25] A line along the northern moonset at minor standstill passes along the length of the Raccoon Creek valley, and close to the center of the Hill Earthwork, a large, perhaps incomplete, circular earthwork, now visible only on aerial photographs.[26]

Finally, we ask whether our hypothesis can account for the size and orientation of perhaps the most puzzling of the major geometric figures at Newark: the Wright Square, shown in figure 12. In the present context the orientation of the Wright Square is puzzling because it bears no obvious relation to astronomical events or other geometrical figures. But when the Wright Square is placed in the context of the observation points H1–H4, there is a plausible and consistent explanation for its orientation. An observer standing at the western vertex of the square (denoted as V_W) sees the moon rise above H4 when it is at the northern extreme of the major standstill. About two weeks

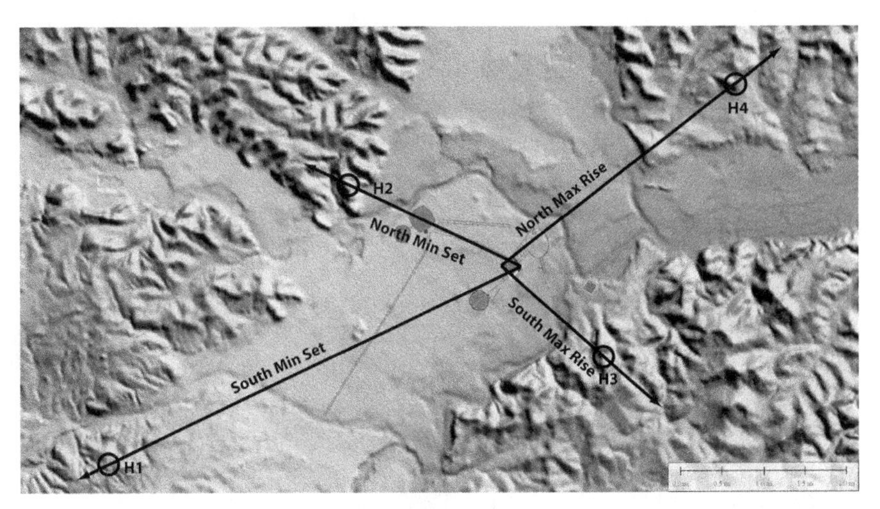

FIGURE 12. This map shows the alignments to four lunar standstills from corners of the Wright Square through the high points H1–H4. Given the abundant additional evidence for intentional alignment to the lunar standstills, these alignments support the possibility that the Wright Square orientation was chosen to achieve this effect.

later the same observer sees the moon rise over H3 at the southern extreme of the major standstill. So as seen from V_W the moon's rise point swings back and forth at the extremes between H4 and H3.

Perhaps the best evidence in support of intentional design in the V_W sightlines would be to find a similar pattern at another vertex. Consider the easternmost vertex of the square (denoted by V_E). An observer at V_E will see the moon set at the northern extreme at the minor standstill along a line passing over H2. Similarly, the V_E observer sees the moon set at the southern extreme at minor standstill along a line passing over H1. These four alignments over H1–H4 have an average accuracy of about 0.3°. The alignments are shown in figure 12. It is noteworthy that the east–west diagonal of the ~931-foot Wright Square is oriented at precisely the angle that minimizes the errors in the aforementioned alignments. If the builders had intended to orient the east–west diagonal for this purpose, they could not have done it more accurately.

Conclusions

After examining the accuracy and precision of the earthworks, the geometrical accuracy of their orientation to lunar standstill alignments, and the encoding of this information into a solar template on the local terrain, we conclude that the Hopewell were engaged in an effort to unify their understanding of the three fundamental cosmological components: earth, sky, and mind. The large number of geometric Hopewell earthworks found in Ohio and nearby shows that the culture invested many generations of effort in experimentation with rudimentary geometric shapes, their perimeters, and their areas. This experimentation culminated in the production of the geometrically exacting earthworks at Newark and near Chillicothe, Ohio. These provide evidence of an understanding and concern with geometrical form, symmetry, and patterned relationships among different figures. The intellectual vitality, effort, persistence, and cultural continuity of the Hopewell planners and builders are well documented by these earthworks on the ground. These are also precisely the qualities that would be inferred from the accurate marking of lunar standstills.

We will attempt to reconstruct what we believe to be the simplest and most straightforward scenario for explaining the geometry and astronomy of the Newark Earthworks and their integration with the local topography. Notice that in many respects this process occurs in exactly the reverse order from our findings at Newark.

The American Indian people occupying the region near present-day Newark and in southern and central Ohio engaged in systematic observation of the surrounding terrain and astronomical events occurring on the local horizon, and they experimented with the symmetries and regularities to be found in geometrical figures. This activity accumulated and recorded knowledge extending over several centuries, most likely reaching back into the Archaic period.[27] At some point after 100 CE or so, two significant discoveries were made at the present site of the Newark Earthworks.

First, local observers noticed that there was a striking correspondence between the motion of the northern extreme moonset between the major and minor standstills and features of the local topography. They saw that this moonset moved back and forth in dramatic fashion between two major valleys, as viewed from the highest elevation in the Newark area (shown in fig. 11). We suppose that the recognition of this relation between the local topography, with its water boundaries,[28] and the motion of the moon, was accorded great importance. These fortuitous correspondences between geometry, topography, and the moon provided the motivation for the Hopewell to launch their great experiment of integrating all three phenomena in the Newark Earthworks in Cherry Valley, at the confluence of Raccoon Creek and the south fork of the Licking River.

Second, planners found that the geometrical octagon construction shown in figure 2 could be approximately aligned with directions to five of the eight lunar standstills as they were seen at Newark. After considerable experimentation they noticed that this alignment could be made even more precise with slight distortions of the figure from the ideal octagonal design. This correspondence between the discoveries of geometrical symmetry and the behavior of the moon, an awe-inspiring and mysterious, perhaps capricious, cosmic actor, was a moment of great significance. It galvanized and sustained interest in geometry and astronomy. The sense of reverence and awe evoked by this discovery was enhanced by the recognition that this correspondence was limited to the Newark area. In modern terms we would say that the fortuitous relation between the octagonal design shown in figure 2 and the lunar standstills was latitude dependent. This close correspondence indeed only exists for a strip of latitudes about twenty-eight miles wide near 40° N. This latitude dependence explains in part why the only other octagon built by the Hopewell near Chillicothe, at a latitude of 39.3°, has a different shape.[29]

This project was ultimately carried out in the following fashion: (1) astronomical observations from local hilltops revealed four prominent elevations (H1–H4) that were connected by solar rise and set lines marking the winter and summer solstices; (2) the directions to the lunar standstills were established by observations conducted from H1–H4 and Coffman Knob; (3) lines along the lunar standstill extreme directions were projected from Coffman Knob and H1–H4 into the valley below; (4) the shapes, locations, and orientations of the large geometric figures in the earthworks were designed to fall along these lunar standstill directions as projected into the valley; (5) the vertices of the Wright Square were then positioned so as to achieve the standstill alignments shown in figure 12.

The accuracy, precision, and intentionality of Hopewell knowledge of large-scale geometric construction and the priority they gave to geometry cannot be disputed. The documented evidence on the ground objectively establishes that fact. The evidence of astronomical knowledge, while similar in accuracy and precision to Hopewell knowledge of geometry, involves an element of inference and abstraction that puts its interpretation in a different methodological category.

While the exacting alignments described can be clearly demonstrated and documented, the evidence that they were intended remains a matter of subjective judgment. Two choices present themselves in the interpretation of the astronomical evidence: (1) the Newark Earthworks were designed to incorporate lunar standstill alignments with a number and accuracy that make the site the most accurate prehistoric lunar observatory known, or (2) the astronomical alignments are fortuitous and random, their supposed intentionality "read into" the site by credulous investigators.

While the possibility that the astronomy at the site is entirely fortuitous can never be absolutely eliminated, we believe it clearly has been eliminated as the preferred interpretation. The hypothesis of a deliberate intent to understand natural regularities by integrating them with geometrical rules discerned by the human mind, we believe, not only is plausible but has no serious competitor. No other hypothesis has been advanced that accounts for so much of the geometrical and architectural design that we find in the earthworks.

Skepticism about the astronomy of the site is generally based on two contentions: (1) there is no precedent in the prehistoric world for disciplined and accurate observation and recording of the lunar standstills, and (2) there is no utilitarian or survival value to motivate an interest in the lunar standstills. These contentions then are used to invoke the axiom that extraordinary claims require extraordinary evidence. Hence the astronomical hypothesis is accorded extraordinary skepticism.

The lunar-oriented Newark Earthworks in the context of the Cherry Valley hilltop solstice stations do constitute extraordinary evidence. No other site offers the same ensemble of geometric constraints on proposed alignments that is found at Newark. No other site encodes with the same accuracy all of the solstice stations and all of the stations of the lunar extreme standstills. No other site so tightly integrates the exacting geometry of its architecture with the local terrain.

There is no precedent for prehistoric earthworks with the combination of scale, geometric accuracy, and precision we find at Newark. The Newark Earthworks are nonetheless there on the ground. They are undisputable proof of unprecedented achievement by their American Indian builders.

Utility is hard to define. The reasons humans inquire are not transparent. Before observers could judge the practical worth of knowing the pattern and period of the lunar standstills, they had to spend substantial effort to discover it. Was it practical to want to decode the moon's baffling behavior and possibly to communicate with the moon? The moon is cryptic and elusive, awesome and mysterious. What would ensue if the builders at Newark offered to the moon their vision of its journey? Did the tradition of standstill observation that Newark epitomizes eventually disappear from memory and record because it was found impractical? This is one of the many things we do not know.

The subject of archaeoastronomy as applied to the Hopewell cultural tradition is still young. We currently believe we have found significant evidence in support of our astronomical hypothesis from our study of the geometry and placement of Hopewell earthworks near Chillicothe. Archaeoastronomical research at the Newark site and at related sites in the Chillicothe region and beyond began in the 1970s and continues today.[30] New survey work combining geophysics with the remote sensing capabilities of LiDAR has already given archaeologists better data about aspects of Adena and Hopewell sites than nineteenth-century surveys.[31] While the evidence continues to be accumulated and analyzed, we believe the question of intentional astronomical alignment is likely to remain a subject of debate and conjecture rather than consensus for some time.[32] However, our best effort to balance appropriate skepticism with three decades of experience in working with the available data and Monte Carlo simulations of randomly constructed geometric figures leads us to conclude that the Newark Earthworks were built to encode simultaneously a knowledge of geometry and regularities perceived in the sky and on the earth below.

An even more important conclusion will enjoy a wide consensus. The construction of the Newark Earthworks stands as a striking example of what human beings can achieve when motivated by ideas with the power to inspire their imagination, discipline, and effort. In that sense the message of the earthworks speaks to us clearly from prehistoric Ohio. Not unlike CERN's Large Hadron Collider, it shows what we can achieve when we seek with determination to comprehend our place between heaven and earth. Pliny would have been impressed. Even Juliet would have been relieved to discover the constant inconstancy of Romeo's blessed moon.

Notes

1. See Lepper's contribution to the present volume.

2. See Hively and Horn, "A New and Extended Case for Lunar (and Solar) Astronomy at the Newark Earthworks"; and Greber et al., "Astronomy and Archaeology at High Bank Works."

3. C. Thomas, *Report on the Mound Explorations of the Bureau of Ethnology,* 440.

4. Fowke, *Archaeological History of Ohio,* 171.

5. See Bernardini, "Hopewell Geometric Earthworks," for detailed labor estimates for the construction of some Hopewell earthworks.

6. This striking view of the symmetry and precision of the Circle-Octagon Earthworks, one never seen by any Hopewell architect or builder, is illustrated in an aerial photo taken by Dache Reeves in 1934. Dache M. Reeves, collection of aerial photographs (Washington, DC: National Anthropological Archives, 1934).

7. The late John Eddy (1931–2009) did a transit survey of the site in 1978 confirming his earlier conjecture, based on the Squier and Davis map, that the Circle-Octagon symmetry axis aligned with the north maximum moonrise extreme (personal communication, 1980).

8. Sims, "Which Way Forward for Archaeoastronomy?," points to the importance of each of these elements in assessing the likelihood that alignments are intended and not fortuitous.

9. See, for example, Ruggles, *Astronomy in Prehistoric Britain and Ireland;* Aveni, *Foundations of New World Cultural Astronomy;* Powell, "The Shapes of Sacred Space: A Proposed System of Geometry Used to Lay Out and Design Maya Art and Architecture and Some Implication Concerning Maya Cosmology"; Sofaer, "The Primary Architecture of the Chacoan Culture" 88–132; Malville, *Chimney Rock*; and Sutcliffe, *Moon Tracks.*

10. See Hively and Horn, "Hopewell Cosmography at Newark and Chillicothe, Ohio."

11. Pliny the Elder, *The Natural History,* II.6.

12. C. Thomas, *Report on the Mound Explorations of the Bureau of Ethnology.*

13. See Hively and Horn, "Geometry and Astronomy in Prehistoric Ohio."

14. See Hively and Horn, "A Statistical Analysis of Lunar Alignments at the Newark Earthworks."

15. See Byers, *The Ohio Hopewell Episode;* and Byers, *Sacred Games, Death, and Renewal in the Ancient Eastern Woodlands.*

16. For detailed discussion, see Hively and Horn, "A New and Extended Case for Lunar (and Solar) Astronomy at the Newark Earthworks."

17. See Hively and Horn, "Geometry and Astronomy in Prehistoric Ohio"; and Hively and Horn, "A Statistical Analysis of Lunar Alignments at the Newark Earthworks."

18. Byers, *Sacred Games, Death, and Renewal in the Ancient Eastern Woodlands,* 435–68, proposes a scenario for the placement of the Circle-Octagon in Cherry Valley that follows our earlier suggestion that the enclosures could be moved in order to avoid the anomaly at wall EF. Among salient questions that should be considered is whether the Great Circle is an Adena-period construction later incorporated in the Hopewell plan for Newark. We will address this issue more closely in forthcoming work. Also see Lepper's contribution to the present volume.

19. See Hively and Horn, "Hopewell Cosmography at Newark and Chillicothe, Ohio."

20. See Hively and Horn, "A Statistical Analysis of Lunar Alignments at the Newark Earthworks"; and Romain and Burks, "LiDAR Imaging of the Great Hopewell Road."

21. See Lepper, "Tracking Ohio's Great Hopewell Road"; and Lepper, "The Great Hopewell Road and the Role of Pilgrimage in the Hopewell Interaction Sphere."

22. See Hively and Horn, "A Statistical Analysis of Lunar Alignments at the Newark Earthworks"; Hively and Horn, "Hopewell Cosmography at Newark and Chillicothe, Ohio"; and Hively and Horn, "A New and Extended Case for Lunar (and Solar) Astronomy at the Newark Earthworks."

23. See Park, *Notes of the Early History of Union Township, Licking County, Ohio.*

24. Squier and Davis, *Ancient Monuments of the Mississippi Valley,* 24, plate 9, no. 1.

25. Lepper, personal communication and his contribution to this volume.

26. See Hooge, "Preserving the Ancient Past in Licking County, Ohio," 179–88.

27. See Brian Hayden and Suzanne Villenueve, "Astronomy in the Upper Paleolithic?," *Cambridge Archaeological Journal* 21, no. 3 (2011): 331–55.

28. See Lepper, "The Newark Earthworks: Monumental Geometry and Astronomy at a Hopewellian Pilgrimage Center," 76.

29. See Hively and Horn, "Hopewellian Geometry and Astronomy at High Bank"; Hively and Horn, "Hopewell Cosmography at Newark and Chillicothe, Ohio"; and Greber et al., "Astronomy and Archaeology at High Bank Works."

30. In addition to numerous articles by Hively and Horn, see, for instance, Greber and Jargiello, "Possible Astronomical Orientations Used in Constructing Some Scioto Hopewell Earthwork Walls"; Greber, "Astronomy and the Patterns of Five Geometric Earthworks in Ross County, Ohio"; Essenpreis and Moseley, "Fort Ancient"; Essenpreis and Duszynski, "Possible Astronomical Alignment at the Fort Ancient Monument"; C. Turner, "A Report on Archaeoastronomical Research at the Hopeton Earthworks, Ross County, Ohio"; Romain, "Hopewell Geometric Enclosures"; Romain, "Summary Report on the Orientations and Alignments of the Ohio Hopewell Geometric Enclosures"; and Mickelson and Lepper, "Archaeoastronomy at the Newark Earthworks."

31. See Romain and Burks, "LiDAR Assessment of the Newark Earthworks"; Romain and Burks, "LiDAR Imaging of the Great Hopewell Road"; and Romain and Burks, "LiDAR Analyses of Prehistoric Earthworks in Ross County, Ohio."

32. See, for instance, Sims, "Which Way Forward for Archaeoastronomy?"

PART III The Newark Earthworks in Cross-Cultural Archaeological Contexts

Nazca, Chaco, and Stonehenge

HELAINE SILVERMAN

An Andeanist's Perspective on the Newark Earthworks

This is the world of exhibition of sacra, symbols of a higher reality; of the dramatization of creation stories; of the appearance of masked and monstrous figures; of the construction of complicated shrines.—Victor Turner, *Process, Performance, and Pilgrimage: A Study in Comparative Symbology*

Most groups . . . engrave their form in some way upon the soil and retrieve their collective remembrances within the spatial framework thus defined.—Maurice Halbwachs, *The Collective Memory*

FROM THE proverbial time immemorial humans have marked their physical environments, imbuing them with stories, subjecting them to particular cosmologies of understanding and ideologies of use, and altering them with visible signs from cave paintings to towering skyscrapers. In so doing, landscapes of meaning are created. Some of these landscapes are of such a monumental nature that their importance in their ancient societies is obvious. Other landscapes may be more elusive for identification in the absence of ethnohistoric records or ethnographic fieldwork among communities retaining some memory of their significance or actual cultural continuity. Recovery of meaning may be so difficult due to the passage and disrupture of time that scholarly interpretation might be more an act of speculation than reconstruction.

One prehistoric monumental landscape that immediately drew the attention of later observers was the earthworks of the Ohio valley. They were so abundant and so large that the pioneers pushing across the North American continent could only marvel and wonder about who had made them and why.[1] Another monumental pre-Columbian landscape—this one in South America—was much less easily perceived though similarly generative of speculation about origin. I refer to the geoglyphs covering the surface of the

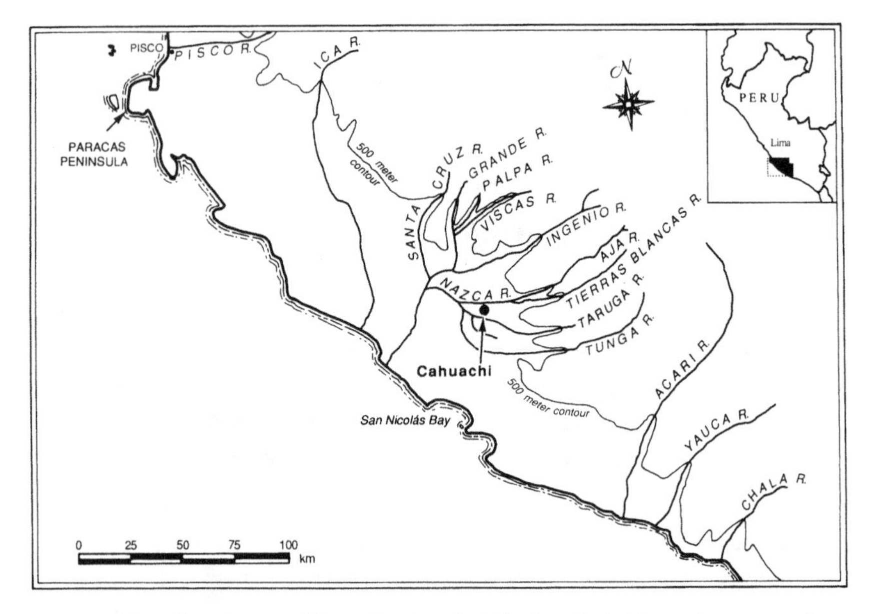

FIGURE 1. Map of south coastal Peru showing the Río Grande de Nazca drainage and Cahuachi, the great early Nasca ceremonial center. The pampa is located between the Ingenio and Nazca valleys.

vast (220 sq km) desert plain, or pampa, in the Río Grande de Nazca drainage in south coastal Peru (fig. 1).

The midwestern US earthworks are a massive *additive* phenomenon that builds bulk on a surface. The Peruvian geoglyphs are *subtractive*—shallow trenches dug into a rocky surface revealing lighter-colored soil and bordered with the removed rock detritus. Though different in technique of manufacture, both sets of ancient remains were coherent, oft-repeated, culturally specific actions that produced large landscapes of a manifestly noninfrastructural, noneconomic nature. These were sacred landscapes whose construction involved religious as well as social and political symbolization. I hope that my Andean perspective can contribute to a greater understanding and appreciation of the Newark Earthworks.

Sacred Landscapes in Nazca and Newark

Sacred landscapes have received extensive treatment by archaeologists for more than thirty years.[2] Ancient Peru figures prominently in these discussions, for the pre-Hispanic Andean world was populated with supernatural beings and sacralized places including mountains, lakes, springs, irrigation canals, boulders, and caves, as well as numina-lodging objects and anthropomorphized forces of nature.[3] The covering of the pampa and surrounding hillsides with geoglyphs made the pampa a sacred landscape and an integral part of Nasca[4] religion in terms of the rituals conducted on it and through transit across it to Cahuachi, the great early Nasca ceremonial and pilgrimage center.[5]

The Newark landscape also had a sacred character. Hively and Horn, for instance, propose a "celestial topography" that produced a Hopewell cosmography on the landscape.[6] The earthworks were sited not just to align with or capture astronomical significance, the locations were chosen "for their significance in their builders' view of the world."[7] Burial mounds at Newark and other Hopewell sites also would have created a sacred landscape, connecting the living to their ancestors and those ancestors to the other, supernatural world.

I believe that the markings on the pampa were the sequential products of individual groups rather than a giant tableau undertaken simultaneously to create comprehensive representation. Thus they were not the "ritual machine" that Lepper and others postulate for Newark.[8]

Pilgrimage in Nazca and Newark

Pilgrimage generates and acknowledges a sacred landscape. It is voluntary travel to and efficaciously quick return from a physically or conceptually distant sacred place at which ritual ceremonies are held. Its archaeological characteristic signatures include evidence of travel, location away from ordinary settlements, monumental architecture, facilities for travelers, places of congregation, offerings, souvenirs, and a catchment area.[9] But even in the absence of these patterns, wherever the data support the reconstruction of a regular and repetitive ingress and exit of distant (supralocal) populations to a ceremonial center for the purpose of participation in ritual activities, regardless of the normal residential density and socioeconomic heterogeneity at that center, pilgrimage is indicated.

A pilgrimage function for the Newark Earthworks has been proposed by Brad Lepper. He correctly observes the obvious: "The monumental size of the [earth]works indicates they were built to serve large numbers of people who gathered periodically here from far and wide."[10] Lepper also identifies a "Great Hopewell Road" between Newark and Chillicothe as a "pilgrim's path." In my own work I have emphasized Cahuachi's correspondence to archaeological and ethnographic indicators of pilgrimage, including pathways across the pampa that lead to Cahuachi or are otherwise part of the Nasca ritual complex.[11]

I am especially intrigued by Lepper's discussion of the nature of the evidence for habitation associated with Newark and other earthworks. He says, "There seems to be little justification for the notion that geometric earthworks were vacant ceremonial centers. Although these structures clearly are not the walls of large urban centers and the interiors of the earthworks sometimes appear to be vacant, when sought, domestic debris commonly is found both outside and inside the embankments."[12] The domestic materials discussed by Lepper are very reminiscent of what I excavated in the plazas of Cahuachi—the premier early Nasca ceremonial center—and that I have accounted for as the result of pilgrimage.[13] Pilgrimage involved the movement of people in and out of the sacred sites. During the course of celebrations at these sites, impermanent housing was created, refuse was generated (of both a domestic and ritual nature), and then these places were vacated until the next round of religious performance. As Lepper is quick to affirm, these sites were not urban or even significantly populated centers. Pilgrimage is a model that can generate the material remains recovered.

A unique modeled Nasca scene (fig. 2) appears to depict a family as a social unit of pilgrimage in early Nasca society.[14] The family would have lived in a village in or adjacent to valley-bottom land. Their house would have been a small wattle-and-daub structure, possibly with cobblestone foundations. They may have made their own ritual paraphernalia (pottery and textiles) for use at Cahuachi, since potters' plates and spindle whorls have been observed at several Nasca habitation sites. They likely met up with other families going to Cahuachi. There is no evidence of non-Nasca people at Cahuachi. Thus, celebration there was highly legible in portable and intangible terms (coreligionists dressed similarly, shared frontal-occipital cranial deformation, and used the same symbolic material culture), just as the sacred space was characterized by monumental legibility.

At Newark and other earthworks, archaeologists have recognized a pattern of earth-filled basket deposition with which the mounds were created. I think these basket loads are the reflection of social units of production in ancient Hopewell society. Regardless of the criticism that Martin Byers's "sacred earth" argument for Newark has received,[15] I do believe that Byers is correct in arguing that from start to finish the construction of the earthworks was a "sacred enterprise"—just as building any Andean ceremonial center was a sacred undertaking. Moreover, as I have argued for the Nazca geoglyphs

FIGURE 2. An early modeled ceramic scene that is interpreted as depicting a Nasca family on pilgrimage. The father plays a large panpipe, and two smaller ones rest on his head covering. The mother carries two more panpipes, one in each hand. Parrots perch on the shoulders of the mother, and another sits on the right shoulder of the daughter. The daughter carries a double-spout-and-bridge bottle in her left hand. The family is accompanied by four trotting dogs; another is held under the left arm of the father. The family may be going to Cahuachi (or another Nasca ceremonial center). Perhaps they are bringing the brilliantly colored birds to sacrifice, panpipes to play, and a fancy ceramic bottle to exchange or ritually consume (i.e., deliberately break). (Tello, "Un Modelo de Escenografía Plástica en el Arte Antiguo Peruano")

(taking my cue from Gary Urton's discussion of *chutas* as the space of the social practice),[16] meaningful social groups were constituted or reiterated in the very act of creating the monument, and this was a ritual activity, indeed a sacred obligation.

The dumping of basket loads of dirt to generate the volume of the Hopewell earthworks suggests an undifferentiated community effort. But "undifferentiated" as I use it here does not close the door to status jockeying at these centers. I agree with the Hopewellianists that Hopewell society did not have powerful hereditary leaders. And I believe the same situation existed in early Nasca society, but in the more sociopolitically complex context of a fully agricultural and densely settled village landscape.[17] The congregation of vast numbers of people at the pilgrimage sites afforded the opportunity for all manner of social interaction, economic exchange, and political negotiation within and between individuals and groups—in addition to all else that may have happened there. Those individuals performing as shamans or priests had positions that could be enhanced or challenged. Headmen must have cast their critical eye on potential challengers. Ambitious males might have been able to advance in these public venues of heightened drama according to whatever cultural criteria were in use.

Pilgrimage is also associated with ritual paraphernalia, including offerings. The modeled Nasca family scene, discussed above, shows the family members carrying objects of high value in ancient Nasca society: a double-spout-and-bridge ceramic bottle and ceramic panpipes. Hundreds of broken panpipes have been recovered at Cahuachi, and the amount of fine, decorated ceramics at the site is legendary, although its surface today is scoured clean by decades of projects and tourists. Llamas and alpacas were sacrificed at Cahuachi.[18] William Duncan Strong's enigmatic "Great Cloth Deposit" was another offering.[19] Lepper observes that at Newark "pilgrims brought offerings from the ends of the Hopewell world that ultimately were buried there."[20] In addition to offerings, human burial itself was appropriate at both of these pilgrimage centers.[21]

Embodied Practices and Ritual Movement in Nazca and Newark

It is likely that the pampa opposite Cahuachi and the pampa behind it were traversed by pilgrims going to Cahuachi from their homes in the valleys to the north and south.[22] Indeed, there is a clear transpampa road running between Cahuachi and the middle Ingenio valley on the north side of the pampa. Many lines of walkable width have been identified on the pampa.[23] These are ritual paths, and, to quote Michel de Certeau, "spatial usage creates the determining conditions of social life. . . . The motions of walking are spatial creations. They link sites one to the other . . . they spatialize. . . . [Walking] is a process of *appropriation* of the topographic system by the pedestrian. . . . It is a spatial *realization* of the site. . . . It implies relationships among distinct positions. . . . The walker . . . *actualizes* some of [the] possibilities and interdictions."[24]

De Certeau's argument is congruous with Simon Coleman and John Elsner's proposition that "diagrams and narratives not only reproduce the topography and experience of the sacred journey, but define and even constitute it. They provide the means for imagining pilgrimage."[25] Walking and movement created the geoglyphs. The performance of social space inherent in the cultural and physical production of the geoglyphs created the space of social practice.[26] Biomorphic figures on the pampa and many of the geometric geoglyphs could have been laid out by particular social groups during the course of pilgrimages to Cahuachi and possibly were themselves—in part—the purpose of pilgrimage. As pilgrims went to or returned from Cahuachi, various geoglyphs could have been important elements of the pilgrim's progress, so to speak.

Movement is also a fundamental aspect of the Newark Earthworks. Lepper presents a compelling case for a "Great Hopewell Road" at Newark.[27] In addition to that road or causeway, he speaks of the walls that connect the enclosure in "an intricate pattern." I see a direct comparison with the Nasca in terms of the ritual movement that was being scripted by the architecture of Newark. Pilgrims (for pilgrimage is what was directing the ancient Ohio people to and from Newark) were moving in meaningful ways, processing out of their ordinary domestic lives to the heightened religious arena at the earthworks themselves.

Importantly, there were only three entrances into the Newark Earthworks, and thus pilgrims were funneled into the site. One can readily imagine the visual and emotional impact of seeing hundreds, perhaps thousands, of elaborately costumed celebrants moving across (today's) Ohio and arriving at this significant place. The management of such density of human bodies must have required organizational planning and skill. By whom? And were the same individuals in charge every year, or were these positions of prestige, like *cargos* that circulated or were competed for? This is why I suggest that pilgrimage is an arena for personal competition as well as religious devotion.

Outside Newark the straightness of the Great Hopewell Road is notable. Its connection to other significant places on the ancient Ohio landscape is comparable to the Nazca roads going across the pampa and to the "ray centers" documented by Anthony Aveni.[28] Like the precisely executed Hopewellian circles and octagons, Newark's roadway was a display of indigenous geometric precision—much like the Nasca people's geometric mastery that produced unerringly straight lines, perfect trapezoids, mazelike spirals, and huge animal figures on the pampa.

Frame, Scale, Perspective, and Experience in Nazca and Newark

Sacred sites have a heterotopic quality that is manifested in their properties of frame, scale, and perspective, which exceed or differ from that of ordinary life. The frame brackets: "it separates from context, focuses attention by screening undesired or irrelevant views, by directing the gaze" and enclosing the desired focus of attention.[29] I believe that the architectural elaboration of Cahuachi and the marking of the pampa (as well as valley hillsides) effectively directed the Nasca gaze at the ritual stage of their own making, notwithstanding the difference between the volumetric architectural places of Cahuachi and the ground-level ritual places of the pampa. I believe that the complementary Andean principles of *kancha* (enclosure) and pampa (open plain) created the ancient Nazca landscape and correspond to the concept of frame.

The immensity of the pampa's geoglyphs must be the result of the ancient Nascas' attempt to transcend the ordinary by means of scale. Gaston Bachelard is clear: "immensity . . . is a conquest of intimacy."[30] The scale of the geoglyphs on the pampa is such that we are clearly witnessing a radical change in landscape perspective. Most of the entire corpus of pampa geoglyphs was not being viewed from an elevated position (e.g., from hilltops) although it was being conceived aerially. The geoglyphs on the pampa "existed as much as a spatial concept as a reality."[31]

The deliberate choice of huge scale must imply civic/communal context. Everyone had the same distorted, ground-level perspective of the geoglyphs (like being a member of the band at halftime on the football field and not being able to visually perceive the moving patterns one is producing). No one person had a privileged or differential viewing position. The visceral experience of the geoglyphs occurred during their elaboration rather than afterward, when they were fixed on the pampa surface. There was a single community of producers and consumers of each of the geoglyphs. Although executed at immense scale, the geoglyphs—whether biomorphic or lineal—were the product of the moving human bodies who created them.

Lepper has summarized the comparable geometrical achievement of the Newark Earthworks.[32] Extraordinarily precise astronomical alignments were embedded in the Newark Earthworks.[33] The immensity of the Newark Earthworks was probably best and most deliberately perceived during their post-construction use. Newark was built to be experienced. Its vast interior spaces can accommodate significant numbers of people. Its causeway conducted people into that space. And Newark itself was connected to other earthworks sites, thus forming part of a dramatic artificially created sacred landscape, all the more remarkable because of the dispersed settlement pattern. The Newark Earthworks and their many sister sites are coherent constructions that were built with careful planning and according to a design.[34] Notwithstanding that the users of the site themselves built it, the sense of moving through and congregating in the great circles and squares, let alone while participating in rituals, including those at night pertaining to the moon, must have been awesome in the most literal sense. The scale and perspective at Newark dwarfed the mundane world physically built by the Hopewells in which they dwelled.

One can only marvel at the achievement of land-marking precision of both groups of prehistoric peoples at Nazca and Newark.

Mapping Memory in Nazca and Newark

In addition to social reproduction, repetitive ritualized acts on a landscape serve "as engines for the creation of time,"[35] for space cannot be separated from time, not in the Andes and not elsewhere. The geoglyphs and earthworks must have "monumented and commemorated the historical linkage" between earlier and later people by means of reference to a sacred place, "a locus of mythic or historic power."[36] In both cases the sociocosmological symbolism of these architectural spaces surely was narrated as it was performed—but for want of a written language we have no record of those oral traditions. In their own time they overcame this lack (that we perceive) through embodied memory. Sacred places are places of memory[37] that are sounder the more fixed in space they are.[38]

The palimpsest of superimposed tracings on the pampa are spatialized memories and a vast mnemonic representation of the social and political relations between the many kin groups forming the Nasca nation.[39] As a map of social and sacred space, the geoglyphs also were a site for intensive interpretation. We cannot know how prior use and meaning of the geoglyphs affected future use and meaning, yet this contingency is expectable. I conceive of each line-making event as an execution or performance "by people whose own knowledge of how to proceed [i]s reworked by their ability to read each situation. The reading involve[s] carrying past experiences forward in the form of expectations that [a]re either confirmed or challenged in each new engagement. The mutuality of each engagement therefore mean[s] something only by reference to the experiences of past engagements and the desires for the future; it is a meaning of memory and of hope."[40] This provokes the question: If the geoglyphs were a map of social and political relations among the Nasca kin groups, what does their effective erasure through superposition of earlier geoglyphs by later ones mean? I do not have a ready answer other than to suggest that the pampa surface was as dynamic as Nasca society itself and that erasure or lack of maintenance may have reflected the life cycles of the groups that made the geoglyphs.

The earthworks of the ancient Newark people and their culturally related groups created a landscape of monuments so physically obtrusive that in their own time their visibility and significance had to have been recalled. The Hopewellians—especially because of their predominantly nonagricultural subsistence base—were constantly moving across the monument-marked landscape as they lived and sustained their ordinary domestic lives. Thus, we must be concerned not only with how this population moved within the great sites at times of ritual but also with how they transected their landscape on a daily basis. I do not know the relevant literature and so would be keen to see temporally fine-grained geographic information system renditions that plot habitation sites and their sight lines to these dramatic constructions. Especially because we are dealing with a nonliterate people, I would guess that oral tradition was tremendously important and that stories were told about this ceremonial landscape that necessarily implicated group memory ranging from myth to last year's celebration.

I end this section with Lepper's eloquent summary of the earthworks: "The giant enclosures were enormous engines of ceremony and ritual intended to do something. They not only symbolized the cosmos, they may have allowed the Hopewell shamans to draw on the energies of the cosmos."[41] The Newark Earthworks were "a gigantic machine or factory in which energies from the three levels of the Eastern Woodland Indian's cosmos—the Upper World of the sky, the watery Underworld, and the Middle World of soil and stone—were drawn together and circulated through conduits of ritual to accomplish some sacred purpose."[42]

Continuity and Discontinuity in Nazca and Newark

The pampa was a field for ritual and social activities from immediately pre-Nasca times to the time of the Inca conquest: geoglyphs continued to be made for more than a thousand years.[43] However, the production of figures, specifically, ended with the demise of recognizably Nasca culture itself.[44] This continuity (geoglyph making) and discontinuity (loss of the biomorphs) in cultural practice on the south coast of Peru is significant in a larger theoretical context. The geoglyphs were embodied actions, the material traces of social and spatial practice. The long history of geoglyph making in Nasca society and afterward can be understood in terms of the interrelated paradigms of practice and materiality as cultural reproduction, a constant regeneration of society through practice, through tangible and perishable rhetoric. The making of geoglyphs did not just manifest Nasca social identity and society; it—along with other behaviors—materialized and reproduced Nascaness.

Reuse (filling in) of the pampa over hundreds of years indicates its long-term manipulability as a dynamic field of representation and re-presentation. In addition, the superpositions may reflect the nature of this sacred space as a negotiated site with contests of power "over the legitimate ownership of sacred symbols. . . . A sacred space . . . is claimed, owned, and operated by people advancing specific interests."[45] By comparison, the end of pilgrimage to Cahuachi resulted in radically different behavior. Construction of mounds ceased. Cahuachi became a cemetery and an appropriate place for the leaving of offerings.[46] The ideological landscape continued to be reworked in such a way that a reading of its dominant significance shifted in emphasis away from Cahuachi to the geoglyphs of the pampa. And it was not reading, per se, that changed but cultural practice as the actual enabling social fulcrum.

Warren De Boer has characterized Newark as a "highly visible, geoglyphic expression . . . of expanded claims to prestige and power."[47] But by 400 CE the Hopewell enclosure tradition had ended. All scholars agree that the Hopewellians were not organized in demographically large, geographically extensive complex polities.[48] The archaeologically recognizable collapse of Hopewell could be due to internal forces such as those I have posited for the early Nasca social formation with its own reified ritualism.[49] Lepper speculates that "the relatively weak power of Hopewell leaders could not withstand the centrifugal forces that sundered their precocious attempts at political integration."[50] I would go further. I would argue that lack of intent to achieve political integration was the underlying problem, perhaps accompanied by lack of Robert Carneiro's important concept of circumscription.[51] In both cases the demands and pressures within society could no longer be handled by the ritual system that produced Cahuachi and the geoglyphs on the one hand and Newark on the other hand.

Conclusions

In this essay I have highlighted several commonalities of the ancient Nasca ceremonial center of Cahuachi and its related geoglyphs with the Newark Earthworks: pilgrimage; performance; movement; physical construction that generates a social topography of meaning and embodies cosmology; marking the landscape; memory; geometric precision through simple construction techniques; the huge scale involved and related issues of visibility and phenomenological experience of place; forces leading to the end of these monumental landscapes.

In this brief conclusion I would emphasize the following. The very *act* of inscription on/of the ground was itself as significant as the final product, and physical building was simultaneously an act of social creation, with work groups and social groups reiterating or reformulating themselves for these projects, reflecting or challenging existing lines of power and authority, and performing rituals that confirmed the fundamental cosmology of each society. Thus, sacred space and sacred place are always culturally specific.

Memory was anchored and social identity was expressed in culturally constructed landscapes that consisted of various kinds of architectural built environments, natural resources, and sacred geographic features. Memory was cued by embodied and geographically situated ritual action; pilgrimage is exemplary of this. Place on these landscapes was generated as space was imbued with meaning by myth, ritual performance, and daily social practice. Earthworking—whether in ancient Peru or at Newark and elsewhere in the Ohio region—expressed cosmology, inscribed social memory and social identity on the landscape, created attachments to particular locales, was a form of social reproduction, and could be a dynamic political arena.

These monuments were extremely complex phenomena in their own times. They continue to inspire archaeological interpretation today.

Notes

1. See, for instance, Lepper and Gill, "The Newark Holy Stones"; and S. Williams, *Fantastic Archaeology.*

2. See, for instance, Bradley, "Sacred Geography"; Carrasco, *To Change Place*; Coe, "Religion and the Rise of Mesoamerican States"; and Townsend, *The Ancient Americas.*

3. Salomon, "The Introductory Essay."

4. As in my previous publications, I orthographically distinguish *Nasca* and *Nazca* (both forms are used in Peru) for the sake of clarity. *Nasca* refers specifically to the

first millennium CE society in the Río Grande de Nazca drainage. *Nazca* refers to that drainage.

5. " Silverman, *Cahuachi in the Ancient Nasca World.*

6. Hively and Horn, "Hopewell Cosmography at Newark and Chillicothe, Ohio."

7. Buikstra and Charles, "Centering the Ancestors," 216, quoted in ibid., 129.

8. Lepper, "The Newark Earthworks," *inter alia.*

9. Silverman, "The Archaeological Identification of an Ancient Peruvian Pilgrimage Center."

10. Lepper, "The Newark Earthworks," 79.

11. Silverman, *Cahuachi in the Ancient Nasca World;* and Silverman, "The Archaeological Identification of an Ancient Peruvian Pilgrimage Center."

12. Lepper, "The Newark Earthworks and the Geometric Enclosures of the Scioto Valley," 236.

13. Silverman, *Cahuachi in the Ancient Nasca World.*

14. Julio C. Tello, "Un Modelo de Escenografía Plástica en el Arte Antiguo Peruano," *Wira Kocha* 1, no. 1 (1931): 87–112.

15. See Lepper, "The Newark Earthworks and the Geometric Enclosures of the Scioto Valley," commenting on Byers, "The Earthwork Enclosures of the Central Ohio Valley."

16. Urton, "Andean Ritual Sweeping and the Nazca Lines."

17. Silverman, *Cahuachi in the Ancient Nasca World.*

18. Valdez, "Alpacas en el Centro Ceremonial Nasca de Cahuachi."

19. Strong, *Paracas, Nazca, and Tiahuanacoid Cultural Relationships in South Coastal Peru.*

20. Lepper, "The Newark Earthworks," 79.

21. Regarding Cahuachi, see Kroeber, Collier, and Carmichael, *The Archaeology and Pottery of Nazca, Peru;* and Silverman, *Cahuachi in the Ancient Nasca World.* Regarding Newark, see Lepper, "The Newark Earthworks."

22. See Silverman, *Cahuachi in the Ancient Nasca World.*

23. Aveni, "The Nazca Lines," 39.

24. De Certeau, "Practices of Space," 129–30; emphasis in original.

25. Coleman and Elsner, *Pilgrimage,* 167.

26. Urton, "Andean Ritual Sweeping and the Nazca Lines."

27. Lepper, "The Newark Earthworks and the Geometric Enclosures of the Scioto Valley," 237–38.

28. Aveni, "Order in the Nazca Lines."

29. Spirn, *The Language of Landscape,* 218–19.

30. Bachelard, *The Poetics of Space,* 195. Also see Anne Friedberg, *Window Shopping,* 22, on the "cult of immensity."

31. Pasztory, "Andean Aesthetics," 63.

32. Lepper, "The Ceremonial Landscape of the Newark Earthworks and the Raccoon Creek Valley," 112–13.

33. Hively and Horn, "Hopewell Cosmography at Newark and Chillicothe, Ohio," *inter alia.*

34. Lepper, "The Ceremonial Landscape of the Newark Earthworks and the Raccoon Creek Valley," 113.

35. Gosden and Lock, "Prehistoric Histories," 6.

36. Lekson, *The Chaco Meridian,* 130.

37. Chidester and Linenthal, introduction to *American Sacred Space,* 22.

38. Bachelard, *The Poetics of Space,* 9.

39. See Urton, "Andean Ritual Sweeping and the Nazca Lines."

40. Barrett, "Chronologies of Landscape," 24–25.

41. Lepper, *Ohio Archaeology,* 165.

42. Lepper, "The Newark Earthworks," 80.

43. See Silverman, "Beyond the Pampa," 452; and Silverman, *Cahuachi in the Ancient Nasca World,* 308.

44. Silverman, "Nasca. Nazca. Continuities and Discontinuities on the South Coast of Peru."

45. Chidester and Linenthal, introduction to *American Sacred Space,* 15.

46. Silverman, *Cahuachi in the Ancient Nasca World.*

47. DeBoer, "Ceremonial Centres from the Cayapas (Esmeraldas, Ecuador) to Chillicothe (Ohio, USA)," 234.

48. See summary in ibid.

49. Silverman, *Cahuachi in the Ancient Nasca World.*

50. Lepper, "The Archaeology of the Newark Earthworks," 134.

51. Carneiro, "A Theory of the Origin of the State."

STEPHEN H. LEKSON

Hopewell and Chaco

The Consequences of Rituality

INTERPRETATIONS and understandings of Hopewell culture invariably focus on ceremony, religion, and ritual. There seems to be no political or economic model that "explains" Hopewell's astonishing monuments, built by hunter-gatherers who may or may not have had fixed villages. This is disturbing; states, not hunter-gatherers, build big monuments. Is Hopewell impossible? No, there is another example of just such an entity in the southwestern United States: Chaco Canyon, the great eleventh-century Pueblo Indian regional center, was a *rituality.*

The idea of a Chacoan "rituality" was first applied to Chaco by Norman Yoffee.[1] Yoffee insisted on the primacy of ritual: "the ritual nature of Chaco cannot be reduced to its being the handmaiden of economic and/or political institutions."[2]

The notion of a Chacoan rituality caught on, underwritten by our notions of modern Pueblo society as fundamentally ritualistic in nature.[3] The exact nature of Chaco's rituality has, for some, settled on pilgrimage: Chaco was a pilgrimage center[4]—of which, more below. That's an important detail: pilgrimage or Puebloan, the basic *rituality* of Chaco has now become widely accepted by southwestern and nonsouthwestern archaeologists. It is the baseline take-home message in public interpretations of the site. Chaco was a marvel, a mystery, a new thing under the sun: a massive urban-scale regional center fueled by ideology.

If Chaco was a rare bird, its rarity was only relative, like the colorful thick-billed parrot, rarely seen in the southernmost mountains of the Southwest, but once quite common in the Sierra Madres of Mexico. (The thick-billed parrot was, almost by definition, rare and endangered north of the border and now, alas, is endangered in its home range in northern Mexico.) Although ritual certainly played a major role, Chaco was not a rituality. Nor was it unique, something never seen in history or ethnology. We've just been reading the wrong histories: Chaco was a garden-variety Mesoamerican altepetl (of which, much more below), the small-scale political/economic system ubiquitous in central Mexico, remarkable only in its location on the far northern frontiers.

Chaco and Hopewell: Separated at Birth?

Hopewell and Chaco are two of the most famous, most popular "mysteries" of ancient North America. Hopewell's gigantic, enigmatic earthworks and Chaco's imposing great houses excite the public (and archaeological) imagination. Chaco, as it turns out, is not quite the mystery we have made it; indeed it is relatively easy to understand. Conversely, Hopewell, without Chaco's validation, becomes even more "mysterious"—that is, more difficult and challenging to understand.

Hopewell's formal geometries recall the precise forms of Chaco Canyon ruins; Hopewell's remarkable geographic reach was approached (but not equaled) by Chaco's extensive "regional system"; Hopewell's likely emphasis of pilgrimage parallels one popular genre of interpretation of Chaco.[5] Pilgrimage, ritual, and ceremony underwrite many academic and most popular understandings of Chaco and its architectural expression. Chaco has become the poster child for rituality in North America and a ready exemplar for Hopewell.

Hopewell and Chaco met early in American archaeology. Lewis Henry Morgan, the late-nineteenth-century "father of American anthropology," linked Hopewell and Chaco in *Houses and House-Forms of American Aborigines,* his last major work.[6] Morgan believed that the San Juan area (Chaco's region) was the source and font for major migrations to Mexico, the Yucatan, and the eastern United States "and that the Mound Builders came originally from the same country [i.e., the San Juan area], is, with our present knowledge, at least a reasonable conclusion."[7] Morgan further suggested that the Hopewell earthworks were substories for raised pueblos, or communal long houses, of these migrants, such as High Bank Pueblo: "The elevated platform of earth as a house-site is an element in Indian architecture . . . which sprang from the defensive and communal principles in living. . . . Such were the pueblos now in ruins upon the Rio Chaco in New Mexico."[8]

Morgan's ideas about Hopewell were quickly superseded by in-situ ceremonial interpretations: the earthworks were not Pueblos. His characterization of Chaco's ruins as "pueblos" carried through (and beyond) the first half of the twentieth century, most notably in the writings of Edgar Hewett, who, by choice and chance, became the most visible Chaco scholar of that era.[9] Hewett's students and colleagues were not so sure; in 1965, Gordon Vivian—a Hewett student who became the resident park archaeologist at Chaco—suggested that the communal pueblo models might not quite work for Chaco.[10] As described below, Pueblo models (which I call "Pueblo space") never vanished from Chaco and still exert great influence, but interpretations of Chaco after Vivian expanded to include a Mesoamerican outpost, chiefly a redistribution center; a pilgrimage center; and finally zeroed in on a unique social form, a "rituality" (which might include pilgrimage and even polity).[11] (I review the history of ideas about Chaco below.) "Rituality" was a term meant to describe something unique, something new under the sun: something we haven't seen in ethnography or history. Chaco reenters the Hopewell interaction sphere as a legitimizing case: if these strange, otherwise unknown things happened at Chaco, then they could have happened in the Scioto valley.

Wesley Bernardini lists parallels, suggesting a "close morphological similarity" of major "great houses" to Hopewell mound groups.[12] "Like Hopewell earthworks, Chaco great houses show little sign of residential use and appear to be largely ceremonial constructions," he writes, and the very largest of Chaco's monuments "approached the scale of individual Hopewell earthworks" in labor.[13] The spatial scales of Chaco's "regional system" and Hopewell's core were also roughly comparable.[14] "Thus, a middle-range society precedent certainly exists for a regional ceremonial system organized around a central ceremonial precinct on the scale encompassed by Hopewell geometric earthworks."[15] And, of course, Hopewell "roads" find ready parallels in Chacoan "roads."[16]

A Chacoan rituality underwrites many Hopewell interpretations. Notably, Ray Hively and Robert Horn, in a recent Hopewell volume, bring Chaco and Hopewell full circle: "At this early stage of the inquiry cultural analogies between the Anasazi and the Hopewell would be as inapt as Morgan's Hopewell pueblos. Yet at present Chaco may be the closest 'architectural' analogy to Ohio Hopewell at and near its core."[17] But what if Chaco was not a rituality, not unique, or untoward? What if Chaco was a minor, garden-variety Postclassic Mesoamerican state, translated into local San Juan idioms? First, Chaco described; then, Chaco decoded.

Chaco Described

Chaco is an otherwise unremarkable sandstone canyon in northwestern New Mexico. The heart of the canyon is a 12 km long stretch with the intermittent Chaco Wash running from east-southeast to west-northwest. The north side of the canyon has towering sandstone cliffs topped by wide "slick rock" terraces; the south side is less dramatic. Chaco is a phenomenon because of its great houses —a dozen huge, monumental, geometrically formal structures built of beautifully crafted masonry. The largest great houses were concentrated in a 2 km diameter "downtown" zone at the center of Chaco Canyon. These include Pueblo Bonito, Pueblo Alto, Chetro Ketl, Pueblo del Arroyo, Kin Kletso, and many other monuments and smaller structures.[18]

Chaco, moreover, extended far beyond the confines of its core at Chaco Canyon, with a 150 km radius inner circle, a 250 km outer reach, and Mesoamerica as its distant trading partner.[19] Chaco was the center of a large regional system, about 80,000 sq km (30,000 sq mi., the size of Ireland), defined by about 150 smaller "outlier" great houses scattered over that vast area, linked back to Chaco by an extensive road networks and line-of-sight signaling systems.

Roads appear much as their name implies: long, straight, wide (typically 9 m) engineered earthen features, leveled and bermed. Roads were meant for masses or formations of people, possibly in pilgrimage or processions or even, perhaps, for troops. The symbolic or monumental aspects of roads were perhaps more important than their utilitarian roles; the dense network of roads in downtown Chaco, for example, included redundant, parallel routes clearly unnecessary for efficient pedestrian use. Roads symbolized, monumentally, political and historical connections between places. Roads ran out from Chaco, like spokes on a wheel. Not all roads led to distant great houses; others led to important natural features.

A complex and extensive line-of-sight communication system, with signal fire stations atop high points, paralleled the road system, allowing information to pass from Chaco to the edges of its region and back again via "repeater" stations, relatively quickly.

Small outlier great houses used the same techniques and design principles as Chaco Canyon great houses, but the outliers were typically about one-twentieth the size of Pueblo Bonito or Chetro Ketl—as if a portion of those giant buildings had been cut away and transplanted up to 250 km away. Almost always, the great house sat amid (and usually above) a scattered community of a score or more commoners' houses, called "unit pueblos." Together, a great house and surrounding unit pueblos formed a village with a total population of perhaps 60 to 655.[20]

The differences were unmistakable. An entire unit pueblo, compressed to its floor area, would fit in a large room at a Chaco great house. Unit pueblos —homes of the people—and great houses—elite residences—constitute one of the clearest examples of stratified housing in archaeology, Minoan in its clarity. In Mesoamerican terms, great houses were the palaces of noble families; unit pueblos were the farmsteads of commoners.[21] Those terms seem untoward for North America, which we have long been told was always simple and "intermediate," but *palaces, nobles,* and *commoners* almost certainly describe the structures and residents of Chaco accurately.

Shortly after 1000 CE, great houses took a canonical turn in form and function: they became monuments, in addition to elite residences. The organic, curved plans of earlier great houses were replaced by precise, formal geometries. Great house plans are typically described by letters: "D"-shaped, "E"-shaped, and so forth. Vast blocks of storage rooms, disproportionate to the relatively small numbers of residents, were added, as were monumental public and official spaces. Most of Chaco's great houses were built along the north side of the canyon in little over a century, from 1020 to about 1125 CE, but each great house has a unique construction history, and several started much earlier.

Great houses were expensive and required a great deal of labor to build. That is, the labor-per-unit measure of floor area or roofed volume far exceeded that for unit pueblos—built and maintained by their resident families. Labor on monumental scales was organized and coordinated for site preparation (leveling and terracing); extensive foundations; massive, artfully coursed masonry walls; timbered roofs and ceilings (hundreds of thousands of large pine beams brought from distant forests); skillful carpentry, which can only be appreciated today from masonry remnants of elaborate wooden stairways, balconies, and porticos; and other monumental features and furniture unique to these remarkable buildings. Among these were colonnades (a Mesoamerican form, found only at Chetro Ketl); unique raised shelf platforms (for storage or sleeping) within rooms at most great houses; and large sandstone disks (approximately 1 m diameter and 30 cm thick) stacked like pancakes as foundations or dedicatory monuments beneath major roof-support posts of great kivas.

Great houses served the dead as well as the living. The earliest part of Pueblo Bonito, the cluster of earliest rooms at the center of the building, became a mausoleum for elite burials. Two high-status middle-aged men—perhaps the buildings' founders—were buried in 850 CE with great wealth in wooden crypts beneath the building's floor.[22] These honored dead defined one aspect of the great house's monumentality. Later construction preserved the early core with its burials, enveloping the older masonry in better-built blocks of rooms. Many more elite deceased were later richly interred in the oldest parts of the building.

In contrast, burials at unit pueblos were typically in middens fronting the homestead, accompanied by a pot or two. Scores of unit pueblos and aggregates of such units—commoner homes—lined the south side of the canyon.

Life at a great house differed in almost every way from life at a unit pueblo. Unit pueblos were built and maintained by the family it housed, with help from close relatives. (During Chaco's era, farmsteads increasingly relied on an organized bulk trade in pottery and foodstuffs.) The few elite families who actually lived in great houses could not build them themselves. Impressive amounts of labor were required to build the huge structure, labor recruited from far beyond Chaco Canyon. Commoners from Chaco's unit pueblo may have done much of the domestic work that kept Pueblo Bonito running. At most great houses, rooms lined with batteries of corn-grinding metates and huge ovens in plazas suggest that teams prepared meals for larger groups—feasts, no doubt, among elites and their followers or in public ceremonies cementing the elite's position in society.

Chaco constituted a city, a regional center that performed services for and transformed its region.[23] More than that, Chaco was a capital city, the seat of political power. Chaco was political and ritual theater, meant to awe and impress.[24] Its population of perhaps twenty-seven hundred was comparable to many second- and third-tier Mesoamerican capital cities.[25] Chaco had an overall urban design. The major great houses and other elements are sited relative to a master axis, a precise north–south line that I have elsewhere termed the Chaco Meridian.[26] That axis was monumented by key great houses atop the cliffs, high above the canyon: Pueblo Alto to the north and Tsin Kletzin to the south. Celestial geomancies shaped Chaco, with alignments to north and to lunar or solsticial points.[27]

Chaco Decoded: The Rituality Demystified?

It would be possible to consider Chaco as a great failure of southwestern archaeology. We know so much about Chaco, yet we agree so little—not at all—about what Chaco was. Historian Daniel Richter, trying to make sense of Chaco, recently complained that "the surviving physical evidence leads archaeologists to wildly different conclusions," most particularly "about the degree to which Chaco Canyon was politically stratified."[28] Severin Fowles recently summarized (part of) the range of variation on Chaco: "There is little agreement on the nature of Chacoan leadership. Some interpretations present the canyon as a pilgrimage center managed in a relatively egalitarian fashion by resident priests. . . . Others claim that at its apex Chaco was a secondary state dominated by an elite who resided in palaces, extracted tribute, kept the masses in place through threat, and occasional practice, of theatrical violence"; he concludes, "Chaco polarizes contemporary scholarship."[29] I think Fowles is quite correct but goes not quite far enough. He omits a third "pole" in Chacoan studies: Chaco as essentially Puebloan.

Interpretations of Chaco varied greatly over the past hundred years. Initially, following Morgan's advice, archaeologists simply assumed that their sites, their parts of Chaco Canyon, were independent farming villages; Pueblo Bonito was one town, Chetro Kelt another.[30] By the mid-twentieth century, it became clear that Chaco was a bit more complicated, more than the sum of its parts; it was a complex, multiethnic settlement "not in the direct [historical] line of the . . . Pueblo continuum as it was exposed at the beginning of the historic period."[31] That prescient, radical reinterpretation exceeded Pueblo space, and it achieved little traction.

Still, Chaco's great houses were anomalous in the eleventh and twelfth centuries, prompting some archaeologists in the 1950s and 1960s to question Chaco's place in the Anasazi sequence. Was the Chaco the result of influence from high civilizations of Mexico? Some concluded that Chaco was the result of Mesoamerican influences.[32] Opinion was sharply divided, and James Judge would later summarize Chacoan thinking of that time as either "Mexicanist" or "indigenist."[33]

The New Archaeology of the 1970s and early 1980s favored local adaptation over diffusion, migration, and extraregional influences. In that intellectual atmosphere, researchers rejected Mesoamerican explanations in favor of the evolution of the Chaco as a "complex cultural ecosystem." New Archaeology posited complex political structures, locally developed but still out of place in a gradual cultural evolution from ancient Anasazi to modern Pueblo. Managerial elites, chiefs, and other political structures went far beyond conventional, egalitarian Pueblo models. Opinion was divided, sometimes bitterly. The most heated debates centered on sites in Arizona; Chaco was (generally, but not universally) accepted as a low-level "complex" society, that is, a "chiefdom" in the terms of those times.[34]

Postprocessual approaches of the 1990s and early 2000s reconfigured Chaco to fit postmodern tastes. Influenced by European revision (and rejection) of Neolithic chiefdoms, southwestern archaeologists emphasized ceremony at Chaco, favoring rituality over polity or economy.[35] Postprocessual approaches also reestablished cultural history and contingency as being at least as important as the evolutionary schemes of New Archaeology. Legal requirements for "culture affiliation" (under the Native American Graves Protection and Repatriation Act) reinforced historical interests. The congruence of postprocessual historicity and legally mandated affiliation studies encouraged an archaeology not unlike cultural history in the 1940s and 1950s; the path of cultural affiliation leading back from Pueblos to Chaco also led forward from Chaco to the modern tribes. Chaco was reaffirmed as profoundly Puebloan while at the same time grandly anomalous.

The range of current interpretations of Chaco is staggering—and a bit embarrassing. Chaco, we are told by eminent authorities, was a canyon of farming villages, much like modern pueblos;[36] or a pilgrimage center;[37] or a militaristic regional capital.[38] These interpretations seem quite distinct, and archaeology should be able to tell them apart. But we can't: there is no consensus, no clear agreement that one interpretation is more likely than another.

It has become a common claim that archaeology's toolkit holds no ethnographic, historical, or theoretical parallels for Chaco (even Pueblo ethnography does not quite cover the spread). We are told repeatedly that no ethnographic or historic model can be found to fit Chaco. Therefore we invent things, uniquities that paper over Chaco's anomalies: "*communitas*, or anti-structure";[39] "rituality";[40] "locus of high devotional expression";[41] "a corporate faceless chiefdom";[42] dual priestly leadership whose "authority was derived from control over ritual and esoteric information";[43] or the leaderless creation of "many thousands of acts of individual decision making to allocate time and energy."[44] These are new things under the sun. But whichever uniquity we chose, all keep Chaco within the Pueblo space.

Chaco, in the end, is declared deeply mysterious—the "mystery of Chaco" suits the manufactured Southwest of Santa Fe and appeals to postmillennial spiritualities of many tourists. An unsatisfactory state of affairs: if after a century of disproportionate expenditure of resources at Chaco we must settle for mystery, there seems scant hope for southwestern archaeology.

Perhaps there's a middle ground. Unlike the stark dichotomy of Mexicanists and indigenists in the 1970s, both ritual and politics are important for understanding Chaco. Few researchers would claim one to the exclusion of the other; it is a matter, rather, of degree. To view Chaco data with both ritual and political emphases is legitimate and appropriate, for the data sustain both interests. But at Chaco (and throughout the Southwest), ritual consistently trumps politics.[45] The bias toward ritual has been attributed to "the dead hand of ethnography,"[46] the still-strong influence of anthropological accounts of Pueblos (Morgan's directive), reinforced by current heritage concerns of those same Pueblos. Pueblos run on ritual; thus Chaco, too, must have been fundamentally ritualist in its structure and operation within the Pueblo space.

Pueblos are powerful attractors. Even in the heady, ahistorical days of New Archaeology, archaeology was still anthropology, or it was nothing. And anthropological New Archaeology projected modern (i.e., ethnographic) Pueblo kinship systems back to fourteenth-century Mogollon sites (Broken K and Carter Ranch). Ever since Morgan (and before and after that scientific detour) Pueblos have been the principal frame of reference for thinking about the ancient Southwest, at least the northern half. Even for Chaco—with its palaces and urbanism—we favor normalizing or minimizing interpretations congruent with our view of the institutions of modern Pueblos, the intellectual Pueblo space. Pilgrimage center or rituality or locus of high devotional expression or faceless chiefdoms seem appropriate interpretations of ancient Chaco, even though modern Pueblos are not and have none of those things.

The past was different. Dramatic changes in material culture and social institutions before and after 1300 CE—a watershed—demonstrate that difference.[47] Those differences were profound. For the ancient past, Pueblo ethnography is not sufficient and indeed may not even be necessary—for some questions, ethnography does more harm than good. It provides the wrong frame of reference.

Pueblo societies developed in reaction to and rejection of Chaco, after 1300 CE.[48] In some ways, modern Pueblos are perhaps the last place to look for Chacoan insights. We need other, independent, non-Puebloan "triangulation points" to define, delimit, and understand Chaco's past. Instead of a Puebloan intellectual space defined from nineteenth-century ethnologies and twentieth-century myths, I suggest that we look at what was happening elsewhere in eleventh-century North America—the ancient Southwest's actual context.

Chaco should be contextualized by its contemporaries, specifically Mesoamerica in the ninth through thirteenth centuries (i.e., the Early and Middle Postclassic periods).[49]

There may be no useful ethnographic or historic models for Chaco in Pueblo space—ethnographic, geographic, or conceptual. But by casting our interpretive nets beyond Pueblo space into Mesoamerica, we find ready models that fit Chaco well and effectively "solve" the mystery Chaco. Mesoamerican models more accurately and effectively represent ancient Chaco than any competing current model. Specifically, Chaco was an altepetl.

In brief: the ubiquitous local polity in Postclassic Mesoamerica was a small unit termed, in Nahua, an altepetl (plural *altepeme*). It surely would have been known to Chaco and other societies in the Southwest.[50]

Most altepeme were rather small.[51] Population ranged from as few as two thousand to as many as forty thousand people—comparable to the population of Chaco's region. An ideal altepetl consisted of a half dozen major noble families, each with its own palace clustered in a central "city." Rulership revolved through the leading noble families.[52] There was a king elected from among the noble families, but the office was not strong, nor did it descend in a dynastic line. Commoners owed tribute in the form of taxes or labor to specific noble families, as did minor nobles to major nobles and so forth. But tribute was not oppressive: a few bushels of corn, a few weeks of labor, occasional military service, and so forth.

The cluster of major great houses at Chaco Canyon is remarkably similar to the central cluster of altepeme. The seven major great houses at Chaco, in this model, represent the altepetl's six to eight major noble families and their palaces. Other buildings represent cadet branches, minor nobility, priesthoods, and so forth, much like the minor great houses at Chaco. The radial divisions of Chaco's region, marked by roads running to scores of small, outlier great houses, parallels the (idealized) radial subdivisions of the altepetl, with each noble family controlling commoners in its piece of the pie. Secondary outlier great houses took care of business in the countryside. There were, of course, differences: Chaco translated Mesoamerican forms into local idioms of architecture, ideology, and cosmology. Most altepeme's central clusters had a pyramid and temple.

The biggest difference between the Chaco polity and the altepetl structure is scale, size. While the total population of Chaco's region fell well within altepetl ranges, Chaco's region (80,000 sq km) was a thousand times larger than the average altepetl, 75 sq km. The alarming difference in spatial scale may reflect fundamental differences in productivity between Chaco and Mesoamerica. Mesoamerican altepeme enjoyed fruitful environments for corn; high productivity supported dense populations in relatively small areas. Chaco's region, in contrast, was bleak. Arable lands were scarce, minimally productive, scattered far and wide. Overall population density was consequently quite low—pockets of settlement ("communities") separated by large stretches of desert.[53] Chaco, perhaps, represents the altepetl political form stretched to its elastic limits, over very difficult terrain.

Thus, Chaco was not a mystery. Chaco was a basic, garden-variety Mesoamerican polity, much like hundreds of other contemporary Mesoamerican polities. It was remarkable only (or mainly) for being the northernmost of those formations, out on the edges of civilization. Or was that remarkable? Cahokia or Cuba might provide other examples if we put them in proper context.

Best Wishes and Hopewell

Hopewell, in contrast, really *is* weird. There *are* new things under the sun from time to time, and Hopewell just might be one. That claim—if we choose to make it—will require assessment from beyond the Hopewell core. It is always useful to think big and to put things into large and larger contexts.

HOHOKAM AND CHACO

There are superficial (that is, surficial) parallels between Hopewell and Chaco, principally in the organization of a symbolic landscape or—better—built environment. The two societies (or aggregates of societies) were, I fear, quite different in social and political structure. Hopewell was not an altepetl. But several potentially important superficial similarities transcend those differences—differences that were merely basic and fundamental—between the two times and places.

Chaco had large earthworks: berms, mounds, and even a few circles comparable in diameter (but not in height) to Newark's. Chaco's were not nearly as massive as Hopewell's. And, in the case of berms and circles, Chaco's earthworks do not appear to represent enclosures. The circles encircle outlier great houses; they exclude rather than enclose—a fine distinction but an important one. Chaco's earthworks and circles defined social spaces around great house occupants and excluded commoners. Long walls running east–west along the mesa top above Pueblo Bonito and Chetro Ketl served the same purpose: the walls were long but low—symbolic rather than practical divisions of space into "Chaco" inside, and "not-Chaco" outside. People approaching Chaco from the north were funneled via roads into small, controlled fates in those long, low walls and presumably had to flash credentials, or sign papers, or pay tithes before making the transition from one space to another. Within Chaco's several circles, we find great houses—but no commoner houses.

Roads, of course, are another attractive and obvious parallel. The so-called Great Hopewell Road was named in part for Chaco's Great North Road, which in turn was named for the famous Roman road in Britain.[54] It is quite likely that Roman, Chacoan, and Hopewell roads were designed for rather different purposes. Roads in the Chacoan world surely conveyed pedestrian traffic, but they also served, importantly, as markers of political/ritual affiliation, and of history—"roads through time" that connected great houses and sites of differing ages.[55] Hopewell roads may have carried similar symbolic loads.[56] There are Mesoamerican examples of lines on the land monumenting social and historical connections. I will return to roads below, in a discussion of Classic period Mesoamerica.

Perhaps the most interesting parallel between Chaco and Hopewell is the very largest: the conception—we might even say "invention"—of space. By "space" I mean an abstract understanding of two or perhaps three dimensions on geographic scales. The projection of lines across space requires a grasp of geometry, perhaps couched as geomancy. Both Hopewell and Chaco lined things up over significant distances.[57] That's a useful tool, when later societies wanted to divide lands, create land laws, and otherwise mark and delimit the earth.

HOPEWELL AND THE WORLD

By looking large and escaping the dense gravities of Pueblo space, it is possible to solve or at least re-solve the "mystery" of Chaco. Its contemporary world provided a fine model for understanding Chaco. Chaco, as it turns out, was interesting but hardly unique—and indeed, not even unusual for its time and place (Postclassic North America). Might that work for Hopewell?

What was Hopewell's context? Typically, Hopewell is considered an anomaly among Middle Woodland populations of the eastern United States. But Hopewell's world was surely larger than that—a world known if not directly experienced. That world was Classic period North America. Hopewell was a close contemporary with Teotihuacan, the great Mesoamerican *ur*-city. Teotihuacan began about 200 BCE, rose rapidly from 1 to 200 CE, dominated central and northern Mesoamerica through 500 CE, and then fell with some violence. It was enormous and largely unprecedented: far larger than other Mesoamerican cities of its time (larger than most European or Asian cities, for that matter). Teotihuacan was a key transformative event in North American prehistory, the eight-hundred-pound gorilla in any reconstruction of those times and after, from the Yucatan to northern Mexico—and perhaps beyond. Teotihuacan's long-distance influences form a recognizable pattern in American archaeology. Ben Nelson has suggested how its fall, around 700 CE, created waves of politicization that eventually lapped up on the margins of the American Southwest.[58]

Teotihuacan famously intrigued with long-distance interventions in Maya, Oaxaca, and other major Mesoamerican civilizations. Less famous, but more relevant to our problem, was Teotihuacan's presence on the northern frontiers of Mesoamerica. Two centers in Zacatecas, in particular, have been proposed as part of Teotihuacan's larger world: Alta Vista and La Quemada.[59] Alta Vista was an early node in the turquoise network linking the Southwest and Mesoamerica, with a clear "trademark" of Teotihuacan, a distinctive pecked circle-cross petroglyph. La Quemada—a city on a hill—was something more and had something more relevant, perhaps, to Hopewell: roads. Teotihuacan had roads; the Avenue of the Dead was a monumental road from 40 to 95 m wide, which formed the central axis of the city and which may have continued south out from the city to a nearby range of low mountains. La Quemada's roads were quite a bit more complex, with a network of raised causeways radiating out from the city on the hill, crisscrossing the small valley in which the city was centered. La Quemada's roads were designed for pedestrian traffic, with broad stairways and ramps; but their configuration suggests something other than pedestrian, quotidian uses.[60] La Quemada's roads may well have inspired Chaco's later road symbolism, but La Quemada was contemporary with Hopewell and much earlier than Chaco.[61] There is nothing about La Quemada that looks like Hopewell and nothing in Hopewell that looks like La Quemada. The two were economically, politically, and socially quite unlike. Yet considerations of Hopewell roads (and other alignments) would do well to consider La Quemada as a paragon of Mesoamerican symbolic landscapes.

Hopewell was not the first monumental tradition in the Mississippi River's greater drainage or in the eastern United States. That honor goes to the Middle Archaic mounds of the lower Mississippi valley (and perhaps the south Atlantic coast). The earliest mound groups date to 3700 to 2700BCE, long before

Olmec or other foundational Mesoamerican civilizations. After an apparent hiatus, monumental construction returned, spectacularly, at Poverty Point, a complex geometric earthwork with concentric polygonal earthworks built between 1400 and 1200 BCE—that is, a few short centuries after Olmec's rise a thousand miles across the Gulf of Mexico. Increasingly, archaeologists are considering the possibilities of a "gulf coast archaeology" spanning the southeastern United States and Mexico—the title of a pioneering collection of papers considering the Gulf of Mexico as a cultural world rather than two separate national archaeologies.[62] Questions of interconnections between eastern North America and Middle America have a long and difficult history;[63] from the perspective of the Southwest, there seems no reason to doubt those connections or to minimize their importance. They may well have been complex and bidirectional: a recent article—on the Southeast, not the Southwest!—in *Science* asks, "Does North America hold the roots of Mesoamerican civilization?"[64]

Those issues—while fundamental to building frames of reference for Hopewell—are beyond the scope of this essay. But surely part of Hopewell's world was the larger region's past, the history to which Hopewell was heir. Despite the hiatus from the end of Poverty Point to Hopewell's beginnings a millennium later, we would be wrong to think that Hopewell's cosmologies and traditions emerged without reference to that earlier monumental center. The astonishing geographic scale of Poverty Point's "connections" (not quite reaching into Mexico) presages Hopewell's, as does the construction of an elaborate symbolic landscape. Curiously, one popular reconstruction of Poverty Point harks back, unintentionally, to Lewis Henry Morgan's vision of High Banks, with rows of houses atop Poverty Point's ring of mounds.[65]

Envoi

Voltaire famously said, "History is a pack of tricks we play on the dead." It is easy for archaeology to "play tricks" on the past. Maya poet-priest-kings turn out to be bloodthirsty, saber-rattling Khans. That dramatic volte-face is a textbook case of lovely theories spoiled by a few new facts, in the Maya case the decipherment of glyphs telling a very different story. How more likely a "Motel of the Mysteries" misfire when no texts remain to correct our aim. (David Macaulay's 1979 farce of a future archaeologist completely misreading a buried mom-and-pop motel should be read yearly by every archaeologist.)

It could happen here: I think it did happen with Chaco. Our narrow frameworks—limited ethnographically to the Southwest, limited conceptually to "intermediate societies"— overrode very compelling evidence for something more. Our misreading of Chaco reflects both an overregard for the Pueblo present and a ritual turn in archaeology, generally. Ritual, it has been taught, was of paramount importance in the past. Perhaps so, perhaps not. Ritual certainly is of paramount importance in the present southwestern archaeology. Southwestern archaeology is chided by non-southwesternists for its fixation on ritual: ritual *uber alles.*

Perhaps aging baby-boomer archaeologists of no fixed religion are getting in touch with their spiritual sides. (There is a strong New Age component to much nonprofessional southwestern archaeology, particularly amateur passions for petroglyphs and archaeoastronomy.) More likely (and less insultingly), waves of theory from across the Atlantic washed up on American shores. Propelled by poststructural and postprocessual disaffections with (mainly American) neoevolutionary social phylogenies, influential British archaeologists insisted that the past was a foreign country, where they did things differently. There was an "otherness" to the past—specifically, something other than modern materialism.

To be sure, there were things in the past that we do not see in the present. Chaco, for example: there are no living examples of an altepetl. But there are many things in the past *that we can see in the past.* Chaco, for example: it was almost certainly at home in Postclassic Mesoamerica. Chaco flourished around 1100 CE; by that time, North America was pretty well integrated, its historical trajectory well advanced.

The coincidence of Poverty Point and Olmec suggest (strongly) a continental dynamic, perhaps a continental *history,* a millennium before Hopewell culture began. We might do well to consider Hopewell and classic Mesoamerica. The stellar life cycle of Teotihuacan—an immense pull of population into a dense urban core, followed by remarkable fusion of cultural energy, and ending with a supernova explosion outward—surely had continental impacts. Chaco tells us something about the formation of secondary states—state-like polities on the edges of empires. Hopewell might tell us something about secondary cosmologies, new ways of thinking born on the edges of civilization.

Notes

1. Yoffee, "The Chaco 'Rituality' Revisited."

2. Yoffee, Fish, and Milner, "Communidades, Ritualities, Chiefdoms," 265–66.

3. E.g., B. Mills, "Recent Research on Chaco"; Ware, "Chaco Social Organization"; and Ware, "Descent Group and Sodality."

4. E.g., Judge, "Chaco Canyon–San Juan Basin."

5. E.g., Bernardini, "Hopewell Geometric Earthworks."

6. L. Morgan, *Houses and House-Life of the American Aborigines.*

7. Ibid., 223.

8. Ibid., 232–33.

9. E.g., Hewett, *The Chaco Canyon and Its Monuments.*

10. Vivian and Mathews, *Kin Kletso.*

11. E.g., Yoffee, "The Chaco 'Rituality' Revisited."

12. Bernardini, "Hopewell Geometric Earthworks," 351.

13. Ibid.

14. Ibid., 352.

15. Ibid.

16. E.g., Lepper, "The Ceremonial Landscape of the Newark Earthworks and the Raccoon Creek Valley."

17. Hively and Horn, "Rejoinder," 207.

18. Lekson, *The Architecture of Chaco Canyon, New Mexico.*

19. Lekson, "Lords of the Great House"; and Lekson, *A History of the Ancient Southwest,* 130–32.

20. Mahoney, "Redefining the Scale of Chacoan Communities."

21. Lekson, "Lords of the Great House."

22. Stephen Plog and Carrie Heitman, "Hierarchy and Social Inequality in the American Southwest, AD 800–1200," *Proceedings of the National Academy of Sciences,* 107, no. 46 (2010): 19,619–26.

23. Monica L. Smith, "Introduction," 7–8.

24. For an experiential analysis, see Van Dyke, *The Chaco Experience.*

25. Michael E. Smith, "Aztec City States"; and Michael E. Smith, *Aztec City-State Capitals.*

26. Lekson, *The Chaco Meridian.*

27. Sofaer, "The Primary Architecture of the Chacoan Culture: A Cosmological Expression."

28. Richter, *Before the Revolution,* 17.

29. Fowles, "A People's History of the American Southwest," 195.

30. E.g., Hewett, *The Chaco Canyon and Its Monuments.*

31. Vivian and Mathews, *Kin Kletso,* 115.

32. E.g., Hayes, "A Survey of Chaco Canyon Archaeology."

33. Judge, "Chaco Canyon–San Juan Basin," 233.

34. E.g., Schelberg, "Analogy, Complexity, and Regionally-Based Perspectives."

35. B. Mills, "Recent Research on Chaco"; Wills, "Political Leadership and the Construction of Chaco Great Houses"; Yoffee, "The Chaco 'Rituality' Revisited"; and Judge and Cordell, "Society and Polity."

36. E.g., Ware, "Chaco Social Organization"; and Ware, "Descent Group and Sodality."

37. E.g., Kantner and Vaughn, "Pilgrimage as Costly Signal: Religiously Motivated Cooperation in Chaco and Nasca," *Journal of Anthropological Archaeology* 31 (2012): 66–82; and Malville and Malville, "Pilgrimage and Periodic Festivals as Processes of Social Integration in Chaco Canyon."

38. E.g., Wilcox, "The Evolution of the Chacoan Polity."

39. Wills, "Political Leadership and the Construction of Chaco Great Houses."

40. Yoffee, "The Chaco 'Rituality' Revisited."

41. Colin Renfrew, "Production and Consumption in a Sacred Economy: The Material Correlates of High Devotional Expression at Chaco Canyon," *American Antiquity* 66, no. 1 (2001): 14–25.

42. Earle, "Economic Support of Chaco Canyon Society."

43. Judge and Cordell, "Society and Polity," 209.

44. Monica L. Smith, "What It Takes to Get Complex," 55.

45. B. Mills, "Recent Research on Chaco."

46. Drennen and Peterson, "Challenges for Comparative Study of Early Complex Societies," 69.

47. Lekson, *A History of the Ancient Southwest,* 192–98; and Glowacki and Van Keuren, *Religious Transformation in the Late Pre-Hispanic Pueblo World.*

48. Lekson, *A History of the Ancient Southwest.*

49. Smith and Berdan, *The Postclassic Mesoamerican World.*

50. My principal references are García and Zambrano, "El Altepetl Colonial y sus Antecendentes Prehispanicos"; Mendoza, "Territorial Structure and Urbanism in Mesoamerica"; Hirth, "The Altepetl and Urban Structure in Prehispanic Mesoamerica"; Hirth, "Incidental Urbanism"; Hodge, "When Is a City-State?"; Lockhart, *The Nahuas after the Conquest;* Michael E. Smith, "Aztec City States"; and Michael E. Smith, *Aztec City-State Capitals.*

51. Hodge, "When Is a City-State?"; Michael E. Smith, "Aztec City States"; and Michael E. Smith, *Aztec City-State Capitals.*

52. Lockhart, *The Nahuas after the Conquest.*

53. John W. Kantner, "Rethinking Chaco as a System"; and Mahoney, "Redefining the Scale of Chacoan Communities."

54. Lepper, "The Great Hopewell Road and the Role of Pilgrimage in the Hopewell Interaction Sphere."

55. Stein, "The Chaco Roads?," cited in Lekson, *The Chaco Meridian.*

56. Lepper, "The Great Hopewell Road and the Role of Pilgrimage in the Hopewell Interaction Sphere."

57. DeBoer, "Strange Sightings on the Scioto"; Hively and Horn, "Rejoinder"; Dennis Doxtater, "Parallel Universes on the Colorado Plateau"; and Lekson, *The Chaco Meridian.*

58. Nelson, "Aggregation, Warfare, and the Spread of the Mesoamerican Tradition."

59. On Alta Vista, see González and Uranga, *Alta Vista*. On Quemada, see Nelson, "Aggregation, Warfare, and the Spread of the Mesoamerican Tradition"; and Nelson, "Chronology and Stratigraphy at La Quemada, Zacatecas, Mexico."

60. Trombold, "Causeways in the Context of Strategic Planning in the La Quemada Region, Zacatecas, Mexico."

61. Nelson, "Chronology and Stratigraphy at La Quemada, Zacatecas, Mexico."

62. White, *Gulf Coast Archaeology*.

63. E.g., Cobb, Maymon, and McGuire, "Feathered, Horned, and Antlered Serpents."

64. Lawler, "America's Lost City."

65. J. Gibson, *The Ancient Mounds of Poverty Point*, fig. 5.11.

TIMOTHY DARVILL

Beyond Newark

Prehistoric Ceremonial Centers and Their Cosmologies

Ceremonies and ritual observance connected to concerns about life, fecundity, well-being, and death are fundamental elements of the human condition and everyday experiences; they are axiomatic to what Martin Heidegger referred to as "dwelling" on the earth and fit within his four-fold concept of "oneness": earth and sky, divinities and mortals.[1] Many aspects of these emotional attachments lie in the domain of intangible heritage—language, music, dance, sacred knowledge, beliefs, representations, cosmologies, and worldviews that peoples and societies hold dear and transmit through oral traditions, participation, pupilage, and performance. But such things also find formal expression in the tangible material world through what Colin Renfrew described as "technologies to cope with the unknown": symbolic and projective—architecture, art, ceremonial monuments, holy objects, sacred places, and special spaces.[2] Embodied in such material culture are the practical realizations of big pictures that serve as mnemonics for beliefs, the means to structure ceremonies and rituals, and, in a very real sense, ways of representing in microcosm the world of particular social realities.

Size is important for ceremonial monuments in a social context. Almost every society provides opportunities for people and their gods to meet together, be it in natural sanctuaries in the landscape, domestic shrines within the house, transportable tabernacles, or towering temples. Most serve local communities and are of commensurate scale. Occasionally, however, much larger ceremonial centers serve extended communities and act as regional foci within sacred landscapes by representing symbols of cosmic, social, and moral order. In prehistoric times such centers typically appear at one of three key moments in the development of social complexity: the emergence of stable agricultural communities, the formation of hierarchical or chiefdom societies, or the coalescence of political units into simple state systems immediately preceding the appearance of urban centers. The Newark Earthworks (fig. 1) represent the physical remains of one such substantial ceremonial center spread over more than 83 ha beside the Licking River in central Ohio.[3] The sheer scale and diversity of Newark's numerous components represented by enclosures, mounds, avenues, platforms, and burial grounds set it apart as one of a handful of important and significant Hopewell ceremonial centers. Much is already known through studies dating back to the mid-nineteenth century,[4] but many questions remain to be answered about how it developed, how it worked, and what the various components meant to those who used it. Some of these can only be answered through new research at the site itself, but comparative perspectives are also potentially useful. Accordingly, this essay looks beyond the Newark Earthworks and the Ohio valley into the wider world of ceremonial centers across time and space in order to provide a broad context through cross-cultural comparisons and mapping possibilities. Working on a wide canvas, I shall explore four themes represented at selected ceremonial centers across Eurasia, Africa, and the Americas: sacred geography, seasonal communal meetings, cosmological structuring, and links between life and death—all of which bear on what can be seen in the archaeological record at Newark and related Hopewell centers.

Sacred Geographies

Ceremonial centers generally both represent and create a sacred geography through a social use of space that is intimately and recognizably linked to beliefs and understandings of sacred and profane worlds. As Paul Wheatley pointed out some years ago, operationally, ceremonial centers were instruments for the creation of political, social, and economic order, but structurally they were symbols of cosmic, spiritual, religious, and moral order.[5] Positioning in the landscape was therefore important, and often gave special meanings to the ceremonial centers of particular communities. A connection with water, especially rivers or lakes, is common. This may have the practical advantage of facilitating communications and access to a site, but water also has many attributes that find expression in beliefs (healing, purification, a liminal zone between worlds, a boundary that restricts the movements of people and spirits). It can also be seen metaphorically as representing the journey of life from source to sea. James Mooney notes that among Cherokee

PLATE 1. Digital reconstruction of the northernmost rising of the moon at the Octagon Earthworks. (John Hancock, Center for the Electronic Reconstruction of Historical and Archaeological Sites, University of Cincinnati)

PLATE 2. Aerial photograph of the Octagon Earthworks in Newark. (Timothy Black)

PLATE 3. Daytime photograph of the moon rising at the Octagon Earthworks. Note that the moon is not rising over the middle of the parallel walls because this is not the northernmost rising of the moon. (Timothy Black)

PLATE 4. Time-lapse photograph of the northernmost rising of the moon at the Octagon Earthworks. (Timothy Black)

PLATE 5. Aerial photograph of the Great Circle Earthworks at Newark in the snow. The Great Circle is one of two remaining features of the Newark Earthworks and is located approximately a mile from the Octagon Earthworks, the other remaining feature. (Timothy Black)

PLATE 6. A golfer prepares to tee off with the moon overhead at the Moundbuilders Country Club within the Octagon Earthworks in Newark. (Timothy Black)

PLATE 7. (*Above*) Photograph of the interior of the Octagon Earthworks. (Timothy Black)

PLATE 8. (*Right*) Wray Figurine, also called the Shaman of Newark, a unique stone carving associated with a burial beneath the largest and most centrally located of the burial mounds at the Newark Earthworks. (Courtesy the estate of Edmund S. Carpenter and the Ohio History Connection)

PLATE 9. Postcard. Golfer at Moundbuilders Country Club. (Licking County Historical Society Archives)

PLATE 10. Postcard. Clubhouse at Moundbuilders Country Club. (Licking County Historical Society Archives)

PLATE 11. An Aztec dancer offering flowers to children at the 2010 Newark Earthworks Day celebration at the Great Circle Earthworks in Newark. Twelve descendants of Aztec people traveled to the Newark Earthworks from Mexico to dance in the Great Circle for this occasion — much as it is believed ancient people traveled to the Newark Earthworks for important occasions two millennia ago. (Timothy Black)

PLATE 12. Three Shawnee youth (members of the Eastern Shawnee Tribe of Oklahoma) waiting to dance at the Octagon Earthworks. Twelve members of the Eastern Shawnee Tribe of Oklahoma traveled to Newark to visit the earthworks and dance for the public in August 2012. (Timothy Black)

PLATE 13. Pilgrimage Walkers in the small circle next to the Octagon Earthworks at Newark, 2010. More than fifty people walked seventy miles from the Mound City Earthworks in Chillicothe, Ohio, to the Newark Earthworks, as it is believed ancient people walked between these sites two millennia ago. Gilly Running and Hunter Garner lead the group around the circle. (Timothy Black)

PLATE 14. Visitors touring the Octagon Earthworks at Newark on one of four annual "open houses" when golf is suspended and the public is invited into the site. Here a group is led through the two parallel walls from the Octagon into the circle. (Timothy Black)

PLATE 15. Visitors listening to Mark Welsh (wearing orange) of the Native American Indian Center of Central Ohio at the Octagon Earthworks on one of four days each year when the public is given access to the Octagon. Mark Welsh, Mary Gordon, and William Weaver (short woman and tall man with white hair next to the sign) acted as tour guides at the Newark Earthworks. (Timothy Black)

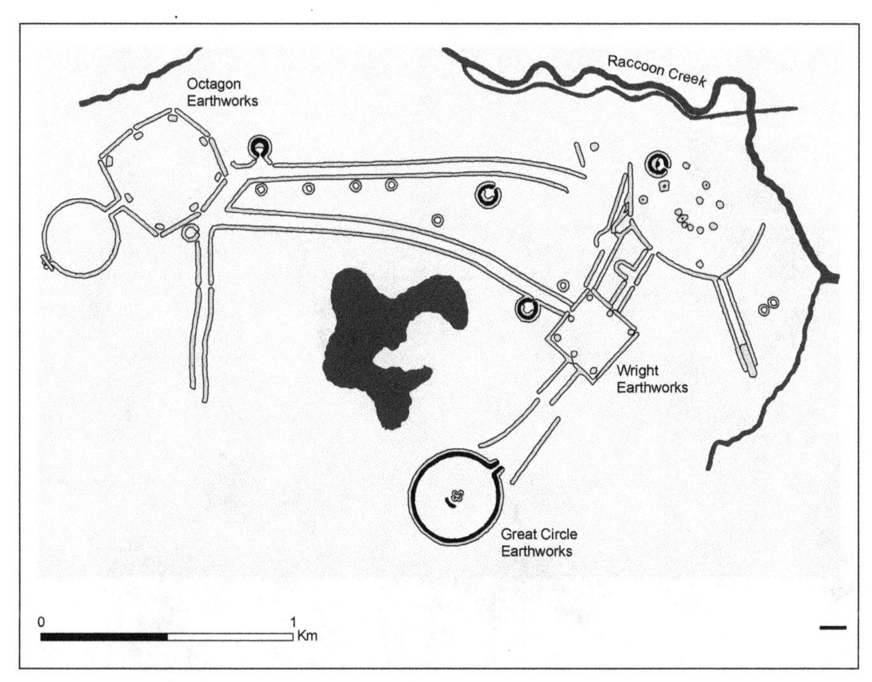

FIGURE 1. Plan of the main component monuments of the Newark Ceremonial Center. (After Squier and Davis, *Ancient Monuments of the Mississippi Valley* [1848], plate 25; drawing by Vanessa Constant)

people every important ceremony contains a prayer to the "Long Person," the formulistic name for the river.[6]

Sometimes the linear form of rivers can perhaps be glimpsed in the elongated enclosures and defined pathways found at nearby ceremonial centers. It has been suggested that at Newark movements between the various enclosures were structured by embanked avenues that link them together and connect the whole complex to the South Fork of the Licking River via the Cherry Valley Ellipse with its burial mounds and enclosures.[7] Similarly, in the quite different cultures of southern Britain in the fourth and third millennia BCE, cursus monuments often run parallel or at right angles to rivers, while avenues link rivers to henges and henge enclosures.[8] At Stonehenge the earthwork avenue linking the stone circle with the River Avon is 2.5 km long with a stone circle at start and finish.[9] Avebury is much the same, with the stone-lined West Kennet Avenue (fig. 2) starting at a stone circle on Overton

FIGURE 2. West Kennet Avenue at Avebury, Wiltshire, United Kingdom, looking southeast with pairs of stones marking the edge of the processional route that originally led from the Sanctuary via the River Kennet to Avebury Henge. (Photograph by Timothy Darvill; copyright reserved)

Hill before passing beside the River Kennet and progressing on to Avebury itself over a total distance of 2.4 km.[10]

Such linearity implies that those passing along the route were arranged in some kind of order that is played out in the timing of their arrivals and departures, the structuring of their movements between nodes within the complex, and their visibility or otherwise to those assembled at the site. Such performances have been documented at the Temple of Heaven (Tian Tan) in Beijing, China, situated south of the Forbidden City (fig. 3). Here emperors of the Ming (1364–1644 CE) and Qing (1644–1911 CE) dynasties worshipped heaven and prayed for good harvests in a complex covering 273 ha. Movements between the main foci—the Circular Mound Altar in the south and the Altar of Prayer for Grain in the north (fig. 4)—were structured along the 360 m long Haiman Road, which forms the central axis of the complex. Changes of clothes and regalia were needed along the way so that some components of the structure, for example the Platform of Changing Clothes, were

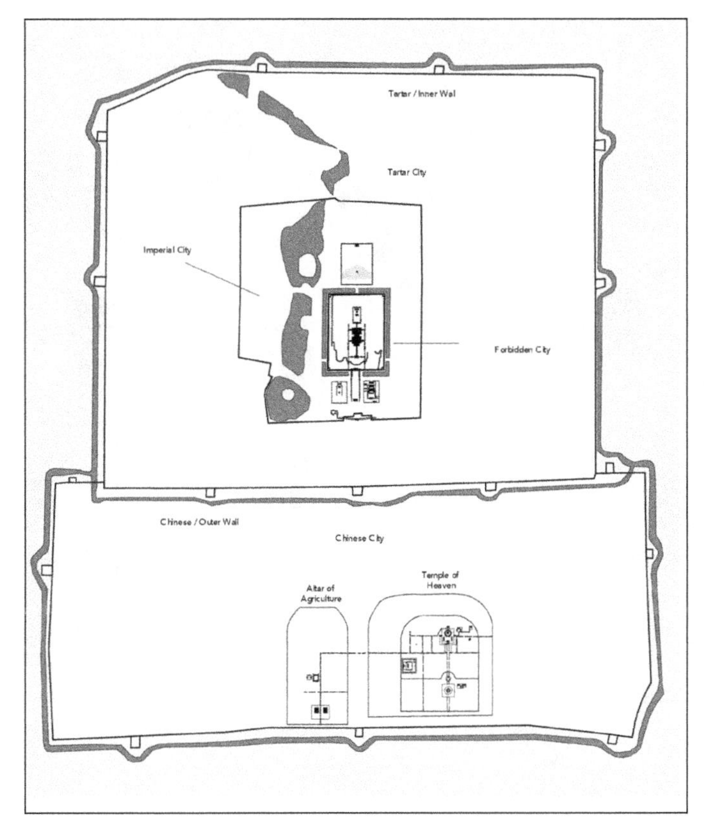

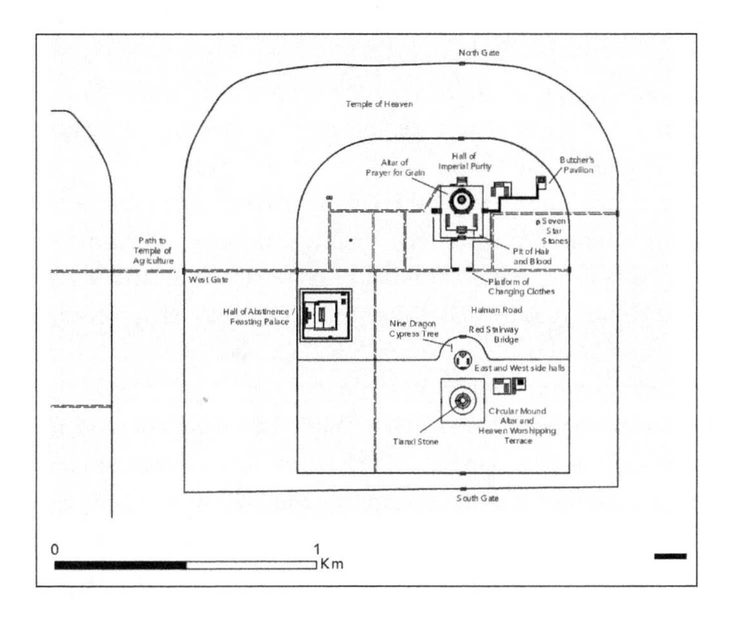

FIGURE 3. Beijing and the Temple of Heaven, China. *Top,* the main component monuments of the Beijing Ceremonial Center, Beijing Shi. *Bottom,* detailed plan of the Temple of Heaven (Tian Tan). (Top after C. P. Fitzgerald, *Ancient China* [London: Elsvier/Phaidon], 1978, 26; bottom after Yang and Lu, *Temple of Heaven,* 96; drawing by Vanessa Constant)

FIGURE 4. Temple of Heaven (Tian Tan), Beijing, China. Hall of Prayer for Good Harvests, looking north. (Photograph by Timothy Darvill; copyright reserved)

principally concerned with transformations in preparation for later stages of the journey.[11]

Circles, squares, and rectangles seem to predominate in the ground plans of individual components at ceremonial centers, perhaps because these geometric shapes are relatively easy to set out and construct. More complicated forms such as the octagons at Newark and High Bank Works in Ohio are not unusual and have the capacity to represent a greater number of relationships and focal points or stations. The use of timber posts or standing stones, separately or in predetermined patterns, is also common, and levels of meaning may be embedded in the source, character, color, and composition of the materials selected, the shape and form of each component, and the arrangement of the whole. Contrasting shapes may imply contrasting components of a conceptual scheme. At the Temple of Heaven in Beijing, heaven itself, following Chinese cosmology, is seen as a circle, while the earth is seen as a square. Thus the symbolic structure of a square juxtaposed with a circle is common in the layout of the whole monument and in the architectural detail of individual components. Details such as shape and form can also be important in the interpretation of meaning, especially binary oppositions that may reflect very fundamental dimensions such as life and death; male and female; day and night; summer and winter; old and young; and, as already seen in the Chinese case, heaven and earth. For the Stonehenge landscape, Mike Parker Pearson and his Madagascan colleague Ramilisonina suggested an east–west division of the local landscape into a domain of the living focused on Durrington Walls and a domain of the ancestors focused on Stonehenge itself.[12] They suggest that such divisions were monumentalized in the use of timber structures in association with the living and stone for the memorialization of the dead. It is a seductive model but one that is founded on questionable cross-cultural structural analogies.[13] Elsewhere I have shown that while Stonehenge was certainly a cemetery in its early years, the great ceremonial complex was transformed through the placing of powerful stone pillars, the so-called bluestones brought more than 220 km from outcrops in the Preseli Hills of southwest Wales.[14] These stones were set up and used within a massive stone structure made from locally available sarsen stones in a fashion that replicated traditional shrines more typically made of timber. The arrangement of the bluestones at Stonehenge represents in microcosm the arrangement of the outcrops in the real landscape from where they came, while the connections between bluestones and water in the Preseli Hills and the traditional association of both with healing and well-being provide a strong reason for the movement of the stones themselves and their subsequent treatment at Stonehenge.[15] Such movements of special materials from a significant source for use in ceremonial contexts at distant centers has been documented in other situations and has been dubbed "shrine franchising" by anthropologist Tim Insoll.[16] In the case of Stonehenge it is especially telling that the earliest available documentary accounts of the site set down in the twelfth century CE, which were presumably derived from oral histories and legends, explain the significance of the stones as having magical powers for healing and locate their source quiet correctly in the far west of Britain.[17]

Seasonality and Communal Gatherings

Some ceremonial centers were also settlements, others simply places where dispersed populations gathered at a particular time; most involve the fission and fusion of populations as numbers swell for significant events and then dwindle as the routines of everyday life demand. At a practical level the availability of subsistence resources may be significant for the timing and duration of gatherings, which thereby mimic seasonal rhythms. Symbolically, such gatherings often revolve around appeasing supernatural forces, vision questing, enlightenment, and attempts to turn favors from deities to secular advantage. The gatherings themselves may thus be destinations for pilgrimage, powwows, trading fairs, sacred games, or a combination of these. Conceptual associations between landscape features, sacred spaces, constellations visible in the heavens, mythologies, and journeys are not uncommon and have been well documented by Linea Sunderstrom in relation to the Black Hills of South Dakota among Lakotas, Cheyennes, Kiowas, and Kiowa-Apache communities.[18]

Something similar may be glimpsed in the ancient world documented by Tacitus, the first-century CE Roman politician and writer. He records that among the Suebi tribe of Germania the oldest and noblest lineage was the Semnones. All the people of this blood gathered together at a set time in a wood hallowed by the auguries of their ancestors. The ceremonies opened with the sacrifice in public of a human victim. It was believed that the nation had been born within the wood and that the god who ruled over them dwelled among the trees.[19] In the case of the ceremonies at the Temple of Heaven in Beijing, the whole population had an interest in successful intersessions with the gods, but those actually involved in the process were limited to the emperor's family and the officiating priests. The emperor was cast as the Son of Heaven, who administered earthly matters on behalf of, and representing, heavenly authority. The ceremonies essentially map a physical and metaphorical journey that started with a period of starvation and ended with sumptuous feasting several days later.[20]

Fixing a meeting place within an otherwise fairly impermanent system has much practical value, but Colin Renfrew has suggested that gathering places were critical for creating human interactions as a fundamental of the human condition.[21] By creating contexts for group-oriented social interaction, often in the form of cosmologically rendered rituals structured through physically constructed monuments, there are opportunities for developing and sharing cognitive understandings of socially constructed realities. It is a view that accords well with studies by the evolutionary psychologist Robin Dunbar, who sees collective talking, laughing, singing, and dancing as key social activities that humankind developed, enhanced, and elaborated because of the intoxicating effects of participation and the emotional enrichment, euphoric state, and sense of well-being that ensued.[22]

FIGURE 5. Temple of Amon-Ra at Luxor, Egypt, looking east along the processional avenue lined by sphinxes with rams' heads, through the courtyard of Nectanebo to the pylon of Ramses II, preceded by two obelisks (one preserved; its missing partner was removed to the Place de la Concorde in Paris in 1836), and two seated colossi representing the Pharaoh. (Photograph by Timothy Darvill; copyright reserved)

Journeying to ceremonial centers and cult places may be as important as arriving. Traditional routes are typically marked by decorated stones, rocks, special trees, or perhaps formal ceremonial roads and spirit paths.[23] Installations may be expected along the routes, as Petersen has shown with the reference to the water systems, khans, mosques, forts, palaces, cemeteries, settlements, and road markings associated with the Islamic Hajj routes of Arabia.[24] Architecturally spectacular are the great sphinx-lined avenues joining temples at Luxor, Egypt (fig. 5), but more typical are the less elaborate structures such as the Great Hopewell Road running southward from Newark for perhaps sixty miles or so to Chillicothe, which may be a formally defined approach route for those attending the ceremonies.[25]

All these dimensions have major implications for the layout and physical structure of the ceremonial center itself. As already noted, elements such as enclosures form arenas for performance, avenues structure movements, and mounds focus attention on particular people or activities. All serve both to include and exclude; they perpetuate social and religious orders. But it is also important to focus on the spaces as well as the structures, especially the large open spaces, as these are likely to be the plazas, dancing grounds, game fields, and campsites that come into use as the occasion builds. In terms of their meaning and significance, big events also demand big attendance.

Studies of contact-period Creek settlement sites in the Southeast of North America shed much light on one kind of ceremonial center: the "square

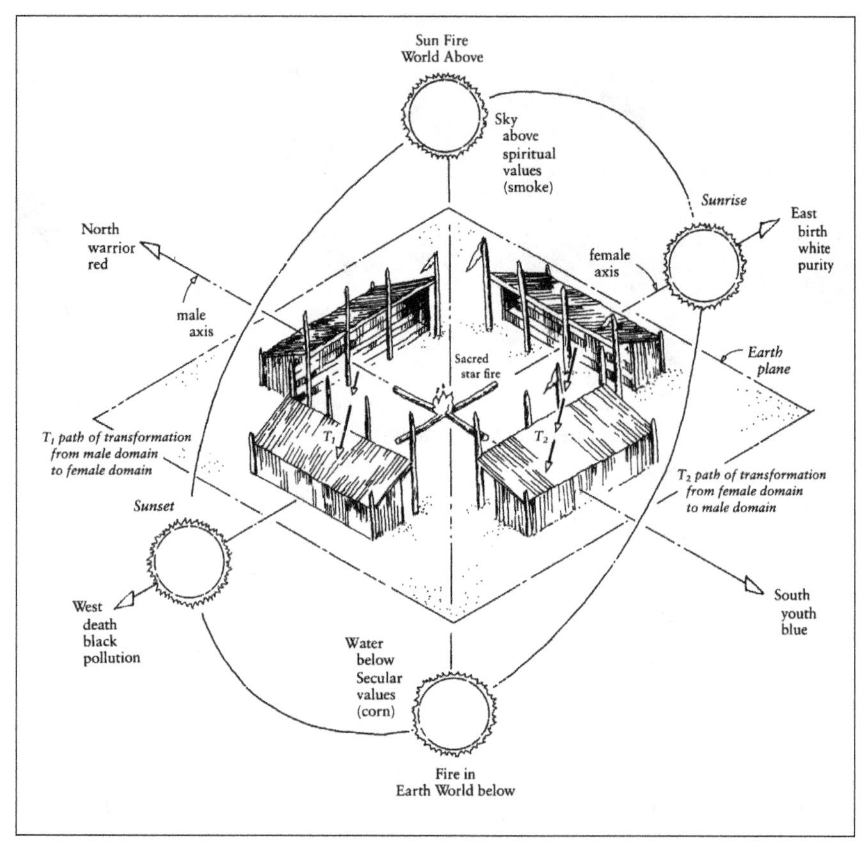

FIGURE 6. Schematic plan of a Creek square ground showing the cosmological symbolism and principal axes. (After Nabokov and Easton, *Native American Architecture,* 110; drawing by Vanessa Constant)

ground,"[26] which is sometimes considered a development of prehistoric plaza sites in the same area.[27] This was the summer location of the sacred fire, and the traditional center of political and religious activities (fig. 6). Four shelters called "clan beds" were situated around a square plaza. In late summer the Green Corn Ceremony was intended to renew ritual and cosmic vitality and reaffirm social structure. On the fourth day of the ceremony the scared fire was rekindled and campfires relit from it. The flame, which represented the sun itself, was believed to renew the vitality of the household, village, and cosmos. Color was symbolic in the square ground. Red clan beds (signifying war) were built astride the male axis, while white beds (peace) were placed on the female axis. A man's seating assignment reflected his change in age and role from boy to warrior to old man. There is also the suggestion that the male passage followed a path from the world of his mother to the world of men and warriorhood and then finally, as an elder, a return to the female domain.

In an archaeological context in Britain there is extensive evidence for seasonal occupation along the Avon valley and around Durrington Walls to the east of Stonehenge during the third millennium BCE.[28] Recent investigations around the great henge enclosure of Durrington Walls suggest that at midwinter this was a large, bustling, and mixed community of possibly thousands of people, a seasonal encampment of revelers awaiting ceremonies at Stonehenge.[29] Hunting, cooking, and eating pigs seem to have occurred,[30] and fires were a conspicuous feature of the small square houses.[31]

The presence at ceremonial sites of repetitious elements is a common pattern and deserves exploration. In some cases there may be a series of similar structures because the very purpose of the event involved gathering essentially repetitious social groups who might each have a focus within the larger communal structure. At Newark, the Octagon with its attached circle is paralleled by the similarly aligned Great Circle and Wright Earthworks, between and around which are numerous smaller embanked or ditched enclosures (fig. 1). The avenues linking these components suggests some kind of progression between the various parts, as discussed above, but in this way of thinking the superficial similarities in the structural form of each element masks the differences. Binary divisions of society along the lines of male/female, day/night, and summer/winter are fairly common and find expression in the use of different sectors of a site. Ternary divisions around a cosmological representation of space in terms of an underworld, earth plane, and sky world or time as being past, present, and future can also be imagined. Close attention to contextual associations is needed here, as particular people, plants, animals, and sprits may be associated with particular domains, while other kinds of being such as birds, trees, water, and shamans may be able to connect or pass between worlds.

Episodic gatherings imply episodic construction and refurbishment. Archaeological evidence suggests that ceremonial centers are dynamic places that develop and change over time as pieces are added and redundant structures abandoned. There is always a temptation to see the plan of a site in its totality as some kind of architectural blueprint, a preconceived design that people gradually moved toward over centuries or millennia. Such a pattern would be exceptional in a prehistoric context. Indeed, Richard Bradley has argued that the design of places was in many ways less important to prehistoric people than the acts of creating them.[32] Thus what we see is not always what was intended but rather what emerged. In this way of looking at sites, the recutting of ditches, the extension of avenues, the reorientation of alignments, the construction of new enclosures, the placing of the next burial mound, and so on arise from social tensions and contests of authority rather than grand designs. In the case of the platform mound at Irene, Georgia, Victor Thompson has argued that the construction of the mound served to cement community relations through widespread participation, but upon completion the top of the mound became exclusive space with access to it controlled both physically and visually.[33]

Material, texture, and color may be important too. The construction of a communal focus can physically embody components from the social world. Mention has already been made of the so-called bluestones at Stonehenge, a range of dolerites, rhyolites, and tuffs that were selected from particular outcrops in west Wales and brought to Stonehenge because of their perceived magical powers.[34] Elsewhere in Europe there are cases where material representing the territories of those building a communal monument are integrated into the structure. On the island of Jersey in the English Channel, for example, the passage grave of La Hougue Bie includes in its construction seven different kinds of stone brought from different outcrops in the north, south, and east. All were variously used to form the walls and roofing of the central burial chamber and approach passage in the early fourth millennium BCE.[35] Somewhat later in date, on the Isle of Man between Britain and Ireland, the great stepped parliament mound known as the Tynwald Hill lay at the center of a Norse open-air meeting place (fig. 7). It is believed to be constructed from soil brought from each of the parishes that it served.[36] Interestingly, the power of the mound is such that people leaving the island sometimes take a tiny piece of the hill with them to reinforce their ancestral connections to the place.

One consequence of building monuments from differently sourced materials is that when new, these structures would have been rather colorful. Miles Russell has suggested that the exposed surface of some Neolithic long barrows in southern Britain would have looked rather like a Battenberg cake of yellows and browns, as materials of contrasting colors were dumped into adjoining fenced bays that structured the body of the mound.[37] At the Thornborough central circle, excavations in 1952 revealed the presence of gypsum on the original surface of the enclosure bank, suggesting that its builders had

FIGURE 7. Tynwald Hill at St. John's, Isle of Man, looking west with the walled processional way opening onto a penannular enclosure surrounding the centrally placed stepped Thing Mound, or parliament hill. Open-air parliaments have been held on the mound since Norse times and continue annually on July 5 to this day. (Photograph by Timothy Darvill; copyright reserved)

deliberately tried to make the monument appear white.[38] It is a practice that chimes with Brad Lepper's recognition that yellowish brown gravel had been deposited on the berm between the ditch and the dark brown bank at the Great Circle at Newark.[39]

Timing and tempo may be fixed into the form and architecture of ceremonial centers. The rondels of central Europe are generally seen as meeting places and cult centers at which there may have been a strong link with the marking of time.[40] But preparing, journeying, attending, and participating takes time, so that the overall duration of gatherings must be seen in terms of weeks or even months. Certain times were no doubt deemed more sacred than others, so being there at the right time was important; in this, astronomy has key a role to play.

Cosmological Structuring

Alignments onto heavenly bodies are embedded into the design and structure of the key components of most ceremonial centers, although not all the components within a single center necessarily have the same orientations and alignments. Many authors have succumbed to the enumeration of all sorts of astronomical alignments for monuments across the globe—stars, constellations, planets, sun, and moon—but always questions have to be asked about the level of precision possible in prehistoric surveying, the quality of the archaeological information relating to the contemporaneity of features used for sighting, and, of course, issues surrounding the visibility and placement of celestial bodies in the sky as seen by potential observers in the past because of changes arising through precession. As Clive Ruggles has admirably demonstrated, the most common alignments are toward key moments in the progress of the sun and moon across the sky.[41] Sun and moon are sometimes seen as deities in their own right, with the sun generally seen as male, while the moon is female (Greek: Apollo/Artemis; Roman: Apollo/Diana; Navajo: Day Traveler/Night Traveler).

At Newark, the axis of the Octagon allows the major north moonrise to be observed from the platform on the southwest side of the Observatory Circle through the opening on the northeast side of the Octagon.[42] This alignment occurs in precise astronomical terms every 235 lunar months or 18.6 solar years, but in practical terms it can be observed for a few days each month for up to a solar year either side of the lunar extremum itself. There is no evidence that key moments in the solar cycle were fixed into the architecture of the Newark Earthworks, although both sun and moon are well represented at other contemporary ceremonial centers in Ohio and at many other centers across the world.[43] Stroke-ornamented pottery culture rondels such as that in Goseck, Germany, in central Europe tend to have entrances that open toward the sunrise and sunset on the winter solstice, whereas the slightly later Lengyel culture enclosures combine solar alignments with markers indicating changes in moon phase.[44]

The principal axis of Stonehenge is famously aligned with the summer and winter solstices, and there is increasing evidence that it was sunset in the southwestern sky at the winter solstice that attracted the most attention.[45] But here too alignments that fix key lunar events are also represented, most notably the four station stones marking the corners of a rectangle whose long northwest-southeast sides align on the major limiting moon rising in the south (full in summer) or setting in the north (full in winter), which happens every 18.6 solar years.[46] Lionel Sims has argued that the builders here were deliberately trying to invest the sun with properties that reflected the moon's religious significance to earlier communities in the area.[47] The significance of the winter solstice may often lie in its association with rebirth and new beginnings. Certainly this was case at Tian Tan, China, where the axis of the site reflects the physical astronomical reality of the winter solstice, while the metaphysical religious purpose of the rituals carried out there focused on ensuring good harvests in the forthcoming year.[48]

In all these instances the relationship between the architecture and the movements of heavenly bodies is fairly general. Ceremonial sites were not scientific observatories for the development of abstract understandings. Rather, they were working models of a socially constructed universe that was understood by each society in its own terms; recurrent and predictable configurations prompted particular activities linked to wider cosmologies. The rhythm of the sky became the rhythm of life. Time and space were no doubt the important considerations in all these cases: being in the right place at the precise moment that worlds collide, deities come close to humans, or doorways between realms open wide to allow movement across other impenetrable divides is central to ritual practices and religious beliefs.

Juxtaposing the movements of the sun and moon is a feature of many early cosmological and calendrical schemes.[49] As a device for representing the passing of time—what Gavin Lucas calls time indication[50]—observing the daily cycle of the sun and marking the two solstices provide secure and verifiable moments that anchor festivals and events to the routine of daily life. Conveniently, it divides the year into two seasons—summer and winter. The addition of lunar observations allows the recognition of lunar months (average 29.53 solar days), and if the lunar extrema are included, then metonic eras each equating to 18.6 solar years can be defined as a basis for time reckoning.[51] At Stonehenge some kind of time-reckoning system was probably built into the architecture of the Sarsen Circle (fig. 8), which contains thirty upright pillars joined at the top by lintels. Notably, stone 11 is about half the size of all the others, yielding 29.5 stones in reality—the average number of solar days in a lunar month.[52]

Calendars were probably only part of the story at Stonehenge and other ceremonial centers. As I have discussed elsewhere, neither the time-indication devices nor the time-reckoning apparatus embedded into the architecture of Stonehenge in themselves provide plausible reasons why the monument was constructed; they were built into its structure to facilitate and program its real purpose.[53] Rather it was the life of the sun attaining its daily passage through the heavens and the underworld that was of greatest interest to those occupying the earth. In a similar way Herman Bender usefully distinguishes

FIGURE 8. Stonehenge, Wiltshire, United Kingdom, showing the Sarsen Circle from the enclosure boundary, looking southwest. (Photograph by Timothy Darvill; copyright reserved)

between the "function" of structures such as medicine wheels in North America that fix alignments in a permanent structure and determine and verify particular celestial events or sequences of events, and their "purpose," which was to help maintain the universal order on behalf of humankind.[54] The beauty of the system is that everyone could recognize the patterns at a general level by locally observing solar and lunar movements against the skyline or in domestic shrines and therefore know when it was time to attend larger-scale gatherings.

Understanding the basic movements of the heavens and the juxtaposition of celestial bodies at key moments is a dimension of traditional sacred knowledge that must have developed over many generations. It is not something that had to be replotted every time it was needed, however, as archaeologically there is evidence for such things being recorded through encoding in material culture. Most well-known is the Nebra Disc from the Mittelberg near Nebra in central eastern Germany, deposited about 1600 BCE in a pit within a walled hilltop enclosure.[55] In its first form this gold-inlayed copper disc shows the sun or full moon, crescent moon, Pleiades, and various other stars encoded to show the leap rule needed to synchronize lunar months with solar years. Later it was modified to include two gold arc-shaped plates indexing the extreme risings and settings of the sun at the summer and winter solstices, and finally a boat symbol was added to reflect the cosmologies of the mid-second millennium BCE in central Europe.[56] Of about the same date is the gold lozenge recovered from the grave of an adult male buried in

Bush Barrow just to the southwest of Stonehenge.[57] Alexander Thom and colleagues also see this as an alidade-type instrument for fixing key epochs within a simple calendar,[58] a view not universally accepted.[59] In North America, star maps were used by Lakota communities and perhaps other groups, but few people knew of them, and even fewer were able to read them and understand their meaning.[60] Such maps may well be relevant to understanding the form, structure, and layout of the Newark Earthworks.

Links between Life and Death

Ceremonial centers often stand on boundaries. Sometimes these are physical boundaries beside rivers or geographically constituted units, sometimes they are social boundaries at the intersection of tribal lands or ground that is shared by adjoining communities, but they may also be cosmic boundaries where the mortals meet the immortals, spirits meet the living, and the quick and the dead are united. Burials are present around most ceremonial centers, as at Stonehenge, Newark, and countless others around the world.[61] However, just as significant as the human remains is the material culture. This represents elements of the prevailing system, contributors to the union, and it is common to find earth, fire, water, and air as familiar themes. Images of shamans such as the Wray Figurine from Newark have an intimate scale about them.[62] Altogether larger are the posts and stones that may represent ancestors or ancestral deities while also providing the sighting points for alignments or mnemonics for calendrical systems. In the case of Stonehenge, I have suggested that the five Sarsen Trilithons in the center of the site could be seen as conjoined deities, pairs of gods, or ancestors born at the same time from a single union who may also represent male/female, day/night, summer/winter oppositions.[63] The Great Trilithon to the southwest, the largest and most prominent, stands astride the principal axis and might cautiously be identified with a pair of deities representing day and night: the sun and moon. In both the Greek and Roman pantheons these might be seen as Apollo (male solar deity) and Artemis (female lunar deity), twins fathered by Zeus and born of Leto. Apollo represents divination, prophecy, healing, music, and causing the fruits of the earth to ripen, while Artemis is goddess of forests, hunting, agriculture, and childbirth; both were associated with the ability to cause sudden death.[64]

At Tan Tian numbers counted as architectural components reflected regular patterns that had meaning to the builders and users of the monuments.

The Hall of Prayer for Good Harvest at the northern end of the complex is constructed on a three-tiered marble base, each tier representing part of the cosmological order indicated by carved reliefs of clouds, phoenixes, and dragons.[65] The roof, covered with blue tiles to represent heaven, is supported by twenty-eight pillars, each identified with one of the recognized constellations; the side halls contain stone tablets representing the gods of the sun, moon, and stars. It was within the square enclosure surrounding the hall that animals were sacrificed in order to create new life, the remains of the ceremonies being ritually burned on stoves before being deposited in the Pit of Hair and Blood in the southeastern corner of the enclosure.

Potentially rather important is the way that prehistoric communities combined fundamental dimensions of their beliefs and worldviews in all aspects of the lived world. Thus with reference to life during the third millennium BCE in the isles of Orkney off the northern coast of Scotland, Colin Richards has shown that the layout of dwelling houses is a reverse image of the ground plan of their tombs and that the same structuring principles were also applied to their ceremonial sites in the form of henges and stone circles.[66] All three articulations to the social use of space drew on the visual imagery of the natural world in their architectural representation.[67] Elsewhere, the distinction between houses for the living and houses for the deities is rather blurred. As Ian Hodder and others have shown with such clarity, at Çatalhöyük, Turkey, the "houses" include remarkable mural painting; reliefs and sculptures of bulls, leopards, and women on the walls; and bodies buried beneath the floors.[68] As anthropology has shown time and time again, the principles that structure people's beliefs also provide the syntax for understanding the worlds they create for themselves, especially the social use of space within domestic contexts.[69] Sadly, few houses occupied by Middle Woodland populations in the Ohio valley are known, but one at Edwin Harness Mound in Ross County excavated in 1976 suggests a two-part structure linked by a short tunnel rather in the manner of the very much larger Octagon Earthwork and Observatory Circle at Newark and the similarly shaped High Bank Works near Chillicothe.[70]

Holistic views of the world do not preclude attempts by the living to steer and control behaviors in other worlds. Mounds, enclosures, fences, boundaries, ditches, and the paraphernalia associated with the technologies to cope with the unknown are all obvious examples. Some are concerned with remembering, prompting emotions and refreshing images and memories of bygone times. But equally there is a technology of forgetting.[71] In Britain there is increasing evidence that the great earthwork enclosures of some henge monuments were the final act of construction, shutting off and separating the powerful ritually charged interiors and memories of the events that took place there from the lived-in world unfolding round about. Certainly this is the case at Durrington Walls and Balfarg and might also apply at Avebury.[72]

Conclusion: Questions of Balance

Across time and space there is much variety in the complexity, construction, purpose, and use of ceremonial centers. Each is unique, but each instances the context in which it was created and represents wider thinking. Belief systems and cosmologies lie at the heart of human action. To understand them requires a social-cosmological interpretation that relates the lives of those involved with the structures to the ongoing process of "becoming" rather than simply "being." A central theme is the idea of balance: balance between life and death, time and space, male and female, young and old, tradition and innovation, the known and the unknown, past and future. Only by understanding something of the bigger picture can individual components of our great ceremonial centers make sense. And this is true not just for the academic understanding of these places but also for their conservation and management. Balance is important here too. Naturally there are practical issues of hegemonic tensions about the ownership of traditions and the need for progress. Assessing archaeological significance and value is incredibly difficult and culturally specific.[73] One possible way forward, however, is linking research and knowledge creation to site management through the development of a research framework that reconciles tensions, defines attainable objectives, and recognizes that new work should not simply perpetuate "scientific knowledge."[74] There are many kinds of "knowledge" that can be developed and explored with reference to ancient sites that individually and collectively allow a bright future for the archaeological remains, exciting prospects for research and investigation, and a recognizable past accessible to the widest possible audience.[75]

Notes

Thanks to Vanessa Constant for the preparation of figures 25, 27, and 30; Wei-Jun Liang for assistance with Chinese texts; Mary Borgia, Brad Lepper, Dan Campbell, and Jeff Gill for discussions about the Newark site and its meaning; and Richard Shiels,

Brad Lepper, Lindsay Jones, and Marti Chaatsmith, organizers of the Newark Earthworks Symposium, for inviting me to speak and contribute to the debate.

1. Heidegger, *Poetry, Language, Thought,* 148–49.

2. Renfrew, *The Emergence of Civilization,* 405.

3. Squier and Davis, *Ancient Monuments of the Mississippi Valley*; Lepper, "The Archaeology of the Newark Earthworks"; and Lepper, *The Newark Earthworks.*

4. Squier and Davis, *Ancient Monuments of the Mississippi Valley.*

5. Wheatley, *The Pivot of the Four Quarters,* 225.

6. Mooney, "The Cherokee River Cult."

7. Lepper, "The Archaeology of the Newark Earthworks"; and Lepper, in this volume.

8. See Brophy, "Water Coincidence? Cursus Monuments and Rivers"; and Roy Loveday, *Inscribed across the Landscape.*

9. Royal Commission on Historical Monuments (England), *Stonehenge and Its Environs,* 11–13; and Pearson et al., "Newhenge," 15–17.

10. Pollard and Reynolds, *Avebury,* 100–105.

11. Yang and Lu, *Temple of Heaven,* 4.

12. Pearson and Ramilisonina, "Stonehenge for the Ancestors."

13. Barrett and Fewster, "Stonehenge"; and Whittle, "People and the Diverse Past."

14. Darvill, *Stonehenge,* 136–40.

15. Darvill and Wainwright, "Stonehenge Excavations 2008"; and Darvill and Wainwright, "The Stones of Stonehenge."

16. Insoll, "Shrine Franchising and the Neolithic in the British Isles."

17. Piggott, "The Sources of Geoffrey of Monmouth II."

18. Sunderstrom, "Mirror of Heaven."

19. Tacitus and Mattingly, *Tacitus on Britain and Germany,* 132–33.

20. Yang and Lu, *Temple of Heaven.*

21. Renfrew, *Prehistory.*

22. Dunbar, *The Human Story.*

23. Loubser, "From Boulder to Mountain and Back Again."

24. Petersen, "The Archaeology of the Syrian and Iraqi Hajj Routes"; and Porter, *Hajj.*

25. Lepper, "The Great Hopewell Road and the Role of Pilgrimage in the Hopewell Interaction Sphere."

26. Walker, "Tribal Towns, Stomp Grounds, and Land"; and Nabokov and Easton, *Native American Architecture,* 109–11.

27. Cf. Swanton, "The Creeks and Mound Builders."

28. Timothy Darvill, *Stonehenge,* 114–16.

29. Parker Pearson, "The Stonehenge Riverside Project," 142.

30. Albarella and Payne, "Neolithic Pigs from Durrington Walls, Wiltshire, England."

31. Parker Pearson, "The Stonehenge Riverside Project," 140.

32. Bradley, *Altering the Earth,* 98–112.

33. Thompson, "The Mississippian Production of Space through Earthen Pyramids and Public Buildings on the Georgia Coast, USA," 455.

34. Darvill, *Stonehenge*, 136–41.

35. Patton, Rodwell, and Finch, *La Hougue Bie, Jersey*, 29.

36. Darvill, "Tynwald Hill and the 'Things' of Power"; A Stranger, *A Six Days' Tour through the Isle of Man*.

37. Russell, *Monuments of the British Neolithic*, 31.

38. N. Thomas, "The Thornborough Circles near Ripon, North Riding," 429.

39. Lepper, *The Newark Earthworks: a Wonder of the Ancient World*, 12.

40. Whittle, *Europe in the Neolithic*, 190.

41. Ruggles, *Astronomy in Prehistoric Britain and Ireland*; and Ruggles, "Sun, Moon, Stars and Stonehenge."

42. Hively and Horn, "Geometry and Astronomy in Prehistoric Ohio"; and Hively and Horn, "A Statistical Study of Lunar Alignments at the Newark Earthworks."

43. C. Turner, "Ohio Hopewell Astraeoastronomy."

44. Bertemes et al., "Die neolithische Kreisgrabenanlage von Goseck, Ldkr. Weißenfels"; Pavúk and Karlovský, "Astronomische Orientierung der spätneolithischen Kreisanlagen in Mitteleuropa"; and Pásztor, Barna, and Roslund, "The Orientation of *Rondels* of the Neolithic Lengyel Culture in Central Europe."

45. Darvill, *Stonehenge*, 143–44.

46. Ruggles, "Astronomy and Stonehenge," 218–21.

47. Sims, "The 'Solarization' of the Moon."

48. Yang and Lu, *Temple of Heaven*.

49. Hodson, *The Place of Astronomy in the Ancient World*.

50. G. Lucas, *The Archaeology of Time*, 71.

51. Ibid.

52. Darvill, *Stonehenge*, 144.

53. Ibid.

54. Bender, "Medicine Wheels or 'Calendar Sites,'" 203.

55. Harald Meller, "Nebra."

56. Cf. Kaul, *Ships on Bronzes*.

57. Hoare, *The Ancient History of Wiltshire*, 202–4; and Ashbee, *The Bronze Age Round Barrow in Britain*, 76–78.

58. Thom, Ker, and Burrows, "The Bush Barrow Gold Lozenge: Is It a Solar and Lunar Calendar for Stonehenge?"

59. Needham and Woodward, "The Clandon Barrow Finery," 22.

60. Goodman, *Lakȟóta Star Knowledge*.

61. On burials at Stonehenge, see Darvill, *Stonehenge*. On burials at Newark, see Lepper in this volume.

62. Lepper, "The Archaeology of the Newark Earthworks," 14.

63. Darvill, *Stonehenge*, 144.

64. Guirand, "Greek Mythology."

65. Yang and Lu, *Temple of Heaven*, 54.

66. Richards, "Monumental Choreography, Architecture and Spatial Representation in Later Neolithic Orkney."

67. Richards, "Monuments as Landscape."

68. Hodder, *Çatalhöyük*.

69. Parker Pearson and Richards, "Ordering the World."

70. Lepper, *Ohio Archaeology*, 132–34.

71. Bradley, *The Past in Prehistoric Societies*, 87–89; and A. Jones, *Memory and Material Culture*.

72. On Wiltshire, see Parker Pearson, "The Stonehenge Riverside Project," 133. On Balfarg, Fife, see A. Gibson, "Dating Balbirnie. On Avebury, see Pollard and Gillings, *Avebury*, 45–46.

73. Mathers, Darvill, and Little, *Heritage of Value, Archaeology of Renown*.

74. Darvill, *Stonehenge World Heritage Site;* Darvill, "*Scientia*, Society and Polydactyl Knowledge."

75. Cajete, *Native Science;* Darvill, "Research Frameworks for World Heritage Sites and the Conceptualization of Archaeological Knowledge"; and C. Turner, "Ohio Hopewell Astraeoastronomy."

PART IV The Newark Earthworks in Interdisciplinary Contexts

Architectural History, Cartography, and Religious Studies

JOHN E. HANCOCK

The Newark Earthworks as "Works" of Architecture

THE CULTURAL TRADITION responsible for building ancient Newark[1] mastered three major types of earthworks (as opposed to "mounds," which for me as an architect is an important distinction): first, geometric enclosures, of which there are many variations, and of which Newark is probably the greatest; second, hilltop enclosures, most spectacularly Fort Ancient and Fort Hill, though there are also many of these, generally a bit better preserved today; and third, complex or figural earthworks, including those at Tremper, Stubbs, Turner, Newark's Eagle, the Tarlton Cross, and certain crescents as at Fort Ancient. Although this volume is about the Newark Earthworks, I hope we don't lose sight entirely of the larger context and typological range of architectural and cultural achievement that will be going forward to UNESCO for inscription.[2]

The earthworks present spatial, perceptual, and cognitive challenges for modern audiences, and these difficulties have fascinated me as an architectural educator. They open up interpretive and experiential issues, and in turn the always-fascinating deep questions about the nature of architecture and its role in the world.[3] Our research and design teams at CERHAS (Center for the Electronic Reconstruction of Historical and Archaeological Sites) at the University of Cincinnati[4] have addressed these challenges in our exhibits and digital publications, leading us to design them with four main "layers" of content, both visual and verbal.

First, there is the problem of orientation.[5] In order to orient people to the location and scope of this monumental ancient place, we connected it visually with the modern city, centered on its courthouse: the camera flies up from courthouse square to reveal the immense, restored complex stretching for miles to the southwest, where the town of Newark now spreads out around the surviving fragments.

Second, there is the problem of visualization. The earthworks are hard to see because of their huge scale, their tree cover, and of course for most of the Newark Earthworks, their destruction. So, for example, we narrate animated eye-level and fly-over tours of the Octagon, cleared of trees and golfers, to reveal clearly its precision, subtle gateways, and visual connections with the distant, undulating horizons.

Third, there is the problem of explanation. Knowledge about the sites is complex and often deeply layered. So we provide stories in locations that encourage audiences to explore and then go in for deeper understanding. For example, after a video clip in which Professors Hively and Horn explain how they discovered the moonrise alignments, another is available to explain and visualize how the moon's rise and set points on the horizon oscillate over 18.6 years, and then another that re-creates how the primary axial moment would have looked in antiquity, from atop the Observatory Mound, as a full moon rose at dusk (plate 1).

Fourth, there is the problem of interpretation. Opening up the capacity of these works to be continuously meaningful, it is necessary to engage various knowledge traditions as well as our own modern frames of reference. The now self-evident primacy of lunar observation for the site's ancient builders is illuminated and enhanced by a Lenape storyteller's account of "Grandmother Moon" and by Delaware Grand Council chief Linda Poolaw's related story of the association of the full moon with the female ancestors:

> The moon is *a woman* . . . we identify her as *"she."* And in some beliefs . . . one of the Canadian sisters has taught me that when our people go, our females, and that moon is full, *they're dancing around the moon.* And that's when we have our ceremony, and we talk to them, and we send them our prayers, so they can take them on to the Creator. But every time I see the full moon, I think about, you know, our ancestors, my mother and all of them, dancing around the moon. And that's a good thought, a good thought.[6]

And in a similar hermeneutic vein, architectural theorist and educator William Taylor recounts the "uncanny" strangeness of a stealth fighter's appearance at the Dayton Air Show and reflects on this aspect of his first experience of the earthworks:

> the *uncanny*—something coming through into the ordinary world that you're *not supposed to see,* that you *aren't supposed to know about*—that

sense of the uncanny. I think the mounds have something of that. Particularly when we saw them, I saw them, on this beautiful, bright, sunny, summer day. But as it got later in the day and the sun started to go down, and the sun angles got lower, and the shadows began to be extended from these objects, when they began to have these atmospheres in them, where you couldn't see distinctly because they were in shadow, you could begin to see what these things might be like when it starts to get dark, or early in the morning, or . . . when there's this layer of fog on the ground, and these things are rising in this improbable way, into ordinary space, as if *this really shouldn't happen*. There's something of the *uncanny* in that.[7]

Three Disciplinary Paradigms

This series of juxtaposed scenes from our public exhibit program presents, in an intentionally nonhierarchical way, various aspects of the site, various categories of meaning, various forms of representation and description, various types of knowledge, and various layers of human cognition and experience. Seventeen years of working on this topic and presenting it to public audiences has afforded us an opportunity to explore, develop, interpret, and present the three main clusters of meaning and interpretation available in consideration of Newark or the other earthworks.

According to our disciplines and traditions, we encounter them via a particular "hermeneutic as."[8] We view them (1) "as" archaeological *sites* that hold knowledge about distant cultures or (2) "as" monumental *works* of architecture or (3) "as" *places* of continuing Native American meaning. (There are other views out there but these are the main ones, it seems to me.) Clearly, each of these has its methods of operation, its criteria for truth, its conventions of knowledge, and its customary modes of representation. What has become clear to me, especially in studying other successful UNESCO World Heritage nominations, is that it will be important to focus on, and perhaps recalibrate, how we are elaborating, balancing, and blending these three categories of interpretive activity.

Now, there's been some tension between (1) and (3), the objective/scientific and the Native/traditional, for a variety of reasons, although that seems to be abating recently. Since I've started advocating for (2), some have suggested to me that maybe a better understanding of the sites "as works of architecture" would actually help to bridge these more epistemologically opposed positions, since it can move between them, around them, and perhaps beyond them, adding value and substance across a wider range of interpretive capacity.

Specifically, elaborating on the sites' identity and capacity "as works of architecture" can help to advance them within the nomination process as "masterpieces of human creative genius" (criterion i), since definitions of these keywords have been provided in UNESCO publications as follows: a masterpiece is "a complete and perfect piece of workmanship, an outstanding example"; "creative" means "inventive, or original; either first in a movement or style, or the peak of one"; and "genius" refers to "a high intellectual/symbolic endowment, a high level of artistic, technical, or technological skills." Moreover, most all inscriptions using criterion i, either primarily or secondarily (as the Ohio Hopewell Nomination Committee has now decided to do) have been "artistic or architectural masterpieces . . . (or) technical and artistic tours-de-force"—in other words, works of architectural design or technical engineering achievements.[9]

To strengthen the Ohio Hopewell nomination in the context of UNESCO, it will be important to explore these values and implications in addition to or even distinct from those more readily associated with "archaeological sites" as such (more prevalent in association with criterion iv): "an outstanding example of a type of building or architectural or technological ensemble or landscape which illustrates (a) significant stage(s) in human history" or, in the example of Cahokia (1982), "which provides an opportunity to study a type of social organization, on which written sources are silent."[10] These values are key to the Hopewell nomination, of course, but obviously have a much lower threshold of evincing "genius, intellectual endowment, or artistic skill" as such.

Works of Architecture

This essay is a sketch of how the earthworks could begin to be understood better "as works of architecture." With recourse to my primary discipline of "architectural theory" (which I define as "reflection on the nature of architecture and its role in the world"), I'll be putting in play some of its disciplinary norms, paradigms, and criteria. I appreciate the interdisciplinary nature of this volume, which explores a wide variety of approaches to the earthworks. Notwithstanding my deep love and respect for my archaeologist friends and colleagues, it is in my view time to explicitly widen the discourse about these places to encompass other kinds of knowledge and experience (within certain limits I'll explain a bit later).

First of all, the interpretation of architecture is inevitably expansive and open ended—if only because buildings are both durable and public; over their lifetimes they stand in front of, and around, a lot of people, and a lot of activities, over a lot of time, in a lot of contexts, and are used, experienced, and valued from a lot of viewpoints. Their reception transcends their origins. Where architecture is concerned, the intentionalist view—the idea that the makers had the only real "truth" about the work—has little applicability.

Architectural interpretations will share with archaeology an interest in the facts of origin, technique, culture, meaning, and use. And they will share with the Native views an interest in the sites' compelling sacred aura, traditional knowledge, or continuities of meaning. But more uniquely, an architectural view will take an interest in form and experience as such. This inclusive, three-way approach explains why, in all of our CERHAS exhibits, we have featured multiple voices and open-ended readings of the works, as well as vivid and immersive (or "experiential") virtual representations.

One way to talk about the broadest capacities of "works of architecture" is to understand them, fundamentally, as "works." They were "worked" (made), they "set to work" (get our attention and cause thought), and are "work-ing" (continuously opening up experiences, meanings, and worlds) in the manner of art "works" as detailed by Martin Heidegger in his essay "The Origin of the Work of Art."[11] There is no more fundamental insight about what "works" do, in the contexts of our lived reality, as distinct from other kinds of things or entities. They show up in our lives and provoke thought and interaction in ways (unlike mere pieces of equipment or processes of nature) that are "always already working" within our activities and interests, our interpretive predispositions and contexts of meaning. Juxtaposing our media research with this conception of "works" has led me to several architectural insights that I'll elaborate on here, within three main areas: engagement, experience, and empathy.

Public Engagement

The first problem with the earthworks is the "I-had-no-idea" phenomenon. Their near disappearance from the public imagination since the mid-nineteenth century means even people who live with them every day hardly know they exist. "Fort Ancient," for most residents of southwest Ohio, is a brown sign along Interstate 71 and nothing more. Brad Lepper may tell you his story of the oblivious car dealer who worked across the street from the Newark Great Circle. Just as surely as most of the earthworks have been plowed down, they had by the early 1990s been largely erased from the public consciousness. So difficult to engage, experientially, they had slipped largely out of view.

When I began to visit the sites, I understood why; they were in various ways invisible: too degraded to find (like the Hopewell Mound Group); too big to see across (like Newark's Great Circle, almost); too forested to show up clearly (like much of Fort Ancient); too complex to be conceived (Fort Ancient again). Then when I began to attend professional archaeology conferences (the only place they were being engaged at all), the only representations in use were "abstract" ones,[12] like what you'll see throughout most of this volume: maps, plans, sections, diagrams, trench grids, more recently those rainbow-colored LiDARs, and so on. While these were essential for the type of knowledge this audience valued, there was really nothing any nonspecialist could engage with or that presented a powerful or significant monument, or a compelling spatial experience.

It was a gradual process by which we discovered the primacy of visualization and iconicity in consideration of these works. To do public education, then, the first step in our work (in the later 1990s) was to try and escape from the abstract and the diagrammatic. So to "make them appear" we created 3-D computer models and began to texture them and animate them, in order to visualize the earthworks anew, to approximate a human experiential context (more on that below).

A few years later, making a poster for the Newark Earthworks Center (plate 1), we inevitably bumped into the idea of the "iconic view" and an "aha" moment about architecture itself: the fact that ancient cultures across the world hold their place in the public imagination by means of an iconic image, of an iconic work of architecture, like the obvious examples we always encounter (Machu Picchu for the Incas, the Great Wall for China, the Parthenon for the Greeks, the Pyramids for Egypt, and so on). The ancient Ohioans had faded so far from public view, perhaps primarily because there was no iconic architectural place or image that lent itself to such vivid visual memory. The Squier and Davis drawings were beautiful but too abstract and really not architectural at all. This "Moonrise" poster was a strong attempt, I think, to give this culture its first "iconic architectural image."

"Engagement" refers to how the work gets attention in the first place, leads audiences into an interpretive or learning process, and then "stays there" in memory—how it becomes the tab on the file folders in people's brains, standing for larger ideas: of the culture, of its architecture, or of specific achievements (like Newark's precision and astronomy, in this case). For this to "work," the "works" need to stand forth in vivid images, visualizations that attain a certain memorable "iconicity" and evoke a concrete human experience of being there, as well as abstract ideas like astronomical precision.

In the search for ways of making iconic visualizations, we encountered the unique difficulties presented by these places, their "uncanny" scale and spatiality.[13] They create space of a kind that is barely comprehensible to us moderns, well beyond, for example, any of Sigfried Giedion's "three space conceptions," which he thought explained everything: object in space, elaborated interior space, or simultaneous space, which he aligned with Egypt (the Pyramids), Rome (the Baths of Caracalla), and modernism (Walter Gropius's Bauhaus at Dessau), respectively.[14]

The earthwork enclosures are of a scale we can barely grasp. This was apparent from our first attempts to make a human avatar walk across our "virtual" Observatory Circle: it just took too long, which is why we abandoned VR virtual reality altogether. It was only when our cameras took to the air, like birds, that we could we show people how the earthworks "appeared in the imaginations of their builders," to use Roger Kennedy's wonderful phrase.[15] Our modern audiences also needed an "imaginative experience": the "conceptual" earthwork needed to be made "perceptual." The "design idea" needed to become a vivid "experience." Besides the bird-flight cameras, this required making the soil a contrasting color, a little "too bright" actually, and as if all freshly laid out at once.

Also, we could not use "scale humans" as clues to the size of these things, since they were smaller than a pixel. Instead we provide the familiar hundred-foot-plus mature deciduous forest edge, which, besides establishing an intuitively sensed scale, also calls attention to the spatiality of the immense clearings, which were undoubtedly monumental in themselves. Details like the haze or the sun reflecting in the rivers (that we moderns recognize from being in airplanes) help place the "conceptual idea" of the work within a "perceptual world" of human life, nature, memory, and meaning—in short, our own "lived world," much of which (apart from the airplanes) we actually share with the earthworks' builders.

The Primacy of Experience

The goal of these virtual representations was to present and interpret such a world to museumgoers as a vivid experience. But now, with the Ancient Ohio Trail[16] we have to curate the actual experience of visitors walking around on the sites: it's as if that avatar now really does have to walk across the Observatory Circle (on a golf-free day, of course). So how do we combine ideas, descriptions, and some new media to make a rich, vivid, and memorable experience out of these places "as architecture"?

Architecturally speaking, then, the Octagon Earthworks at Newark are the preeminent example in the world of a precise geometric earthen monument and one of the true wonders of the ancient world. The axial Circle-Avenue-Octagon combination is over half a mile long and composed of exact geometric figures, generated and unified by a common unit of measure. The 1,054-foot-diameter circle is perfect to within less than 24 inches.

The wall heights of the Octagon are perfectly uniform and at approximately human eye level, helping them to suggest gigantic "artificial horizons" for people within or moving among them. The uniform but slightly taller gateway mounds inside the Octagon help ensure this effect by visually "blocking" the eight openings seamlessly except from certain angles. The whole arrangement sits on a perfectly level protected river terrace.

The spatial effect of being within these giant, perfect figures is uncanny, as noted by Professor Taylor: they are so large that their boundaries become vague as an experience of "enclosure" and yet the key lines and points remain discernable. In the climate conditions of Ohio these huge spaces often seem to fill with mist or frost or other "atmospheres" that cause the forms to present themselves in surprising ways. The amazing precision of the wall profiles and gateway designs combined with the vast scale create a completely unique landscape experience.

To interpret these sites for visitors along the Ancient Ohio Trail, it will be necessary to develop new theories of spatiality and to teach new ways of perceiving monumental architectural space. The idea, for example, that this is an architecture that we are "among": even though all spatial relations are prepositional (over, under, in, out, near, with, beyond, etc.), the "among" relation is definitely an unusual one. Or the idea that, in these spaces, a special kind of attentiveness must be focused on a narrow band at the horizon, with its subtle shifts of depth and parallax (much as canoe trippers learn to read the details of distant lakeshores in Minnesota).

Or the idea that it is the surface of the earth itself that is swelling into precision shapes, not objects placed upon it, altering a fundamental understanding of over and under, or earth and sky, or more concretely in terms of known Native traditions, the beneath, surface, and upper worlds are interacting in unusual ways. We're also working on a new concept for waterway arrival, kayaking caravans, if you will, to try to change the whole orientation.

For new media techniques, we're developing "augmented reality," which puts on your portable screen a view of a restored timber building, or a gallery of artifacts from a particular mound, or the moon rising at a certain point. And we're creating "fly-ups" to connect the on-the-ground view with the overall conceptual view from above, like lighting the burner in a balloon and gently rising into omniscience. Appreciation for this architecture will require new tools, in connection with the real on-the-ground experience, by which we can connect perceptions with conceptions, and then with now.

Human Empathy

To conclude, then, what is the ultimate goal of heritage travel, or of all historical and humanities education, for that matter? By cultivating the experience of the earthworks as "works," we can better understand their builders as our fellow human beings, living together in communities and having aspirations and needs and their own ideas, hopes, tragedies, and so on: like us, on the earth, within a world, and under the sky, as Heidegger would say.[17]

As works, the earthworks are already "working" toward us by projecting shared human volition, aspiration, meaning, knowledge, craft, social unity, and orders of various kinds. In short, a "world"—a full context of life and being, where ritual practices were planned, enacted, and remembered; where community life was sustained; where ancestors were remembered and natural phenomena both felt and explained and where the rain fell and the haze lifted. The work points to, mimics, frames, and foregrounds the surrounding landscape; it marks an act of clearing; it both anticipates and remembers an act of gathering, two of the most primordial human gestures. It also reaches for eternity in its form and its material.

Such a description is not merely "soft and poetic" but rather an "accurate" description of how both the makers and ourselves are humans living in a comprehensive existential world, in which meanings and experiences don't divide cleanly among disciplinary categories or advance into "objectivity" except under specific conditions. The earthworks are not mere "objects of study" or mere "subjective experiences" but "works"—enabling a world of meaning to appear and, at best, a "fusion of horizons" to occur between their ancient builders and ourselves, as hermeneutic philosopher Hans-Georg Gadamer would say.[18] Moreover, in their specific ways of "standing, rising, opening, closing, and extending" they sustain their "capacity" as "existential" phenomena, as Christian Norberg-Schulz would say.[19]

So in our interpretive horizon, the works have room for deep tribal traditions, accurate scientific data, detailed formal analyses, and rich metaphors about "ritual machines,"[20] or even stealth bombers. And when you ask me how we will know when we have opened up the space of interpretation too far (think crop circles or aliens), I will say that mainly, we have only to attend to two core principles: we must respect the core scientific facts about who, what, when, and how, and we must respect the fundamental humanity and wide-ranging intelligence of those who built them.

We should also respect our own encounters with these places, and their capacity to teach us new things about ourselves. That is, the ongoing "working" of the "earth-works."

Notes

1. I will limit my use of the term "Hopewell" as much as possible and only to mean an imprecise "era" rather than a culture or a people. I am mostly persuaded by arguments against naming distinguished ancient societies after modern, unrelated places or landowners, notwithstanding the persistent strength of this archaeological convention.

2. The Ohio UNESCO nomination effort is now under way on a sequential strategy: first the Hopewell Ceremonial Earthworks nomination will be prepared to go forward, and then the nomination for Serpent Mound.

3. This is my operational definition of the "theory of architecture," the primary discipline in which I have taught and published throughout my career.

4. Writers, researchers, designers, architects, and content advisors at CERHAS at the University of Cincinnati, since our beginnings on this topic in 1997, too numerous to name, under the leadership of this author. CERHAS outcomes include interactive digital exhibits at all the major Ohio earthwork sites, plus the Ohio History Connection in Columbus; the CD-ROM publication *EarthWorks: Virtual Explorations of the Ancient Ohio Valley;* and most recently the Ancient Ohio Trail website, www.ancientohiotrail.org (see note 16).

5. At the symposium, I played several video scenes from the Newark Earthworks exhibit to illustrate this layering as it would be experienced by the audience; here I provide descriptions and quotes.

6. Poolaw, "Moon."

7. Taylor, "The Uncanny."

8. Hermeneutics, the study of interpretation as such, observes that any encounter, prior to any objective or intentional thought, is already built upon a deep and ready-made, historically and contextually conditioned understanding of what kind of thing is before us; we take any thing "as" what we already "know" it to be, before we start building any knowledge about it.

9. Jokilehto et al., *What Is OUV?*, 18–21.

10. Ibid.

11. Martin Heidegger, "The Origin of the Work of Art," in *Poetry, Language, Thought.*

12. It may be worth observing that the word "abstract" means, at its Latin roots, a "pulling away from," and this helps us understand what is really going on here: we are being distanced from the thing itself, the lived experience, the work itself as it is actually encountered.

13. The term "spatiality" denotes space as perceived and felt qualitatively through the activities, movements, memories, and sensations of embodied humans, as distinct from "space" as an infinite quantitative abstraction. See Norberg-Schulz, *Genius Loci,* 3–23.

14. Sigfried Giedion, "Introduction." in *Architecture and the Phenomena of Transition* (Cambridge, MA: Harvard University Press, 1971).

15. Roger G. Kennedy, videotaped communication, Chillicothe, Ohio, Nov. 1999 (CERHAS Archives).

16. A heritage tourism website collaboratively developed by the Newark Earthworks Center at The Ohio State University and CERHAS at the University of Cincinnati, with many other statewide partners, and funded by the National Endowment for the Humanities, 2008–14, http://www.ancientohiotrail.org.

17. Heidegger, "The Origin of the Work of Art."

18. Gadamer discusses interpretive "horizons" in *Truth and Method,* 302–7.

19. Norberg-Schulz, *Genius Loci.*

20. Lepper, "Processions."

THOMAS BARRIE

The Newark Earthworks as a Liminal Place

A Comparative Analysis of Hopewell-Period Burial Rituals and Mounds with a Particular Emphasis on House Symbolism

Critical acumen is exerted in vain to uncover the past; the *past* cannot be *presented;* we cannot know what we are not. But one veil hangs over past, present and future, and it is the providence of the historian to find out not what was, but what is. Where a battle has been fought, you will find nothing but the bones of men and beasts; where the battle is being fought, there are hearts beating. We will sit on the mound and muse, and not try to make these skeletons stand on their legs again. Does Nature remember, think you, that they *were* men, or not rather that they *are* bones?—Henry David Thoreau, *A Week on the Concord and Merrimack Rivers*

THIS ESSAY discusses house symbolism in Hopewell funerary practices with the goal of providing broadened perspectives regarding the Newark Earthworks and Hopewell culture burial mounds. Even though much has been written about the Hopewells' social structures, ritual practices, and material artifacts, relatively little attention has been directed to their domestic architecture and, in particular, its appropriation and application in the funerary practices that were central to their culture. This essay argues that even though Hopewell sacred sites such as Newark materialized a range of symbolic agendas, they were primarily positioned as liminal places believed to provide physical and metaphysical connections to what otherwise were inaccessible, and that domestic symbolism was central to this agenda. Pertinent comparative scholarly areas serve to contextualize the subject, including the essential roles of sacred architecture and places, the ontological significance of home, and the symbolic potency of funerary architecture and traditions of house tombs. It concludes with the position that prehistoric sacred architecture and sites retain a certain vocality of symbolic content, though often in cipher, which may reveal the significance and meaning (however diverse and diffuse) they hold for us today.

Because material artifacts, like language, are subject to multiple interpretations, this essay utilizes aspects of hermeneutic retrieval to analyze a broad array of direct and related material and scholarly evidence from a contemporary perspective. This scholarly approach and methodology recognize that it is in between the diverse perspectives regarding Hopewellian culture and its artifacts that some dependable and useful perspectives and knowledge might be accessed. This is an ambitious undertaking in a short essay but one that may be justified by its goal of contextualizing particular aspects of Hopewell funerary architecture in a manner that renders them accessible. Moreover, prehistoric sites present significant challenges to interpretations of their meaning and use where one must rely on archeological findings, restored sites, contemporary scholarship, and in some cases, ethnographic comparisons. At best it is a speculative endeavor that mediates incomplete physical evidence, the absence of historical records, and a discontinuity of ritual use. Our interpretations, therefore, must rely in part on comparative analysis of extant or restored sites and dependable examples of architectural manifestations of cultural and religious archetypes.

Given the history of erroneous postulations and conclusions regarding Hopewell funerary architecture, we proceed cautiously, recognizing and (in some cases) incorporating our own personal, cultural, and political presumptions. We can with some confidence identify the prejudices of the scholars and quasi-historians of the past; it is more challenging to recognize our own. From a hermeneutical perspective an appropriate question is, What do places like the Newark Earthworks "mean" to "us" today? In which ways do they still "speak" to us? Theory is put in service of just such questions as a means to frame and contextualize material evidence in a manner that allows them to "speak for themselves." Theory from this position is applied not to explain the available evidence by predetermined criteria but to contextualize it in a manner that may render it more accessible. According to Christopher Carr and D. Troy Case, this can entail studying ancient cultures "from an actor's point of view," but (quoting Sherry Ortner), "this does not imply that we must get 'into people's heads.' What it means, very simply, is that culture is the product of acting social beings trying to make sense of the world in which they find themselves, and if *we* are to make sense of a culture, we must situate ourselves in the position from which it was constructed."[1] In the context of hermeneutic retrieval, the capacity of the built environment generally, and Hopewell sites in particular, to communicate, to elucidate and enlighten, retain a certain kind of veracity. The communicative capacity of ancient sites depends in part on a shared language of enduring characteristics of human culture and its material artifacts, a subject to which I now turn.

The Essential Roles of Sacred Architecture and Places

Architecture, of all of the arts, perhaps including even religion itself, has been a potent means to embody and express ontological content, and its creation can be essentially understood as a symbolic (in all of its aspects) activity. It is rarely neutral, especially in significant state- or religious-commissioned works, and is often an embodiment of predominant cultural beliefs and imperatives. Significant architectural works can be understood as cultural artifacts that served as potent communicative media, which contained and expressed symbolic, mythological, doctrinal, sociopolitical, and, in some cases, historical content.[2] This is not to suggest that all architecture neatly fits this loose definition but that the preeminent works of established cultures, particularly in the case of the cult and ritual architectural sites of predominantly religious cultures, can be fruitfully contextualized in this manner.

We can productively frame issues germane to Hopewell funerary architecture by outlining pertinent and related roles that the built environment (in its most diverse definitions) has been asked to play. First, architecture, and in particular sacred architecture, has served as a potent medium for the materializations of symbolic content. Second, its spaces and path sequences have served as settings for individual and communal rituals. The architecture in this aspect provides the setting for ritual enactments, which consequently complete its symbolic content. It is a "player" in ritual enactments, similar to the roles "played" by the priesthood and participants. Third, sacred architecture has been positioned and articulated as a place of mediation and is believed to comprise a potent and portentous place where connections to lost knowledge, revered ancestors, and the gods are possible. Similar to holy figures, shamans, scripture, ritual, and prayer, the sacred setting was (and in many cases still is) believed to have the power to mediate between the corporal world and what otherwise would be inaccessible. Just as shamans "serve as mediators" between a culture, clan, or tribe and "the gods, celestial or infernal, greater or lesser,"[3] the architecture was similarly positioned (for those who choose or are anointed to "play" along) as a potent intermediary.

Lastly, to extend the analogy, sacred architecture distinguishes itself from other forms of building as "elect" and powerful places, believed to have specialized powers.

The Ontological Significance of Home

Domestic shelter was arguably the first architecture and building type. It may have begun as a simple structure, but eventually, like most human endeavors associated with surviving in an unstable and unpredictable world, it assumed symbolic and ontological significance. In Marcus Vitruvius Pollio's mythical account of the origins of the first "dwelling house," its development is conflated with the establishment of language, political discourse, and civilization. For Vitruvius, home and house are not mere shelter but both emblematic and catalytic of culture.[4] From an ontological position, the home not only inculcated safety and stability but explicated the world surrounding it.

Traditionally, the home was not only the place where individual and family lives were centered but a place upon which society as a whole depended. It is not surprising, therefore, that the ontological roles and content of house and home have been appropriated and translated into symbolic materializations. From the earliest recorded texts, the temple has been conceptualized as the "house" of the god(s). Solomon's Temple is described in 1 Kings as the "house of the Lord" and built according to exact specifications to ensure that the deity would "dwell" there. Hindu temples are similarly conceptualized and materialized. The Jewish temple is often described as the "house of God," continuing a long tradition of an earthly habitation for the divine, and Christian cathedrals were called "dominus dei" or "God's house."[5] In this manner the material or metaphorical house, in many scriptural and formal contexts, occupies an in-between zone, believed to be dynamically positioned as a place where realms formerly inaccessible may be revealed to connect humans and the divine.

The Symbolic Potency of Funerary Architecture

Generally, sacred places, either believed to be revealed by the gods or consecrated in the hope that they would appear, were specific places clearly positioned in relation to their surroundings. A common agenda of these fundamental placemaking acts was the creation of eternal places reserved for the gods and rituals of the clans, tribes, or cultures that created them. The gods, in all their manifestations, symbolized eternal life, and from the earliest building programs, architecture was put in the service of embodying and representing their continuity.

In the case of deified rulers, funerary architecture assumed the role of sustaining the power of rulers beyond the grave and mediating between humans and their god-rulers. This is apparent at the early and extensive burial mound–building programs in prehistoric Japan and Korea. It was at the vast and sophisticated necropolises of pharaonic Egypt, however, that this type was clearly established. The early funerary complex and step pyramid at the Mortuary Temple of King Zoser at Saqqara was an extensive amalgam of surface and formal symbolizations of environmental setting and political hegemony. The deification of Zoser and establishment of his eternal status were made possible through enduring stone, geometry, and symbolism, all of which communicated the continuity of the ruler and the political union of northern and southern Egypt associated with his reign. Death and place are inextricably bound to the human experience and pose enduring ontological questions regarding corporality and continuity.

Funerary architecture is arguably the most symbolically potent and diverse of sacred building types, which often symbolized and communicated cultural and religious content and served as a mediator between the known and the unknown, the living the dead. Funerary architecture plays multiple and multivalent roles (with inflections to each culture) of mediating between life and death—permanence and decay—humans and their gods. At Saqqara, the architecture materialized the cultural components embodied by the ruler while serving as a medium by which the ruler and all he represented could be accessed. In this context, funerary architecture not only memorializes the dead but provides liminal places where contact with the dead (and all they represented) was believed possible.

Traditions of House Tombs

The conceptualization and materialization of burial mounds and house tombs from other prehistoric cultures may provide other fruitful contexts to understand the meaning and significance of Hopewell sites—in both their own time and ours. Just as symbolizations and materializations of home were appropriated by the temple, they were similarly applied to the tomb. According to Michel Ragon, "from the earliest civilizations . . . the tomb has appeared as the obverse of the house."[6] The Assyro-Babylonians called graves "houses and residences," and ancient Egyptians referred to tombs as "eternal habitations" or "residences for eternity," which included many of the accouterments of living within their sepulchral domiciles. Etruscan tombs even more explicitly re-created the house in stone, their underground chambers including doors, staircases, servants' quarters, beds with stone pillows, kitchen implements, and pitched roofs. In some cases erotic frescos were painted on the walls—scenes of sensual pleasures to enliven the quiet rooms of the tomb (and promising a certain kind of eternal existence). For the Gallo-Romans death was merely a continuation of life in which one simply forsakes one dwelling for another, and Lycian rock-cut tombs were rendered in a manner that replicated the wooden construction of the homes for the living.

The Neolithic peoples of northern Europe produced monumental funerary constructions that indicate the existence of beliefs and rituals focused on the dead (and death itself), concretized and facilitated through stone architecture.[7] Aubrey Burl suggests that early funerary sites in the British Isles utilized wooden charnel houses, which eventually transformed into megalithic ones. Burl suggests that the amount of labor required to construct megalithic barrows indicates that they were not merely tombs but places built to "harbour powerful forces" and materialize the "compulsions and dreads that drove people to these efforts."[8] The West Kennet Long Barrow, located on the highest point of a ridge in the gentle hills of southwestern England, stretches more than 300 feet along the ridge in an east–west orientation. The east-facing entrance was defined by a transverse row of sarsen stones and led to a series of corbeled chambers that held remains. Burl suggests that they "resembled the rectangular houses of the Neolithic although larger, visualized as dwellings for the dead whose spirits would watch over the living family." However, Rodney Castleden argues that they appear to have served purposes beyond the simple commemoration of the dead and were not "primarily graves, but cenotaphs. They were not monuments to the dead, but to death itself, and they can be seen as magic gateways through which life could be started anew and where the living and the dead could meet."[9]

American Indian burial customs comprised a diverse though related range of practices.[10] According to Martin Byers, our contemporary understandings of Indian cultures suggest that their views on death (and life) produced culturally distinct rites and artifacts. Charnel houses in particular were used throughout the eastern part of North America, where multistage rituals, which may have included inhumation followed by excarnation, cremation, and a more final deposition, were centered (what Byers describes as the "laying-in" process, in which the body was first placed in a charnel house, followed after an interval of time by removal and either cremation or deposition.)[11] In Hopewell culture the house tomb took on specific functions, symbolizations, and meanings. It served as a temporary abode for the corpses and spirits of the dead, while the practice of destroying the charnel houses before entombing them in an earthen mound made them temporary abodes themselves.

Hopewell Charnel Houses and Burial Mounds

Aubrey Burl has suggested parallels between the mound-building cultures of North America and Neolithic burial rituals and sites. He states, "With such banked rings and with their conical burial mounds they resemble quite closely the middle Neolithic societies of eastern England where round barrows were already being built. It is turning to circular forms, perhaps in imitation of round timber houses, even of wigwams, that distinguishes the easterners from the people of Wessex."[12]

Burl's formal comparisons may be of limited relevance, but his four stages of Neolithic burial rituals and artifacts can be applied to expand our understandings of Hopewell sites. According to Burl, first the dead were buried in their own houses, then in barrows, which then led to the construction of burial mounds. In the fourth stage of stone circles, however, bodies were no longer necessary for the commemoration of the dead. Communion with ancestors was an important component of many of these sites, and the symbolic agenda of many of the Neolithic necropolises was to "transform collective burials into spectacular and indestructible monuments" where "communication with the ancestors is ritually assured."[13] In particular, some megalithic monuments in western Europe embodied the dead, the enduring stone symbolizing the continuity of the rulers and their people.[14]

We do not definitively know the types of funerary rituals that were performed at Hopewell funerary sites, even though there is substantive archeological evidence of their materials and contents. We do know that wooden charnel houses, some with multiple rooms, served as ritual centers and repositories for cremations and talismanic artifacts. N'omi Greber and Katharine Ruhl provide a detailed catalog of a mound at the Hopewell group that included two or three wooden structures (presumably built over time). At Mound City there is evidence of numerous wooden charnel houses. At mound 13 (often referred to as the Mica mound), two sequential structures

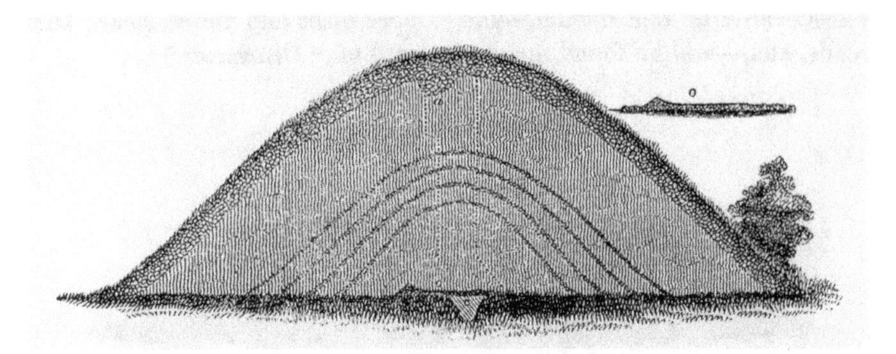

FIGURE 1. The mounds, covered in dark earth or gravel, would have been distinct from their surroundings as befitted their function and symbolic content. Section through mound 7, Mound City. (Squier and Davis, *Ancient Monuments of the Mississippi Valley*)

were erected, the second measuring 40' × 42.5'. It has even been posited that the rectangular form of the earthwork enclosure itself at Mound City symbolized the base of a Hopewell house.

At Hopewell sites, the house was eventually covered by an earthen mound, sometimes burned beforehand—replacing impermanent flesh and materials subject to rot and decay with more permanent earth, sand, and gravel (fig. 1). The mounds would have looked very different when first built, covered with dark earth or gravel, distinct from their surroundings as befitted their function and symbolic content. Carr and Case describe various functions in the extensive use of charnel houses by tribes of southeastern North America. They were places for internment, to "process the dead," to connect with the dead through food offerings, to feast "in honor of de-fleshing," to store ritual objects, and to hold the "sacred fire."[15] Carr and Case outline multistage funerary rites, which may have included participants from diverse geographical locations (similar to the periodic rites of the Huron and Algonquian tribes). The burial or cremation and storage of remains in these ceremonial houses concluded with their transformation into an earthen mound, a culminating event that the authors suggest may have been part of important periodic ceremonial events. Charnel house funerary practices were widespread among the Native peoples of eastern North America, and in many instances they were referred to as "temples."[16]

Not all Hopewell burials occurred within earthwork enclosures, which raises the question, Were only a select number of important individuals afforded this honor—elite burials for personages playing roles in life and in

FIGURE 2. Charnel houses provided settings for shamanic rituals performed for the benefit of the living and the dead. (National Park Service, Harper's Ferry Center Commissioned Art Collection, Louis S. Glanzman)

death for the benefit of the community?[17] At the Eagle Mound of the Fairgrounds Circle at Newark, there is evidence of human remains and one or two large charnel houses over which an effigy mound was constructed. It has been suggested that the wooden charnel houses replicated typical Hopewell houses and provided settings for shamanic rituals performed for the benefit of the dead and the living. Perhaps shamans, dressed as supernatural animals, would perform rites that mediated between the dead and the living (and all they represented) within the dark interior of the house. In this context the common term "charnel house" should be questioned. Perhaps "ritual house" is a more accurate term to describe a setting for a diverse series of rituals, some (or most) of which were focused on the dead.

Shamanic rituals found panculturally share themes of the passage from one world (of the living) to another (of the dead, or "undead" enduring spirits) and depend on the guidance of one who has the ability to mediate between these worlds (fig. 2). Specifically, in elite burials there may have been rituals to commemorate those who died and to ensure the passage of their souls to the next world (thus perhaps the significance of fire to release their souls) and consequently to ensure the health and continuity of the tribe and culture. Many were aligned with positions of the moon—the nocturnal celestial object most closely associated with death and the "mirror world" of the land of eternal spirits, which may have served to reinforce their roles as portals to the land of the dead and the spirits of dead ancestors.[18] Many have argued that the houses (along with the earthwork enclosure's only opening) faced in the direction of significant periodic moonrise events. And so, in one aspect, the Hopewell period houses of the dead were positioned as liminal places with the ability to mediate between worlds. The house consequently was not a place to return to, but to depart from. Home, in this context became a vessel, ferrying the dead to another world while also serving as a place where they might be accessed.

Conclusions

Any conclusions regarding sites such as Newark need to be tempered by recognition of our culturally based prejudices regarding ancient sites. Given the history of erroneous postulations and positions regarding Hopewell culture funerary architecture, we should proceed cautiously in recognition of our own personal, cultural, and political presumptions, prejudices, and proclivities. Recognizing and incorporating our own (in Hans-Georg Gadamer's words) "preunderstandings" can serve to distinguish the "meanings" of Newark and what it "means" to us today—effectively bridging between the past and the present. The plenitude of meanings Newark may have today are not unitary—and are certainly not ideals—but they can perhaps provide accessible contemporary perspectives. Perhaps the most dependable foundation for contextualizing Newark is the enormity of our environmental setting (over which we have little control) and the mystery of death (and what may lie beyond it). These are contexts that retain potency. If we ignore, minimize, or objectivize them, we risk truncating our ability to provide contemporary interpretations of both ancient sites and these perennial conditions. With this in mind I now proceed with a summary of some aspects of funerary architecture in general and conclusions regarding Hopewell culture mortuary sites specifically.

PERMANENCE

Throughout the history of architecture, the earth was molded and transformed as a means to articulate the enduring questions of *who* and *where* we are. Funerary rites have often included the preservation of the body or spirit through eviscerating, embalming, and other means, as if to symbolize the endurance of the soul by the preservation of the body. This was certainly an important component of the complex funerary rituals and practices of pharaonic Egypt. Egyptian funerary practices included colossal architectural monuments that were essentially cenotaphs for the dead (but eternal) pharaoh and all that he represented. Hopewell funerary architecture may be similarly understood as memorializations of the dead and even death itself, translated into the permanence of earth and stone, as a means to symbolize and signify the endurance not only of the dead but of the family, the clan, and the tribe.

MARKING OF TERRITORY

The (intended) permanent constructions of funerary architecture may provide houses for the spirits of ancestors, but as complex symbol systems, they also served to mark the territory of the tribe, clan, or culture that built them. One of the roles of religion has been to bind together the clan, tribe, culture, or nation in support of hierarchical and territorial agendas, which is why religion, culture, and place are so inextricably connected. Architecture typically served these unified and multifarious cultural, religious, and territorial agendas. Even the most primordial of architectures, the burial mound, served to reinforce the continuity of the clan and its territorial claims. Even though it has been argued that the Hopewell culture was more interested in religion than competing with its neighbors,[19] claimed territories and built artifacts for ritual purposes constitute acts of appropriation. In this context the houses of the dead marked a sacred precinct but also appropriated the larger territory that it (in part) represented.

Carr and Case have suggested that precincts such as Newark were ritual centers, with only a small population of priests and caretakers living nearby (what Olaf Prufer termed the "vacant ceremonial center—dispersed agricultural hamlet hypothesis").[20] Consequently they can be contextualized as places reserved for episodic gatherings of a more regional character. Archeological evidence suggests that they often served as mortuaries, places for ritual building activities, craft centers, places for feasts, and initiatory sites.[21] As centers of the Hopewellian world, they materially and ritually appropriated large and even cosmic territories. Similar to the Algonquian and Huron Feasts of the Dead, the ritual gatherings at Newark would have served, in part, as a means to concretize Hopewellian religious and ritual hegemony. Moreover, even though singular Hopewell burial mounds and earthworks may have established central ritual precincts, their broad geographic distribution in the Ohio River basin could also have served as material means to consecrate (or claim) larger territories.

PLACE OF VENERATION

In funerary architecture the dead are symbolically materialized (and preserved) by means of the medium of built forms, spaces, and materials. The objectification of the dead consequently presents the artifact as a place of veneration, which takes two principal forms. In the first, the ancestors or rulers may be venerated directly and as specific personages (as at Saqqara, where a bust of Zoser was placed in a *serdab* adjacent to the pyramid). In the second, the dead are represented more generally and abstractly, with the funerary site consecrating the powers embodied or materialized by the architecture. This latter type may be understood as similar to effigy figures (such as Serpent Mound), which venerated not the animals they represented but the powers they symbolized. In this context Eagle Mound can be understood as aligning with Burl's fourth stage of funerary architecture, as more of a cenotaph or a monument to death itself.

In many American Indian cultures the dead were believed to be active participants in the activities of the tribe. They may have changed to spirit forms, but like animal spirits, they could be accessed though rituals performed at specific locations. According to Carr and Case, "Whether spiritual beings are a subtle part of the objective reality or projections of the imaginations of the unconscious mind onto objective reality does not matter. In either case, the person experiencing the spiritual being acts in relation to it and to other humans as though it were real."[22] Hopewell funerary sites and rituals can be understood in this context as places where the spirits of ancestors were embodied by the architecture. The materializations and rites at Newark acted not as memorials but as settings for acts of veneration in the belief that the blessings and knowledge of the dead and their eternal spirits could be accessed and received.

COSMIC MODELS

We know that the colossal earthworks and enclosures of the Hopewell culture were not only funerary sites and in some cases did not contain any bodies or remains. It is unclear whether the house or houses built at Eagle Mound served as charnel houses, but if not, they most likely represented them. Consequently, the house and the geometric organizations and celestial orientations of the earthen enclosures may have served as cosmograms—as both replications of and connections to the cosmos. There are numerous examples of American Indian domestic architecture serving, at least in part, as cosmic models. From the Navaho hogan, with its orientation to the cardinal directions, to the Plains Indians' teepees, with their celestial directionality, the domicile served to align with and appropriate cosmic contexts. The Creek Square Ground enclosure, in particular, was known as the "big house" and served as a cosmic microcosm of the Creek world.[23]

It may be pertinent to recall that Avebury did not contain any burials within its earth and stone enclosures but served as the setting for funerary processions and rituals. Similarly, the Fairgrounds Circle at Newark may have performed multiple roles—as a place for funerary rituals but also, by means of its celestial alignments, as a ritual calendar, a place of cultural and cosmic elucidation, and even as a center of the Hopewellian "world." The charnel houses and their earthen enclosures would have served in part to represent what the Hopewell understood to be their cosmic contexts and their relationship with them.

WORLD RENEWAL

With its multifarious symbolizations and ritual functions, Newark can also be understood (from an ethnographic position) as a place where life and death and, by extension, fertility and death were intertwined. Similar to ritual itself, in which sacred time is accessed, the "eternal return" (to use Mircea Eliade's term) of the gods and powers of primordial time, Newark may be conceptualized as a sacred center where timeless powers were believed to be present. Similar to Burl's assessment of Neolithic funerary sites, Newark would have served as a power place and "cult centre(s) for the protection of the living." Just as funerary crematory rites transformed (and renewed) the soul through fire, smoke, and shamanic rituals, the sacred enclosures themselves may have served as places of periodic individual, communal, and cultural renewal.

Myths from around the world share a common theme of events of world renewal, often preceded by destruction. American Indian mythology generally shares this enduring theme, with the Earth Diver myth and its descriptions of the founding of the world from an aquatic abyss perhaps the most pertinent to our inquiry.[24] There is evidence that Newark was subject to periodic floods. If so, we can imagine water surrounding the Newark earthworks, with Eagle Mound rising above the flood like the "first place" of so many creation myths.[25] As in ancient Egypt, flooding and renewal would have become conflated, with the earthworks symbolically establishing the Hopewell as demiurges, actively participating in the periodic renewal of the world (and bearing the responsibility that this entailed).

PLACE OF CONNECTION

An enduring aspect of sacred architecture is the role it was often assigned of serving as an intermediary, with the potential or capacity to access what otherwise would be inaccessible. From Hindu temples to Christian cathedrals, sacred places were typically precisely located, planned, and materialized as a means to house or connect with the deity or deities. In the case of funerary architecture, the tomb, monument, or cenotaph often served as a potent setting where the worlds of the living and the dead connected. Similar to the ability of shamans to access the world of the dead, the architecture itself was believed to hold certain powers and serve as a portal or threshold to the otherwise inaccessible—lost knowledge, dead ancestors, and the gods.

In Eliade's words, "the pre-eminently shamanic technique is the passage from one cosmic region to another—from center to the sky or from earth to the underworld."[26] Just as the shaman served as a metaphysical bridge between worlds, Hopewell funerary architecture may have served a similar role. As in the conjoining of worlds symbolized by the Serpent Bridge in Ojibwa mythology, the charnel house and its surrounding earthworks were liminal places, straddling worlds and connecting the known with the unknown. Religious rituals share common characteristic rites of separation and liminality, where one spiritually, symbolically, and physically occupies the in-between. Carr and Case discuss the relationship between the ghost water barriers of American Indian beliefs and the borrow ditches of Hopewell earthworks. According to them, these may have been filled with water as a material means to ward off malevolent ghosts, to represent the "world axis in cross section," and to underline the enclosure as a bridge to the dead.[27]

American Indian funerary rituals typically emphasized the process over any one specific event, which included separation, liminality, and reincorporation. The three-phase mortuary programs of the Hopewell, similar to the better-known rituals of the Algonquian and Huron, included burial or cremation in the charnel house, storage within its sacred enclosure, and lastly its transformation, often through fire, into a burial mound.[28] These rituals, perhaps enacted over time, but especially during propitious times established by the celestial calendar, were rites of passage that guided the dead to their changed status and guaranteed that they would continue to participate and aid the tribe. The charnel house, oriented in the direction of the spirit or mirror world, served as a potent transistor with the power to effect these connections and transformations.

CLOSING THOUGHTS ON THE CONTEMPORARY RELEVANCE OF NEWARK

Archeological evidence of Hopewell sites suggests sustained practices centered on the dead and perhaps death itself. However, there may have been additional displacements as the burial mounds came to serve as environmental and social foci. It is clear they were places that brought people together through the shared condition of death and connected them through ritual with the goal (perhaps) of deeper understandings of its meaning. The architecture memorialized the continuity of life in a discontinuous world while straddling the two worlds. In this manner it created an enduring material place while simultaneously symbolizing an immaterial threshold that connected the living, the dead, and the cosmos of which they were a part.[29]

Gadamer provides accessible and applicable perspectives when he insists that works of art, including architecture, are intrinsic to the perennial human endeavor of structuring our place in the world. For Gadamer, art exists not only within its formal, aesthetic, and symbolic parameters but also within a larger context of which it is a part. The work of art "is perhaps the only way that is granted to us finite beings to relate to what we call eternity."[30] In all, material artifacts such as Newark's act as mediators between who we are now and who we were, with the implicit hope that we may rediscover lost perspectives, understandings, or even knowledge. Places such as Newark act as mediators between present understandings and aspirations and occluded historical positions, knowledge, and understandings, with the capacity to disclose perspectives that may have contemporary relevance. We may have a multitude of positions, each reflecting a particular ideology, but implicit in all is the hope that who we are has a greater promise for understanding, based on who we were. Even though Hopewell sacred sites outlive their creators, they continue to retain their communicative efficacy in various forms. They speak to us, sometimes obliquely, other times eloquently, in the enduring, transhistorical and pancultural language of the built environment.

Notes

1. Carr and Case, "The Gathering of Hopewell," 44–45. Emphasis in the original.

2. Veselay, *Architecture in the Age of Divided Representation*, asserts, "What the book is to literacy, architecture is to culture as a whole" (8), and suggests that the power of architecture to explicate a culture equals or exceeds textual or historical evidence.

3. Eliade, *Shamanism*, 8.

4. Vitruvius, *The Ten Books on Architecture,* 38.
5. L. Jones, *The Hermeneutics of Sacred Architecture,* 2:100.
6. Ragon, *The Space of Death,* 26.
7. Castleden, *The Stonehenge People,* 5.
8. Burl, *Prehistoric Avebury,* 95–96.
9. Castleden, *The Stonehenge People,* 185.
10. See, for example, an early-nineteenth-century study, completed in 1878 but first published in 1988 by H. C. Yarrow and V. LaMonte Smith as *North American Indian Burial Customs.*
11. Byers, *The Ohio Hopewell Episode.*
12. Burl, *Prehistoric Avebury,* 127–28.
13. Eliade, *Shamanism,* 123.
14. L. Jones, *The Hermeneutics of Sacred Architecture,* 2:159.
15. Carr, "Scioto Hopewell Ritual Gatherings," 468.
16. Romain, *Mysteries of the Hopewell,* 158.
17. There is little evidence to assume the existence of social or political hierarchies. For example, both genders were represented in mound sites. See Milner, *The Moundbuilders,* 91–92.
18. Romain, *Mysteries of the Hopewell,* 157.
19. Carr and Case, "The Gathering of Hopewell," 42–45.
20. Carr, "Historical Insights into the Directions and Limitations of Recent Research on Hopewell," 59.
21. Carr, "Scioto Hopewell Ritual Gatherings," 463.
22. Carr and Case, "The Gathering of Hopewell," 42.
23. Nabokov and Easton, *Native American Architecture,* 111.
24. See Carr, "Scioto Hopewell Ritual Gatherings," 475; and Romain, *Mysteries of the Hopewell,* 191–93.
25. Romain, *Mysteries of the Hopewell,* has even suggested that the mounds represented the "mountains" found in southeastern Ohio.
26. Eliade, *Shamanism,* 259.
27. Carr and Case, "The Gathering of Hopewell," 44.
28. Carr, "Scioto Hopewell Ritual Gatherings," 470.
29. As Castleden, *The Stonehenge People,* suggests, burial mounds were enduring symbols of "people living in a half-tamed landscape, seeking to forge a covenant with continuity with their short-lived ancestors and with unknown generations to come" (190).
30. Hans-Georg Gadamer, *The Relevance of the Beautiful and Other Essays,* 45.

MARGARET WICKENS PEARCE

The Cartographic Legacy of the Newark Earthworks

BEFORE THE town surveyor platted the city of Newark, Ohio, before the public land surveyors, before the surveyors of the Western Reserve, an ancient team of surveyors assembled at the confluence of Raccoon Creek and the Licking River to measure and mark the landscape. The time was somewhere between 100 and 400CE.[1] They were not the first to shape this landscape; already the valley was a place of mounds, habitations, agricultural fields, and roads and pathways built by previous generations, to which the surveyors' work seems closely tied. Yet the design they planned and initiated was far larger and more complex than anything that had come before, and it would remain so for the rest of the Woodland era. The pace at which they worked is not known, but the geometric precision and cohesiveness of their design suggests completion within a generation. When they were finished, they had created a place that generations of Indigenous peoples from across North America would visit for ceremony and study.

Perceived through the lens of Western scholarship, the Newark Earthworks are material culture for archaeological investigation—the evidence of cultural, astronomical, mathematical, and engineering knowledge during the Middle Woodland period. The most recent studies have significantly expanded our understanding and appreciation of the Earthworks' contributions to these sciences.[2] Mapping has been central to outsiders' perceptions of the Earthworks' identity since the beginning, and the history of that cartographic representation by European and American mapmakers is a chapter of North American map history. Perceived through an Indigenous lens, the Earthworks appear differently. Their physical spaces and networks, and the creation and embodiment over time of these networks, become an arc of intersecting Indigenous traditions from the Middle Woodland period to the present, an Indigenous people's cartographic legacy that is itself a part of North American map history.

In this essay I ask you to perceive the Earthworks through an Indigenous lens, to consider the legacy of Indigenous cartographies here, that it may open the door to what is possible. In that inquiry, we will turn that Indigenous lens on ourselves, to perceive that respect, responsibility, and reciprocity must lie at the heart, so that the Earthworks as a place to *take from* becomes a place to *contribute to.*[3]

Indigenous and Western Cartographies in North America

To understand the ways in which the Earthworks are part of map history requires us to consider first the differences between Western and Indigenous map traditions. In the United States, the mental image that we associate with the word "map" most likely resembles something from the Western tradition: a screen display depicting the road network of a city with driving directions, a wall map displaying the coastlines and countries of the world, or a color reproduction of a nineteenth-century European explorer's map. Our conception also likely includes a scale bar, a north arrow, and perhaps a legend to complete the picture. These images arise despite the fact that we encounter different kinds of cartographies in our lives, including transit maps, star charts, satellite maps, store locators, and flight path diagrams, maps that "break the rules" of cartographic conventions.

Yet somehow the rules remain central, for within the diverse and fluid category of Western cartography, there remain a number of assumptions about what is considered the natural, accurate, and sophisticated way to represent space in a map, and what is considered unnatural, incorrect, or crude. For example, it is assumed that a map should depict space with a uniform, measureable scale within the limits of what is possible at that size and that to introduce scale shifts in a map would be considered a deviation or distortion. It is also assumed that the best way to control uniform scale and measure is with the use of organizing rules such as coordinate systems, which locate and interrelate places by x-y coordinates on a global grid, and with map projections, which mathematically rearrange those grids to emphasize qualities of distance, direction, area, or shape. It is assumed that the most accurate angle from which to represent geographical features is from either a decentered, positionless, orthogonal perspective (the "view from above"), or from a converging, linear perspective, as in the oblique or "bird's eye" views that place the reader on an imaginary hillside above the landscape.[4]

Some conventions of Western cartography are so naturalized that it may seem unusual to state them. For example, Western cartography assumes that a "map" is a product, an object produced for a market, intended for interpretation by a map reader, not a mapmaker, and for placement on a wall or in a drawer, library, or online, where it is archived.[5] The production of this map may have involved the contributions of many people in different locations, yet the map itself is intended to be independent, a scientific document or expressive form whose meanings can be interpreted by its reader apart from the process of its production. Finally, Western cartography assumes that terrestrial, celestial, and religious spaces are separate, and to include such spaces in the same map would be unscientific or "cosmographic." It assumes that space and time are separate and that time cannot shift the spatial coordinates of a feature in the map; to do so would be considered unconventional.

Indigenous cartography, on the other hand, is based on different assumptions. It assumes that the map is a process, that mapping is an ongoing series of located or situated events among people and places, and that maps themselves are but artifacts of or mnemonics for that process, not intended to carry all the meanings of the process as a whole.[6]

It assumes that the map will represent the celestial, terrestrial, social, and spiritual spaces as influences in or dimensions of an inseparable, connected space. What is above is mapped to what is below (on the land, and below the surface of the land). In North America, Plains peoples are known for mapping traditions that emphasize connectivity between the celestial and terrestrial. Celestial geography is a map of terrestrial geography and vice versa, and each is the home of the spirits. Plains cartography may embed the terrestrial in the celestial, by connecting constellations in ways that mirror geographical features on the ground, or encoding celestial geographies into the built and physical landscapes. In Lakota tradition, the geographical locations in the Black Hills are tied to corresponding sites in the geography of the stars, navigating the routes and stories of the place, so each constellation is observed from its corresponding location in the Black Hills. In this way, the Black Hills are permanently stored in and tied to the sky. To map the sky into the earth is to map the earth into the sky, "mirroring" and connecting above and below in a continuous geographical dialogue.[7]

Indigenous cartography assumes the map is not one archivable object but something dispersed, embodied, and archived across sites in the landscape and within the stories and teachings of a community. An Indigenous map in a drawer is only the fragmented evidence that mapping has occurred. It assumes that the particular logic of that community's worldview is re-created in the logic and structure of the cartography itself. For example, the strength of a political alliance influences the degree to which a feature is centrally or marginally situated or its proximity to or distance from other features in a map.

It assumes that space and time are a single, inextricable concept, the spatiotemporal, as for example in the painted cloth *lienzos* of Mesoamerica, where places are located based on a combination of terrestrial location and genealogical histories. As Barbara Mundy writes, Indigenous people "saw space as so deeply connected to time, be it historical or calendrical, that the two could not be rent apart. Thus, in the secular sphere the line between 'map' and 'history' is a blurry one, as is the line between 'map' and 'almanac-calendar' in the sacred sphere."[8] Indeed, embodiment is itself a spatiotemporal dimension of the Indigenous map, for to stand where one's ancestors stood, to view the landscape from the angle and direction of the ancestors and to hear their stories is to fuse ancestral and historical time with the present, as well as with intentions for future activities at the place.[9]

Indigenous maps often use devices to locate a place beyond the horizon, to the earth and the universe, somewhat analogous to coordinate systems. In Navajo cartographies, this device is built of story routes, rocks, and rock crystals.[10] For Indigenous peoples in the Andes, the device is a network of (mostly straight) radiating lines called *ze'ques,* along which are natural or built sacred sites called *wak'as* (such as caves, temples, and springs), which connect places, encode social ties and water rights, and align to astronomical events such as the summer solstice, and are reencoded in the *khipus,* as mobile versions of the map.[11]

In sum, Indigenous cartographies are processual and embodied and archived in landscape and community. They connect above and below, sacred and profane, and time and space. They consider the locations of political, social, and spiritual interrelationships as they also consider locations of the spatiotemporal. Beyond these generalities, they are as differentiated as spoken language itself. What, then, is known about the traditions of people at the Earthworks?

The Language of Southeastern Indigenous Cartography

The logical direction to look is the Indigenous cartography of the Southeast. Though this region may at first seem outside the cultural sphere of the Ohio valley, Creek migration stories record that they came to their land after following a white path from the West,[12] and their oral tradition further documents a legacy of pilgrimages to mounds in Ohio, an association of mounds with celestial observations, and a sacred tradition associated with the moon.[13] Other southeastern Native communities record early migrations out of this region as well.[14] It is likely, then, that there is cartographic common ground.

The published literature about southeastern Indigenous cartography is based on scholars' analyses of eight maps.[15] The first is a deerskin robe that made its way from Virginia to England sometime in the first half of the seventeenth century. As it is believed that a Powhatan chief either sold or gave this robe to an English visitor, possibly from the chief known as Wahunsunacawh or Powhatan himself, the robe is known as "Powhatan's mantle."[16] The map is composed of thirty-four circles sewn with shell beads and arranged around three central figures, one human and two feline, the latter symmetrically arranged on either side of the human.

A second map is an account of a 1607 mapping ceremony witnessed by John Smith, who recounts the construction of three concentric circles of cornmeal around a central fire and a pile of sticks placed outside. His interpretation was that the first ring of corn is the Indian's country, the second ring is the edge of land and sea, the sticks are England, and the fourth ring is the edge of the world.[17]

Two maps were gifts to Governor Francis Nicholson of South Carolina: a 1721 Nassau map that delineates the political and geographical interconnections and relationships of thirteen Native polities and their relationship to two English polities, and a 1723 Chickasaw map that expands the political geography north and west to take in about 700,000 square miles.[18] In the 1721 map, Nassaus dominate as the central circle in a confederacy of eleven nations, with outside connections to the Cherokee and Chickasaw nations (as circles) on the upper edge, Virginia Colony as a square on the bottom right, and Charlestown as a grid and sailboat on the left edge. Running horizontally through the map is a prominent double line connecting Nasaw to Charleston and labeled "The English Path to Nasaw." In the 1723 map, four large circles represent the relative locations of the confederated Chickasaw, Cherokee, Choctaw, and Creek nations, and a fifth circle on the eastern edge represents South Carolina. A network of lines representing rivers and paths connects these circles; along the south edge a line represents the coastline, and to the northeast and west, the map depicts the Seneca and connects to watersheds in what is known now as Minnesota, Kansas, and Texas.[19]

Two maps were drawn on spy missions in 1737, during the war of the Natchez and Chickasaw against the French, one by Chickasaw chief Mingo Ouma for the Alabama chief Captain of Pakana, and one by the Captain of Pakana himself.[20] In both maps, polities are represented as circles. Ouma's map positions the Chickasaws at the center with ties to peripheral southeast Native nations, and it connects outward to the English to the south and Hurons to the north, with south at the top. The Captain of Pakana's map is a large-scale layout of the Chickasaw villages.

The final two maps were painted sometime before the late eighteenth century, a Choctaw map of a war expedition and a Creek map of the events surrounding a military ambush, redrawn together by Bernard Romans.[21] These maps were each posted at the location where the events occurred and are thought to have been drawn on bark.

All eight maps come from colonial transactions, and with the exception of the robe, all eight are copies of the originals. Yet even in these compromised translations, we can perceive a distinct cartographic language unlike any other in North America.

All but one of the maps are guided by a specific cartographic language of circles, lines, and pictographs that encode physical factors such as locations and distances and human factors such as cultural identity and difference, political affiliation, and demographic proportion. Circles dominate this language. Circles represent villages, as is conventional in Western cartography, but they also aggregate villages having cultural or political unity and encode alliance.[22] Kay Galloway elaborates that, in southeastern Indigenous terms, the circle represents the fire or "community bound by political, genealogical, and ceremonial ties."[23] In both the 1721 Nassau map and the 1607 ceremony map, colonial polities are mapped with a square or grid.

The ties between circles and squares are expressed with line symbols. Lines represent travel routes (by river or footpath or both) and cultural or political alliances, and Gregory Waselkov concludes that lines generally explain the quality of communication between groups. If the communication route is open—that is, if the diplomatic or cultural relations are strong—the path, too, is strong and the line is solid. If relations are weak or antagonistic, the line is broken or disconnected; that is, it is extended from one group in the direction of the other, but stops short of connection to that group. This is true even if a river or footpath runs between the two communities; the connection cannot be traveled, or not in a good way, and so does not exist.[24]

In three of the maps, all circle and line symbols are composed of two parallel lines. In the 1721 Nassau map, the circles of cohesive groups are drawn with single lines, and communication routes between groups are drawn with double lines. The categories remain flexible, however, for to code a line as symbolizing a footpath and not a relationship precludes the line from encoding multiple meanings.

Pictographic images are part of all but the 1737 spy maps. These images include humans and other animals, boats, and the implements of war and hunting. In every case, they are depicted in profile, in contrast to the circles, squares, and lines, which combine conceptual shape with aerial perspective. It is interesting to note that the two oldest maps, the robe and the ceremony map, are composed of circles and pictographic images only. All significant spatial relationships are conveyed by the arrangement of the circles and images themselves. Perhaps, in the telling of these maps, the oratory or embodiment that activates the map is also the component that makes those connections clear.

In two maps, color communicates qualitative differences in geographical information. In the 1737 Chickasaw map, red signifies enemy, and black signifies ally.[25] Of course, Western cartography also conveys meaning through color, and this is an English copy of a Native map. But to attribute this color differentiation to circumstance or possibly only the influence of a Westerner's edits would be to overlook the symbolic meaning of color in southeastern Native culture.[26] Mingo Ouma's map includes a written narrative for each feature.[27] Five of Ouma's descriptions use "white" to signify a benevolent quality or a feature associated with the Chickasaws' allies, and the Chickasaw Nation is "white within" although "the space surrounding it is nothing but blood." Paths labeled *S* are "white paths that lead to their friends," and paths labeled *V* are "hunting paths of the Alabamas, white." Paths labeled *Q* are warpaths, and although the mapmaker has drawn them with broken or incomplete lines, "they hope that they will become white."

Quantitative information is encoded in the Choctaw map, the ceremony map, and the Nassau and Chickasaw maps. The Choctaw map uses a countable series of repeating identical bars or pictograms to record the battle's numerical information, including a tally of the number of leaders and the number of second in commands, the number killed, and the total number of people in the war party.[28] And in one version of Smith's account of the ceremony map, we see this same counting technique embodied. Smith describes the actions of one of the chiefs, who begins building the map after completing one of the songs: "At the end of the song, he laid downe 5 or 3 graines of wheat [maize seeds or kernels] and so continued counting his songs by the graines, til 3 times they incirculed the fire, then they divided the graines by certain numbers with little stickes, laying downe at the ende of every song a little sticke."[29]

The Nassau and Chickasaw mapmakers depict quantities by varying the sizes of circles in order to symbolize population differences between the villages. As in the Western tradition of graduated circle mapping, the use of circle sizes to denote hierarchy or rank without specific number allows the mapmaker to generalize a quantitative aspect in such a way as to combine it with a second, ranked dimension of information. In this case of southeastern Indigenous cartography, the second ranked dimension is the relative status of that polity's political power or significance.[30]

The language of southeastern Indigenous cartography also includes a rule system for how these symbols will interrelate. One of these rules is to shift scale and direction when it would be impossible to maintain uniformly measured distances and angles and antithetical to the purpose of the map to incorporate these qualities in the first place.[31] Southeastern Indigenous cartography necessarily distorts scale and direction so that circle symbols can be located based on a combination of physical location, relative political centrality or peripherality, status of confederation or autonomy, and identity as ally or enemy in a way that allows connecting lines to represent both the actual routes of footpaths or river paths and communication quality between them.

In sum, southeastern Indigenous cartography shares the general characteristics of Indigenous mapping traditions while also revealing a unique graphic language of its own. This language is recognizable by a visual aesthetic of circles and lines, connected or disconnected along symmetrical networks, punctuated by pictographic images in profile, flexibly encoded, and expressing both numerical and narrative concepts. From this vantage point of a general knowledge of Indigenous cartography and some fluency in southeastern traditions specifically, we can begin an exploration of the many ways in which the Newark Earthworks belong to the Indigenous map history of the Ohio valley.

Born of Mapping

The observations and survey measurements by Hively and Horn[32] and Morgan[33] provide a basis from which we can piece together the way in which these Indigenous land surveyors worked. Hively and Horn explain that they combined a keen mathematical sense with sophisticated field techniques in order to achieve “a remarkable degree of symmetry, precision, and geometrical harmony” in their work. Measured angles and lengths were used with precision and according to an overall logic; for example, the unit of measure derived from the Observatory Circle, 321.3 meters, formed the basis for mapping the Octagon. This mathematical relationship between octagon and circle was then repeated in the circle and octagon at High Bank Works.[34]

The final design included six near-perfect geometrical shapes: four circles, one octagon, and one square. These discrete shapes were connected in a specific pattern through a built network of diagonal corridors constructed from parallel mounded walls that curved in some places and elsewhere ran straight. This central network of about 10.4 square kilometers was, in turn, aligned and oriented to connect with other earthworks and water sources in the valleys of Raccoon Creek and the South Fork of the Licking River.[35]

The surveyors worked with precision. At the Great Circle, a comparison of diameters indicated length deviations of 10 meters in a total circumference of 365.9 meters; at the Observatory Circle, diameters deviate only 0.8 meters in a total circumference of 1008.6 meters.[36] Though the tools may have been as simple as peg and twine, the overall symmetries and structures of the network indicate that these Indigenous surveyors were adept in more advanced calculations and mapping techniques.[37] At the Octagon, corners align with lunar positions, other Earthworks mounds and corridors, and other, more distant structures in the landscape, beyond the horizon. These measures would have required astronomical observations from the valley floor or from the longer sightlines obtainable from surrounding hills or features, then transferred to the valley floor.[38] In their examination of the latter possibility, Hively and Horn found that the surveyors could have taken measures either from three adjacent hills or from other features, including the Octagon; the central earthwork network; surrounding mound features including the Salisbury Square, the Ellipse, and Alligator Mound; and other, preexisting mound structures in the valley.

The overall size of the Earthworks is itself a testament to the skill with which the surveyors worked. It is one thing to stand holding a peg, on a relatively flat field, and ask a friend to walk around you with a long piece of rope and mark the arcs that become the diameters for circles and the right-angle intersections for squares and octagons. It is another matter entirely to stand on a hill, take a bearing, and then find a way of transposing that bearing at another location on the valley floor. Today we would achieve that bearing mark at a distance through cell phone communication; before cell phones, we used walkie talkies. With what tools were these bearings marked? Were they signaled by sign language from hilltop surveyor to valley surveyor, and if so, expressed as an intersecting landmark or an abstract angle identified by a quantity? Or were they noted and then inscribed onto bark, hand, shell, gesture, or the surface of the mind?

Any land survey is also limited by accumulating error. The first few distances and angles are simple to map, but as human error inevitably limits measurement, be those measures embodied or recorded with tools, errors accumulate with each discrete measurement. A large-scale symmetrical network quickly becomes asymmetrical when compounding error stretches or contracts the distances and angles between located objects. Today, we keep accumulating error in check through the use of ground control, that is, frequent adjustment of our newly surveyed locations to the previously located and known coordinates of control points and backsights. Did the surveyors use parts of this landscape for ground control? If so, which parts? The other mound structures? Or was it checked by astronomical readings?

A final indicator of the level of skill involved in the survey of the Earthworks is the placement and orientation of the network itself, which was designed to fit the valley and region in a particular way. The network connects to and incorporates the topography, water, and other earthworks of the valley with specific bearings.[39] To achieve this required a scope of vision beyond the merely local, a vision, rather, for the intersection of these two river valleys, for the ways in which these rivers connected to the larger fabric of the Licking River watershed and the Licking to the Muskingum watershed, and at a much wider scope, for the ways in which the Muskingum related westward to the surveyed landscapes of the Hocking and Scioto watersheds, all three flowing south into the Ohio River valley.

Brought Alive by Mapping

Human hands then continued to build the map upward and downward, bringing it into three dimensions. Guided by the surveyed measures, wherever and however marked, they placed soil, clay, and stone into the layered embankments, walls, ditches, and enclosures that raised and lowered the earthworks above and below the surface of the land. In so doing, they rendered the flat map visible and tangible to the human eye, and they created small and large places for people to gather.

People arrived from all over to use those spaces. In this journey and arrival, they contributed the next stage of the mapping process, that of embodiment. Through embodiment, they retraced the outlines of the map by foot as they walked and by eye as they followed the sightlines and comprehended the terrestrial and astronomical geographies presented to them there. Through embodiment, they tied the disparate sections of the map together by direct experience, and in so doing, forged an empathic bond with the geography and knowledge of the map.

A central purpose of the Earthworks is clearly created by the Octagon and Observatory Circle, whose structure charts the locations of the 18.6-year lunar cycle. At the Observatory Mound, for example, and the platforms at the Octagon openings, direct observations of the positions of the lunar moonrises and moonsets can be made.[40] To stand on the Observatory Mound and look up the corridor into the Octagon is to inhabit that part of the map connecting earth to the northern maximum moonrise. Another purpose of the Earthworks is expressed by the siting of the Great Hopewell Road, whose parallel embankments form a straight corridor now traced to 14 kilometers in length and, if extended, would logically end at the place now known as High Banks Works in Chillicothe.[41] The corridor thus maps the earthworks at High Banks into the earthworks at Newark. As with other corridors in the Earthworks, the archaeological and historical analyses of the road strongly suggest that it was walked in a sacred way, different and separate from everyday footpaths. Walking the road, these ancient pilgrims rendered the three-dimensional map into themselves, embodying the map by inhabiting it.

By tracing the flat map into raised and recessed spaces and incorporating soils and stone in different combinations at each site, the earthworks and indentations also seem to express symbolic axes of connection to the vertical and horizontal, thus joining the vertical and horizontal mapping functions of the lunar sightlines and corridor to High Banks. Vertical and horizontal mapping elements orient the visitor with respect to the universe by establishing axes of connection between the local and the universal, just as coordinate systems connect local to global in Western cartographies. Lepper reminds us that one quality associated with Middle Woodland peoples and their descendants is a unified cosmography based on the inextricability of water and earth and the vertical and horizontal dimensions of experience. To reveal this "cosmographic" mapping more fully, then, we must look beyond mound shapes and attend equally to where and how water, recessed spaces, and the pattern and proportion of materials are incorporated into the framework as a whole.

Underwater spaces and recessed spaces that may or may not have held water also form part of the Earthworks' geometry. The Earthworks are intertwined with water, sited at the confluence of two rivers and enveloping a small lake that once dominated the center of the main complex, between the Octagon and Observatory Circle and the Great Circle. In the Great Circle, a ditch 2 to 4 meters deep runs the length of the circle's interior.[42] On an adjacent hill, an oval enclosure of about 150 meters includes an exterior ditch. That the site also provides a comprehensive view of the Earthworks has led scholars to conclude that it, too, is of Middle Woodland origin.[43]

Colors appear in the Earthworks at particular locations. At the Great Circle, the layering of soils in the circular embankment is brown on the exterior and yellow on the interior. Lepper writes that although "color is likely to have carried great symbolic meaning for the Hopewell people, as it did for later Eastern Woodland tribes," it is unclear whether these color arrangements had symbolic significance. At the great house foundation in the central mound, however, the layered placement of black, yellow, and red clays strongly suggests symbolic arrangement. "The use of soils of varying color," Lepper concludes, "reinforces the sense of a highly dynamic, ritualized, and symbolically charged architecture."[44]

Different types of minerals and rocks, in various shapes and arrangements, also inhabit particular locations in the Earthworks. A copper ax, a bear tooth, and a stone figure of a person transforming into a bear are known to have been placed in the central burial mound at the Ellipse, and large quantities of mica were located in the smaller mounds. Two copper objects, one crescent and one beaver, were located in the central mound of the Great Circle. An arrangement of flint and different types of stones were located at Salisbury Square, and a quantity of Flint Ridge flint was arranged in a cone at the southeast corner.[45] Some of these materials came from the local area, but some of them were brought from long distances. The flint was dug locally at Flint Ridge. Mica was brought from what is now western North Carolina, copper from northern Michigan and Wisconsin, and obsidian from Yellowstone. The presence of these items in particular locations has been interpreted as representing the artifacts of ceremonial offerings from distant visitors. Do their locations also tell us something about spatial, social, or political relations?

In these ways, the Earthworks function as a map when remapped by its visitors: to read the map is to participate in the map, to embody and thus reinscribe the map, in a continuously regenerative process.[46] The embodied Earthworks map the 18.6-year lunar cycle with the spaces and sightlines of the Octagon. They inscribe the landscape with interconnected spaces and paths between spaces. They connect the above and below, or skyworld and underworld, mathematically through astronomical measurements and symbolically through a network of raised and recessed spaces made of earth and water. And along a southwestern diagonal corridor, they point us to the terrestrial world at High Banks.

These actions are made possible by the structure of the Earthworks, itself mapped by Indigenous surveyors using techniques for foresight, backsight, error adjustment, linear and angular measure, and large-scale collaborative coordination. This network of measured and surveyed geometric relationships is built of spaces of different shapes, predominantly circles, but also ellipses, squares, lines, and a crescent, and of different sizes, from small, discrete shapes to locate a sightline or enclose a group, to areas and lines so large that their full extent cannot be understood from any one position on the ground. The network disperses itself across the landscape but is also integral to that landscape, constrained by its scale and conforming to its shape. It was created with different soil, rock, and mineral types to support varying structural demands and create a design aesthetic through color and pattern. Through this framework of earth, sky, and water, the network measures that which is above and below while also encoding, connecting, and mirroring these worlds on the earth's surface.

What, then, can we conclude about Indigenous cartography at Newark? That it was a process begun by measured marks, rendered into three dimensions above and below the earth through embankments and ditches, rendered to a wider scale by objects, rendered into the sky by sightlines from and between embankments and corridors, and rendered within by tracing, reading, and experiencing the sightlines. Each stage of this mapping process tied all parts of the map together in its own way, yet only when they functioned together was the map made whole, activating its ability to map a local place to its location in the bodies of its visitors, its location on the North American landmass, and its location in the universe.

The kinship between the Earthworks' cartography and other Indigenous cartographies is evident. Like other Indigenous cartographies, the Newark Earthworks emphasize maps as the process or act of mapping and not the products resulting from that process. The Earthworks connect the terrestrial, celestial, and spiritual worlds as part of the same network. The Earthworks are both read and activated through embodiment and dispersed in the landscape through embankments, corridors, and ditches.

The parallels between the visual aesthetic of the Earthworks and that of southeastern Indigenous cartography are striking. Does that shared aesthetic indicate a shared cartographic language as well? It requires a leap of the imagination to ask this question. But there are many gaps in our understanding of what the Earthworks do and the reasons for their forms and structures, questions that begin where the archaeological and LiDAR records end. What might our knowledge of southeastern cartography, and Indigenous cartography generally, contribute to those questions? In other words, if we were to translate the silences in the Earthworks using the language of southeastern cartography as the dictionary, what would that translation say?

Translating the Visual Language of the Earthworks

The embankments of the Earthworks are dominated by circles of varying sizes, although the meanings of most of these circles are not yet known. If we consult the southeastern cartographic dictionary, we would interpret these circles as signifying discrete polities of varying social and cultural hierarchies and population sizes. The circles' sizes and positions relative to each other and to the Earthworks as a whole would tell us the rankings of these communities in the hierarchies.

There are, in addition, two prominent squares: Wright Square and Salisbury Square. In translation from southeastern cartographies, which depict qualitative difference through a shift from circle to square, we would interpret these squares as representing qualitatively different polities or concepts from their circular neighbors and functioning as spaces from which to measure or observe celestial geographies.

By the same logic, if some circles represent polities, it would follow that some corridors, in addition to mapping celestial or lunar alignments, also map the quality of communication between polities. It is notable that the corridor embankments are designed as double lines built in parallel. Some of the corridors do not connect, such as, for example, the corridors in the network around the Ellipse and Wright Square. Although two of these obviously end at cliff edges, four other corridors have rounded, disconnected ends. As it is acceptable in southeastern cartography for map symbols to have multiple functionalities or responsibilities, a corridor might map celestial or lunar alignments or the strength of communication between polities or both. If these meanings were not visually differentiated in the embankments, if the embankments were otherwise visually identical and differentiated only when embodied (when the map was "read" or explained, to each other and to themselves) during specific moments in a ceremony, this too would be conventional to the southeastern tradition.

If we were to attempt an actual translation of the geography represented among these circles, squares, and lines to discern which communities might be represented, we would keep in mind an aspect of southeastern cartographic logic, that scale and direction may shift within a single mapped space in order to accommodate the representation of political relationships and hierarchies, as well as the size and shape constraints of the medium (in this case, the valley itself).

In the particular case of the corridors, it seems necessary to also consider the alternative possibility that the corridors and small circles may be more similar to another Indigenous cartographic device, that is, a locational device resonant of the Andean system of *ze'ques* and *wak'as*. If visitors arrived at the Earthworks from distant communities, a map that included their home in its coordinate system would orient them in the map with respect to their homeland and the homelands of the other visitors. On the other hand, if the embankments of the Earthworks are primarily for the portrayal of celestial geographies, the corridors and circles would then logically constitute a coordinate system for the stars.

We can imagine further translations if we consider the other known characteristics of southeastern cartography and compare them to the characteristics of the Earthworks. For example, southeastern cartography uses color to connote specific concepts, as in the color white connoting an ally or something good and the color red connoting an enemy or something associated with war. Knowing this, it would be useful to examine the geography of soil colors at the Earthworks, as well as the geography of the colors of rocks, minerals, and carved objects placed within specific embankments.

Southeastern cartographic language expresses number through the use of repeating forms or collections of objects. There are no visual manifestations of this technique in the embankments and ditches. But is it manifested in that other dimension of the mapping process, the survey of angles and locations? One of the gaps in our knowledge is understanding how surveyors transferred their calculated bearings from hillsides to the valley floor. If the

Earthworks surveyors also expressed number through repeating forms or objects, we know something about how numbers were recorded and moved and how they could have been signaled from hillside to valley, perhaps with a discrete hand-held device having countable sections, as in a *khipu,* or with repeating symbols painted on bark.

These informed and imaginative ruminations open up exciting new pathways for investigation. Already, scholars have gained many insights into how people created the Earthworks by putting themselves in the shoes of those Middle Woodland surveyors, imagining the logical locations for survey setups, and hypothesizing and testing the kinds of tools they might have used to create a perfect circle. I suggest that we can also gain insight into the Earthworks' identity by thinking in the visual language of Indigenous cartographers, to imagine what kinds of symbols would be logical to use for different features and what kind of visual language would logically and aesthetically connect those features into a map.

AS DISCUSSED BY Tom Bremer and others in this volume, in the fall of 2009 an alliance of community members from Ohio and elsewhere, Native and non-Native, retraced the ancient route of the Great Hopewell Road from High Banks to the Octagon in the seven-day "Walk with the Ancients." They perceived it to be their collective responsibility to render the road visible once again: to map the landscape of the road to the landscape of Ohio residents' daily experience, by walking. Two cartographies intertwined, through embodiment. Like the Great Hopewell Road, the Earthworks are also our collective responsibility. Shall we golf on it? Or shall we listen to and activate what it is asking us to do?

Imagine you are heading west down Main Street, away from Old Town in Newark and across the river. Imagine you cut south on Union Street. Right about where you cross the railroad tracks, you have entered the parallel cartography of the Earthworks. You're on Union, and you're also on the Ellipse. This is the cartographic archive. You're in it, and you are it. If you continue south down Union, take a right on Hancock Street, and head west across the interstate, you've walked through the Ellipse and are proceeding into Wright Square. Walk through the square, taking care to walk around two of the mounds inside the square. Climb over the embankment on the other side, and leave the square. Now, in the cartography of the city of Newark, you're at the corner of Hancock and South Williams, and ahead of you to the west is the intersection with 21st Street. In the cartography of the Earthworks, that intersection is a body of water, a lake.

If you follow the water's edge southward, it will take you down to the Great Circle. If you follow it northward, it will take you up to where you can join the processional pathway between Wright Square and the Octagon.

From here, the route is up to you.

What will you map next?

Notes

1. Lepper, "The Ceremonial Landscape of the Newark Earthworks and the Raccoon Creek Valley," 114–15.

2. See, for instance, Hively and Horn, "Hopewell Cosmography at Newark and Chillicothe, Ohio"; Hancock, "The Earthworks Hermeneutically Considered"; and Lepper, "The Ceremonial Landscape of the Newark Earthworks and the Raccoon Creek Valley."

3. See L. Smith, *Decolonizing Methodologies*, 15–16, 120.

4. See Pickles, *A History of Spaces*, 75–106.

5. See ibid. and Rundstrom, "Mapping, Postmodernism, Indigenous People."

6. Rundstrom, "Mapping, Postmodernism, Indigenous People."

7. See Lewis, "Maps, Mapmaking, and Map Use by Native North Americans," 125; and Rice-Rollins, "The Cartographic Heritage of the Lakota Sioux," 42–43.

8. Barbara E. Mundy, "Mesoamerican Cartography," 193.

9. See Basso, *Wisdom Sits in Places*, 89; and Pearce, "The Last Piece Is You," in "Cartography and Narratives," special issue, *Cartographic Journal* 51, no. 2 (May 2014): 12.

10. See Kelley and Harris, "Traditional Navajo Maps and Wayfinding."

11. See Zuidema, *The Ceque System of Cuzco.*

12. Albert S. Gatschet, *A Migration Legend of the Creek Indians, with a Linguistic, Historic and Ethnographic Introduction*, 250.

13. See Lepper, "The Ceremonial Landscape of the Newark Earthworks and the Raccoon Creek Valley," 119.

14. See ibid.; and Galloway, *Choctaw Genesis, 1500–1700*, 327.

15. See Galloway, "Debriefing Explorers"; Waselkov, "Indian Maps of the Colonial Southeast: Archaeological Implications and Prospects" (1998); and Waselkov, "Indian Maps of the Colonial Southeast" (2006).

16. See Waselkov, "Indian Maps of the Colonial Southeast" (2006), 455–57.

17. See Waselkov, "Indian Maps of the Colonial Southeast: Archaeological Implications and Prospects" (1998), 207; and Lewis, "Maps, Mapmaking, and Map Use by Native North Americans," 69.

18. See Galloway, "Debriefing Explorers," 224–26; and Waselkov, "Indian Maps of the Colonial Southeast" (2006), 469–81.

19. Waselkov, "Indian Maps of the Colonial Southeast" (2006), 481.
20. Ibid., 481–86.
21. Galloway, "Debriefing Explorers," 227–28.
22. Waselkov, "Indian Maps of the Colonial Southeast" (2006), 445.
23. Galloway, "Debriefing Explorers," 224.
24. Waselkov, "Indian Maps of the Colonial Southeast" (2006), 445.
25. Ibid.
26. E.g., Lepper, "The Ceremonial Landscape of the Newark Earthworks and the Raccoon Creek Valley," 119.
27. Waselkov, "Indian Maps of the Colonial Southeast: Archaeological Implications and Prospects" (1998), 209.
28. Galloway, "Debriefing Explorers," 227–28.
29. Quoted in Waselkov, "Indian Maps of the Colonial Southeast: Archaeological Implications and Prospects" (1998), 209.
30. Waselkov, "Indian Maps of the Colonial Southeast" (2006), 447–49.
31. Ibid., 474.
32. See, for instance, Hively and Horn, "Geometry and Astronomy in Prehistoric Ohio"; and Hively and Horn, "Hopewell Cosmography at Newark and Chillicothe, Ohio."
33. R. Morgan, "Ohio's Prehistoric 'Engineers.'"
34. See Hively and Horn, "Geometry and Astronomy in Prehistoric Ohio," S9.
35. See ibid., S1.
36. Ibid., S10, S8.
37. R. Morgan, "Ohio's Prehistoric 'Engineers,'" 3.
38. Hively and Horn, "Hopewell Cosmography at Newark and Chillicothe, Ohio," 132.
39. See Hively and Horn, this volume.
40. See ibid.; and Lepper, "The Ceremonial Landscape of the Newark Earthworks and the Raccoon Creek Valley."
41. See Lepper, "The Great Hopewell Road and the Role of Pilgrimage in the Hopewell Interaction Sphere."
42. Ibid., 101, 105.
43. Ibid., 105.
44. Lepper, "The Newark Earthworks: Monumental Geometry and Astronomy at a Hopewellian Pilgrimage Center," 78.
45. Ibid., 77–79.
46. For groundbreaking work on the generative and performative nature of all mapping, see Kitchin, Perkins, and Dodge, "Thinking about Maps."

THOMAS S. BREMER

The Modern Religiosity of the Newark Earthworks

FASCINATION WITH ancient earthworks in Ohio goes back at least to the time that Euro-American explorers and settlers first entered the Ohio valley. English colonists in America knew of Indian burial mounds, and the cultural elites of the new American nation following the Revolution found evidence in these mysterious structures of a glorious antiquity and premonitions of future greatness for their civic ambitions.[1] Indeed, fantasies of cultural greatness danced atop these silent mounds. But Ohio's ancient earthworks also recall a history of loss and longing. Disappeared and entirely forgotten were not only the people and cultures who had produced these monuments but also the meanings, purposes, and uses the earthworks served in ancient times. A gulf of cultural differences lie between today's modern people and the imaginative genius—neglected, forgotten, and irretrievably vanished—that originally produced the earthworks found throughout Ohio and much of eastern North America.

The distance between contemporary American society and the ancient societies that built the earthwork structures has encouraged speculations about the meanings and purposes of these monuments. Not surprisingly, nearly all theories about their origins assume a religious dimension underlying their importance for the people who constructed them. Religion serves such discourses as a ready-made explanation of the mysterious; it also imbues such cultural achievements with heightened significance, regarding them as representations of ultimate value for the cultures that produced them. Such assumptions allow contemporary people to claim an intimate understanding of ancient people with insights into their highest ideals expressed in sacred architecture that has lasted a millennium or more.

Rather than adding to the speculations about the ancient people who left the earthworks on the Ohio landscape, this essay contemplates what the modern fascination with them might reveal about our own highest ideals and most cherished values. In particular, I aim to explore, at least tentatively, the manner in which these ancient places have in fact attained religious significance for contemporary people as sites of modern religiosity.

A Meaningful Modernity

Addressing the question of whether the Newark Earthworks are in any way "religious" for contemporary people requires a shift in how we think of religion. Rather than regarding religions as the characteristics and traditions of particular religious systems that communities of adherents follow, we might instead consider religion according to a particular view of modern subjectivity constituted by a secular/religious opposition, an approach that allows a different sort of analysis of the cultural and social histories of ancient artifacts such as the Newark Earthworks and of the people who built them. Presuming that a religious component underlies modern subjectivity turns attention to the question of why modern people attribute a religious character to these antiquities and what motivates modern people to engage in their own religio-aesthetic relationships with the artifacts of former cultures. In following the methodological recommendations of religious historian Ann Taves, this sort of scholarly investigation can focus on the processes by which modern people deem the Newark Earthworks special, sacred, and religious.[2] For purposes of this analysis, then, I suggest employing the term "religion" to indicate the processes, discourses, and attributions by which modern people regard their lives, both individually and collectively, as meaningful.[3] Religion in this way constitutes modern subjectivity as strategies of meaningfulness.[4]

Modern people, of course, engage in myriad practices, habits, discourses, ways of prioritizing their values and attachments, and many other strategies to make life meaningful. Without insisting on a particular approach that best demonstrates the religious underpinnings of modern subjectivity, I have returned time and again to a specific set of modern practices and discourses that have proven exceptionally useful for exploring religion in modern life; these practices and discourses are most commonly denoted as "tourism." In its distinctively modern form, tourism exemplifies the circumstances of modern subjectivity, thus offering especially productive opportunities for asking critical questions about religion, modernity, and strategies of meaningfulness.

I regard tourism as a peculiarly modern set of travel practices aimed at producing aesthetically pleasing experiences. Historically, the emergence of tourism in the Western world as a particular mode of travel corresponds to the emergence of the secular as a distinct category in opposition to the religious. Although usually associated with the secular, modern tourism retains a religious element, sometimes explicit, but most often merely implied in the cultural logic that motivates modern people to engage in journeys of pleasure and edification. In particular, a religious component underlies tourists' desire for aesthetically meaningful experiences. Thus, by engaging in practices and discourses that characterize tourism, modern subjects produce and sustain aesthetically meaningful identities through the conventions of market capitalism.

Two characteristics of tourism make it especially helpful for thinking about religiosity in modern societies. First, in their tendency to aestheticize, commodify, and consume virtually every corner of the natural world and all aspects of human cultures, tourist practices and discourses exemplify modernity; tourists are, in short, exemplary modern subjects.[5] Second, participation in the practices and discourses of modern tourism reveals in many instances the reflexive nature of modernity's implicit reliance on religion. In addition to, or more often in collusion with, the secular desires of tourists for pleasure, relaxation, escape, adventure, or edification, modern touristic travel nearly always involves a religious propensity for creating and sustaining meaningful selves, both individual and collective. Tourism, then, serves as a useful point of entry into the entanglements of the secular/religious opposition that generates modern subjectivity.

The interpretive prism of tourism can reveal particularly illuminating moments in the history of touristic practices and discourses involving the Newark Earthworks, allowing us to explore the ways that these ancient earthworks are very modern places of religiosity for people today. In particular, we can ask how the Newark Earthworks and other mound sites contribute to the creation and sustenance of meaningful selves for modern people. This is seen prominently in the history of modern interpretations of the Newark Earthworks.

Modern people have struggled over how to interpret the significance of ancient Native American earthworks in North America since they first stumbled upon them in the eighteenth century. By the end of the nineteenth century, Europeans and Euro-Americans remained puzzled and bewildered by the architectural artifacts of advanced civilizations in ancient America. British traveler Lindesay Brine, writing in 1894, found it remarkable that after nearly four centuries of investigations and speculations about civilization in

Mesoamerica, "nothing satisfactory has yet been ascertained which explains the manner in which that civilization could have arisen amongst those exceptionally instructed races." Brine himself had explored sites in Mexico, Ohio, and elsewhere during an extended journey in the 1870s in an attempt "to solve the difficult and complex problem of this Indian advance towards higher conditions of life."[6] Most puzzling to him and most other Europeans who gave much thought to the Native cultures of the Americas was how people who were obviously not modern could produce cultural artifacts that rivaled the sophistication of modern people. In particular, modern Westerners presumed that unfamiliar indigenous cultural practices and values regarded as "religious" defined the primitive Other in contrast to the rational and predictable values of modernity. Yet belying such simplistic characterizations of the nonmodern Native was the confounding evidence of sophisticated and advanced civilizations, especially in the monumental architecture that Westerners found throughout much of the Americas.

This conundrum of primitive people building remarkably sophisticated structures persisted in the nineteenth century when Lindesay Brine visited Ohio. His examinations of the geometrical precision of the earthworks in Newark led him to surmise that their "plans and measurements are evidences of the existence of mental capacities which were far in advance of those of the present Indian races."[7] He found it difficult to imagine "that an accurately designed work of this shape and magnitude could have been planned by Indians," causing him to conclude that the earthworks "may have been raised by native tribes, acting under the direction of European surveyors, or of men who had received a mathematical education."[8] In support of this hypothesis of ancient Indians building earthworks under the supervision of Europeans, Brine relates local stories he heard in Ohio, some attributed to Native American oral traditions, that white people inhabited the land long before Christopher Columbus initiated European settlement in the Americas.[9] On the other hand, Brine concludes, "Although it has to be admitted that all theories as to the Mound Builders must be necessarily indeterminate, . . . nothing has been found amongst the ornaments or weapons that were placed in their burial mounds, which supports the hypothesis that they were different in race or intelligence from the tribes that surrounded them."[10]

This question of who built the Ohio mounds and what purposes they served for the builders and other ancient peoples has persisted throughout the history of modern interpretations of the earthworks. Nearly without exception, modern speculations have assumed that these structures had significant religious meanings for ancient people. The possibility of a purely secular purpose has rarely occurred to the modern imagination. This perpetuates a long tradition of modern Westerners interpreting the nonmodern Other based on a foundational assumption that such remarkable cultural achievements could only be attempted for religious aims. The point of such interpretations, however, is not necessarily to "understand" Native peoples and their cultural traditions objectively but in many cases simply to assert the superiority of the rational modern self, as seen in Brine's conclusions. The notion that nonmodern people would pursue such advanced cultural achievements for purely secular, rational purposes would contradict a fundamental opposition between modern and primitive that underlies modernity. Thus, interpretations of the Ohio mound sites often have more to do with the subjective concerns of the modern interpreters than with the ancient concerns of the people who built and used them.

Meaningful Ritualizations

Interpretations of places like the Newark Earthworks are closely associated with the practices performed there; in fact, processes of producing and sustaining meaningful selves often involve the ritualization of particular places. Ritual theorist Catherine Bell uses the term ritualization "to draw attention to the way in which certain social actions strategically distinguish themselves in relation to other actions. In a very preliminary sense," Bell continues, "ritualization is a way of acting that is designed and orchestrated to distinguish and privilege what is being done in comparison to other, usually more quotidian activities."[11] In other words, ritualization as Bell understands it involves distinctive behaviors, usually conventionalized and strategically engaged for specific purposes, that gain rhetorical force in being set apart from other ways of acting. In this regard, the conventional behaviors of tourist visitation, as an exemplary practice of modernity, amount to ritualization of particular destinations, not unlike the ritualizations of pilgrimage sites with the conventional behaviors of devotional visitors. At the Newark Earthworks, visitors participate enthusiastically in the rituals of tourist visitation: they visit the site museum at the Great Circle, they purchase souvenirs and postcards, and most of them walk the site, often capturing their experiences with pictures. As tourists, visitors at the Newark Earthworks perform conventional practices of modern subjectivity.

Touristic engagement with the Newark Earthworks follows long-established conventions that Westerners have used to ritualize these and countless other sites in North America from the earliest days of European settlement on the continent. From the time of the European Renaissance onward, and especially during the heyday of the European Grand Tour in the seventeenth and eighteenth centuries, modern Westerners have made particular destinations special through travel practices undertaken for recreational and educational purposes. These practices include discourses aimed at aestheticizing certain places and activities, heightening their appeal to visitors, and making them desirable tourist destinations. This in turn serves to commodify these destinations, thus transforming them into highly profitable attractions for those who own and control the sites, as well as for the many purveyors of goods and services that facilitate and support travel there. Transportation and communication services, food, lodging, banking, tour guides, purveyors of souvenirs, and myriad other commercial interests all benefit from the touristic ritualization of desirable destinations. At the Newark Earthworks, travelers rely on many of these services, namely hotel accommodations, restaurants, airlines, and rental cars; Internet sites that describe and promote the Newark Earthworks as a desirable destination; photography equipment and services; and the site museum and gift shop at the Great Circle. Participation in the ritualized conventions of touristic visitation perpetuates the processes of aestheticization and commodification of earthwork sites that mutually benefit visitors and commercial interests alike.

The modern ritualization of the Newark Earthworks occurred in different forms in the nineteenth century. The Great Circle, for instance, besides its appeal for tourists as an unusual instance of ancient architecture, has been the site of state and county agricultural fairs; for several decades it included a track for horse and sulky races and later for bicycle, motorcycle, and car races. The grounds served for a time as a military training camp, and they were the site of a popular summer resort with a hotel and restaurant, as well as an amusement park that included a Ferris wheel, roller coaster, casino, boating and fishing, a bowling alley, and other entertainments. Since 1925 the Great Circle has been the main feature of Moundbuilders Park in Heath, Ohio, with a museum that opened on the site in 1971.[12] These sorts of developments served to ritualize sites as appealing attractions for the conventional practices of tourist visitors.

One recent example of ritualization at the Newark Earthworks illustrates especially well how touristic practices can facilitate processes of religious meaningfulness in sustaining modern subjectivity. In an event dubbed "Walk with the Ancients," about thirty walkers trekked roughly seventy miles across central Ohio in October 2009 to follow, more or less, the ancient Hopewell ceremonial road that archaeologists believe may have stretched from the earthworks in Newark to those near Chillicothe, Ohio. Officials at The Ohio State University's Newark Earthworks Center organized the walk, which they characterized as a "pilgrimage," in conjunction with a Newark Earthworks Day, a celebration that also included an art exhibit, musical performances, and an academic symposium about the earthworks.[13]

Leading the group of walkers across the Ohio landscape were several Native American ritual specialists headed by Gilly Running, a member of the Sicangu Lakota tribe of South Dakota, who served as the group's "spiritual leader." The group first gathered at the Hopewell Culture National Historic Park outside of Chillicothe, where they "sang a sacred song led by Gilly Running, . . . a song we would all sing together in strength and unison by the time our time together had come to an end," according to ethnohistorian Vincent Stanzione, an invited participant who completed the entire walk. The group of walkers then began with "a ritual procession" as they entered the park's mound enclosure "to sing and pray asking the Great Spirit, Grandfather God, the Ancient One to watch over us on our way."[14] After the procession they set out on their northward trek through the mostly rural landscape toward the Newark Earthworks, led by "Native American Indian staffs," festooned with colorful banners and feathers, "that opened the road as they sang out to the four directions."[15] At night the walkers camped, and in most locations hosting groups prepared meals for them and joined in nightly programs related to their walk or to Ohio's earthworks. All along the way, Gilly Running led them in songs and prayers (see plate 13).

After a week of traversing the rural Ohio landscape, the walkers entered the more urban precinct of Newark and its surrounding communities, where a small crowd joined the pilgrims for the final procession to the Octagon Earthworks. Led by their sacred staff, the group entered the enclosure of ancient mounds for a final round of ceremonial drumming, prayers, and offerings. Joining them there was Chief Glenna Wallace, leader of the Eastern Shawnee Tribe, one of the groups that had been removed from Ohio in 1832.

Many of the participants regarded their experience of the Walk with the Ancients as religious, especially if we think of religion in terms of meaningfulness. In fact, several indicated that the meaningful nature of the experience had a transformative effect on their lives. During the walkers' panel discussion at the Newark Earthworks Day celebration on the day after completing their walk, Tom Krupp stated that his experience of walking with others across the Ohio countryside was "something I will never forget." On the same panel, Joan Stoufer emphasized a connection to each other and to ancestors; she characterized walking as an act of prayer that transformed the entire group through the power of love. The connection for Bob Neinast, known to his fellow walkers as "Barefoot Bob," was to the ground itself; he had walked the entire distance without shoes, claiming that barefoot pilgrimage enhances a connection to "Mother Earth."[16] Some months afterward, Vincent Stanzione recalled the transformative effect of the walk as participants realized "that we were integrating something new into our lives by allowing the sacred road and our experience on that road to transform us in whatever way that destiny seemed to be taking us."[17] That experience of "the sacred road" also had a transformative impact on Bob Pond, a self-described "religious skeptic" who nevertheless regarded the Walk with the Ancients as a religious experience. He recalled in particular how he became immersed in the constant drumming and singing, which, he said, "gave me humility." Some eighteen months afterward, Pond described the walk as "a highlight of my life so far."[18]

As many of the walkers grappled with finding language adequate for expressing the profundity of their experience, a common religious strand became evident. For the most part, they found meaningfulness through their experience of otherness. For many of them, this otherness became manifest in the image of "the Ancients," most often regarded as spiritually powerful beings extant in Native American religious practices and present in the land itself. Several of the walkers claimed a communal connection to these Ancients; this connection became possible, according to Stanzione, by "practicing our own eclectic kind of walking religion."[19] This walking religion not only allowed an intimate attachment to the nebulous figure of "the Ancients," more concretely it strengthened communal bonds among the walkers. Together they walked, Stanzione recalls, "as Mother-Fathers and Sister-Brothers of this sacred earth, Sons and Daughters, Grandsons and Granddaughters of this hallowed land."[20] Immersed in the native traditions of singing and drumming, led by the sacred staffs, accompanied by a Native American spiritual leader, walkers felt a meaningful, transformative bond with the Ancients in the Ohio landscape.

But the divine experience of walking with the ancients, with its "mystical feeling that seemed to fill the air,"[21] could not last. On the final day of the walk, things "quickly turned toxic" as the route brought them into the Newark area and took the walkers along a busy highway.[22] The modern reality of the Great Hopewell Road did not match their romanticized expectations of its sacred route. According to Stanzione, "You could feel the pilgrimage falling apart as we entered the rather ugly world of low-cost/maximum-square-footage commercial design and architecture."[23] Their sacred journey skidded to a disappointing conclusion as their religious experience of walking with the Ancients collided with the secular realities of modern America.

The last day of the Walk with the Ancients also revealed the limits of Euro-American engagements with the Native Other. In the morning they passed a Fraternal Order of Elks lodge building with a large mural of Mount Rushmore painted on the side of it. The walkers stopped to pose for a group photograph in front of this image of the monumental icon of American triumph carved in the Black Hills, the sacred land of the Lakota known as Pahá Sápa. This angered the Native American lead walker, who resented interrupting their journey to recognize Mount Rushmore, which many Native peoples regard as a despicable desecration of their holy land. An uncomfortable rift fractured the communal bonds of the group.[24] In this moment the immense gulf between non-Native walkers and the Native American participants became evident. The history of colonial imaginings of conquered peoples and Native resistance to conquest and subjugation, a history that remains integral to modern subjectivity, came crashing down on their divine journey.

Yet most if not all of the walkers came away unscathed by the conflicts and disappointments of the last day. Their recollections at the Newark Earthworks Day meeting on the day after the ceremonial conclusion to the Walk with the Ancients, as well as comments in the coming months, overwhelmingly reflect the profound meaningfulness of the experience for most of the participants. It was, in short, a religious undertaking for many who completed the journey, even religious skeptics like Bob Pond.

At the same time, however, the walk was also a thoroughly touristic event; indeed, the walkers remained modern tourists throughout their experience. The organizers and participants relied on touristic travel practices that serve to commodify tourist experiences. Most obvious among its touristic features was the extent of planning that went into packaging the travel experience. According to Richard Shiels of the Newark Earthworks Center, it took more than a year of planning to work out the multitude of logistical details for the weeklong trip.[25] These included arranging for the overnight stops and the various programs that the walkers enjoyed each evening; producing colorful and informative maps; providing support vans to shuttle the walkers to campsites and meals and to carry their gear while they walked; and arranging for people to drive the vans, prepare food, and take care of other needs of the walkers. The organizers also invested considerable efforts in creating an enticing product that would attract walkers and marketing it to the public, which included a series of essays about the Newark Earthworks and pilgrimage traditions that appeared in the local newspaper.

Creating an enticing package to attract walkers to the event included emphasizing the religious character of the walk. From the beginning, organizers characterized the journey as a "pilgrimage," and the invited participants in the walk emphasized this theme. In particular, Vincent Stanzione shared his vast knowledge of indigenous pilgrimage traditions among the Maya people of Central America, and there was deliberate inclusion of Native American spiritual practices throughout the walk led by the Lakota spiritual guide Gilly Running. Besides bringing attention to the Newark Earthworks as an ancient pilgrimage destination, organizers aimed to produce meaningful experiences for contemporary walkers.

Their attempts to produce meaningful experiences amounted to a commodification of the Newark Earthworks as an appealing travel destination; this process of commodity production relied largely on appeals to modern aesthetic discourses common in touristic contexts. Most prominent of these was the touristic emphasis on authenticity. Tourists often seek, and tourist providers attempt to supply, authentic experiences. For tourist visitors, the appeal of the Newark Earthworks consists in their status as authentic artifacts of an imagined past, authenticated by scientific research as well as by cultural traditions of Native peoples. Organizers of the Walk with the Ancients emphasized this appeal by insinuating the promise of an authentic connection to a transtemporal community, the so-called Ancients. In an essay published in the *Newark Advocate* a month before the walk, Richard Shiels asks, "How better to connect with the ancient people who shaped the landscape where we live? We will walk where they walked and celebrate where they celebrated."[26] This implied promise was bolstered with a direct connection to contemporary communities. A real Native American spiritual leader guided participants in authentic practices of drumming, singing, prayers, and offerings, allowing non-Native walkers to experience the otherness of indigenous spirituality. Thus, the aesthetic of authenticity constituted a fundamental element in the touristic commodity of walking with the "Ancients."

Auspicious Sites of Modern Meaningfulness

The touristic quality underlying the meaningful interpretation and experience of the Newark Earthworks does not in any way diminish their importance. On the contrary, as an exemplary practice of modernity, tourism typifies the complex entanglements of the religious and the secular in modern subjectivity. In the Walk with the Ancients, religion erupts in secular discourses even as the secular impinges upon the religious. It reveals a particular way that modern people are religious: by engaging capitalist strategies of commodification and consumption to produce meaningful, transformative experiences, participants were pursuing a profound moment of modern religiosity.

This is not to say, however, that engaging touristic travel practices to create and sustain meaningful selves always involves commodification. Religious meaningfulness can also be found in other uses of modern leisurely travel, utilizing the conventions of tourism to generate meanings that escape commercial manipulations. A particularly poignant example became apparent at the conclusion of the 2009 Newark Earthworks Symposium, following the Walk with the Ancients. In the keynote presentation of the symposium, Chief Glenna Wallace, leader of the Eastern Shawnee Tribe of Oklahoma, told an emotionally powerful and compelling personal story of recovering a displaced self, both for herself personally and for the Shawnee people collectively. She spoke of hardships and cruelties that stretch back at least to the forced removal of her ancestors from Ohio in the 1830s; the Shawnee people, she related, have endured nearly two centuries of conquest, subjugation, and colonization in a place foreign to their ancestral traditions. By the time of her own childhood, her family members were almost entirely estranged from their Native American heritage; their indigenous past had been erased almost completely. Chief Wallace experienced extreme hardship during childhood when her family lived as migrant farm workers for a number of years. Displaying exceptional talent as a student, however, Chief Wallace found success in school and was able to enter college, eventually earning an educational specialist degree (EdS), leading to a long academic career at Crowder College in Neosho, Missouri. She also found her way back to her ancestral community among the Eastern Shawnee Tribe. In 2006 she became the first woman to serve as the tribe's chief.

As the leader of her people, Chief Wallace arranged a tribal excursion to Ohio in 2007, not to reclaim their homeland, but as tourists. They chartered buses and embarked on a tour of the significant sites of their tribal past, including the birthplace of the great Shawnee leader Tecumseh; the place where the tribe made its first overnight stop on the Trail of Tears after being removed from their homes in 1832; and finally, Wapatomica in Logan County, Ohio, the former village of their ancestors that had lain abandoned for more than two centuries. Now owned by the state and closed to public access, it had been the political center of the Shawnee people in the eighteenth century and the site of several intertribal councils. The arrival of the Eastern Shawnee Tribe led by Chief Wallace marked the first time Shawnee people had been to Wapatomica since an American militia burned the village to the ground in 1786.[27]

The return of the Eastern Shawnee Tribe to Ohio certainly relied on touristic practices; riding on chartered buses, staying in local hotels, and eating in local restaurants, as well as such familiar tourist pastimes as shopping for souvenirs and taking pictures were all part of the trip to their historical homeland. But more important, their journey involved, as Chief Wallace put it, a search for stories. After describing the poignant and emotional experience of returning to Wapatomica, Chief Wallace concluded, "We went in search of, and have found, ourselves." It was certainly a transformative moment for her personally and for her community; the touristic experience of discovering the places of their history reconfigured the meaningfulness of their sense of self, specifically in relation to the places of Ohio.

The Newark Earthworks, however, have not figured prominently in the historically meaningful connection that the Shawnee have explored in Ohio. Certainly, Chief Wallace has been an enthusiastic supporter of efforts to preserve the earthworks and to educate the public about them and their history. In this way, she has participated in the touristic practices that commodify earthwork sites as appealing travel destinations. But to suggest that she represents an authentic historical link to the mound sites seems to miss the point of why the Shawnee people have returned to Ohio. Their initial excursion was not merely heritage tourism in the sense of a recreational but edifying trip involving what Barbara Kirshenblatt-Gimblett has described as "the transvaluation of the obsolete, the mistaken, the outmoded, the dead, and the defunct."[28] The Eastern Shawnee people, at least as Chief Wallace describes their experience, seem less interested in consuming a commodified past and more intent on reclaiming their own story grounded in particular places. Foremost among the places of their story is Wapatomica, a most uncommodified, nearly hidden place that remains entirely absent from tourist itineraries. In contrast, the Newark Earthworks, well established on tourist maps of Ohio, hardly rate any mention at all in the Shawnee reclamation of their displaced past.

To the extent that the Eastern Shawnee Tribe are modern people, they cannot escape the tensions and discourses that define modernity. As a result, they are deeply implicated in the forces and conventions of capitalist economies; likewise, they are as subject to discourses of the religious and the secular as anyone else. This is not to say, however, that all moderns are the same, or even recognizably similar. Modern subjectivity relies on a diversity of experiences, perspectives, and situated understandings. There are many ways of narrating the meaningful stories of the modern self. For the Eastern Shawnee Tribe, a history of displacement and colonial subjugation has made for a very different meaningful engagement with the places of the Ohio topography than how descendents of European people have regarded Ohio. Chief Wallace's pilgrimage to Wapatomica means something far different to her and her Shawnee compatriots than the meanings that the pilgrim walkers encountered as they entered the Octagon earthwork structure in Newark at the end of the Walk with the Ancients.

The Newark Earthworks, like many tourist destinations, serve as sites of modern meaningfulness. They rank as special places, what at least some people regard as sacred places, in their role as sites where modern subjects can create and sustain meaningful selves, both individually and collectively, through the practices of modernity. Earthworks in Ohio have served modern people in this way for two centuries, and they continue to generate meaningful understandings of selfhood for diverse individuals and communities even today. They are, in short, auspicious places of modern meaningfulness and thereby remain a site of modern religiosity.

Notes

This essay has benefited greatly from the help and support of numerous people. Above all, I must acknowledge the debt I owe to the convener of this project, Lindsay Jones, who was my first teacher of religious studies at The Ohio State University. I am also grateful for the differing perspectives of participants in the symposium "The Newark Earthworks and World Heritage: One Site, Many Contexts," held May 1–4, 2011, in Granville, Ohio. In addition, this work has profited from the helpful comments and criticisms of Melanie Bremer.

1. See Sayre, "The Mound Builders and the Imagination of American Antiquity in Jefferson, Bartram, and Chateaubriand," 225–26.

2. Ann Taves, *Religious Experience Reconsidered*, xiii.

3. Taves discusses work by Wayne Proudfoot, Phillip Shaver, Bernard Spilka, and Lee A. Kirkpatrick regarding "meaning-belief systems" in ibid., 100–102.

4. I am not proposing a definition of religion but instead only suggesting a particular way to think about religion in complex modern societies.

5. The argument that tourists serve as exemplary modern subjects appears in Dean MacCannell, *The Tourist,* 11–13. Judith Adler also notes how touristic practices participate in discourses on modernity in "Origins of Sightseeing." I have previously argued that its engagement with modern global capitalism makes tourism "an exemplary practice of modernity" in Bremer, "A Touristic Spirit in Places of Religion," 38.

6. Brine, *Travels amongst American Indians,* v, vii.

7. Ibid., 69–70.

8. Ibid., 98–99.

9. Ibid., 94–95.

10. Ibid., 103.

11. Bell, *Ritual Theory, Ritual Practice,* 74.

12. Ohio Historical Society, *Newark Earthworks Historic Site Management Plan,* appendix I, "A Brief History of the Newark Earthworks," Ohio Historical Society, http://ohsweb.ohiohistory.org/places/c08/pdf/Appendix1_History.pdf.

13. This trek became an annual event, continuing in 2010 and 2011 with more emphasis on educational goals.

14. Vincent James Stanzione, "My Walk with the Ancients," unpublished document, 2010, 3–4. Organizers of the walk had invited Stanzione, an ethnohistorian living in Guatemala, to participate as an expert on walking as ritual practice based on his experiences with Maya people in Central America. I thank him for his generosity in making his essay available to me.

15. Ibid., 7.

16. The comments by walkers are recorded in my personal notes of the panel discussion at the Newark Earthworks Day event, Oct. 17, 2009, on the campus of The Ohio State University at Newark.

17. Stanzione, "My Walk with the Ancients," 49.

18. Bob Pond talked to me about his experience on the Walk with the Ancients during the 2011 Newark Earthworks Symposium on May 2, 2011, at the Robbins Hunter Museum in Granville, Ohio.

19. Stanzione, "My Walk with the Ancients," 24.

20. Ibid., 45.

21. Ibid., 51.

22. Ibid., 65.

23. Ibid., 66.

24. Ibid., 65.

25. Josh Jarman, "In Ancient Footsteps: 'Walk with the Ancients' Involves Hikers in American Indian Spirituality, Community," *Columbus Dispatch,* Oct. 16, 2009, B2.

26. Richard Shiels, "Newark Was a Place of Pilgrimage," *Newark Advocate,* Sept. 4, 2009.

27. Information about Wapatomica is available in a report about the Eastern Shawnee Tribe's third visit to the site in July 2010 for the dedication of a new flagpole and monument. See http://ohsweb.ohiohistory.org/enews/0710b-nf.shtml. For a newspaper account of the dedication, see Joel E. Mast, "New Pole, Monument Mark Shawnee Home," *Bellefontaine (OH) Examiner,* July 20, 2010.

28. Kirshenblatt-Gimblett, *Destination Culture,* 149.

PART V: The Newark Earthworks in the Context of Indigenous Rights and Identity

American and International Frames

MARTI L. CHAATSMITH

Native (Re)Investments in Ohio

Evictions, Earthworks Preservation, and Tribal Stewardship

TWO THOUSAND YEARS AGO, ancestors of contemporary American Indians created the Newark Earthworks amid bountiful woodlands surrounded by rushing creeks and wetlands. The natural landscape provided the inspiration for people to plan and build a massive complex on the scale of the physical world around them. From this act of community and ritual emerged a map reflecting their spiritual and social world. When completed, maintained, and landscaped as the architects envisioned, the Newark Earthworks complex was surely experienced as a wonderful place: a place of anticipation, mystery, medicine, magic, and grandeur, perhaps especially during the three years around lunar standstills.

While archaeological methodology, research, and documentation provided most of what is known about the Newark Earthworks to date, archaeology alone has not been sufficient for understanding the earthworks. The interpretation of Newark and other sites requires a balance among what has been learned from the archaeological perspective, the physical geography, social histories of the Ohio valley, and what has yet to be learned from tribal histories and indigenous traditions that have persisted in some form and originated long before European contact. The Newark Earthworks have been designated as monuments, as an example of "outstanding universal value" and human achievement. What has yet to be understood are the complexity and nuances of the peoples' cultures; how their experiences and appreciation of life in the Eastern Woodlands and their familiarity with the sky above led to earthen architecture with which they conveyed knowledge from one generation to the next. American Indian governments and people whose tribes have significant historical connections to the Ohio valley and earthen architecture in the Eastern Woodlands and have a deep knowledge of their traditions should be participants and leaders in developing a more complete understanding of the Newark Earthworks.

American Indian Perspectives and the Interpretation of the Newark Earthworks

The iconography and architecture associated with the Newark Earthworks celebrate metamorphosis, transformation, and duality. The "Wray Figurine" is a stone carving of a person holding in his or her lap a head adorned with regalia of Mississippian culture sculpted in the act of shape-shifting into a bear, perhaps using "bear medicine" for healing. Other artifacts included mirror images of animals and geometric shapes etched or carved into copper and stone, perhaps for tattoos or textile prints. The architects of the Newark Earthworks constructed earthen boundaries signifying entryways between the natural world outside and the enclosed spaces within; the act of crossing boundaries through an entryway meant moving from one state of being into another. Low earthen walls encircled the entire complex. Tall seamless curves were configured in precise geometric shapes with specific points of entry. At the Octagon, large rectangular barrier mounds at each of the eight entryways obscured the view inside the octagonal space and provided two entryways. There may have been structures in addition to the earthen architecture, perhaps made of wood, tall grasses, or water plants. Water held special significance: waterways transported people to the site, creeks protected the complex on three sides, slate slabs placed within deep furrows at the base of high walls simulated a waterway inside the Great Circle, and a natural pond was maintained in the center of the entire complex. Ponds and a water-filled perimeter just inside the Great Circle earthworks calmly reflect earth and sky. People went to the Newark Earthworks in the dark of night to observe the moon and the night sky and in the daytime to participate in communal activities.

Examine the Newark Earthworks survey maps from the mid-1800s (see fig. 1 in Bradley T. Lepper's essay in this volume, for example).[1] The earthen walkways bordered by low earthen walls led to individual earthworks, and people moved around the complex in specific and meaningful ways; each of the four enormous geometric enclosures represented different shapes, and each had a different purpose. The 50-acre Octagon connected to a 20-acre circle was an astronomical observatory that marked the lunar standstill every 18 years and 291 days, two different shapes connected by an unbroken earthen walkway. The ovoid enclosure, called the Ellipse, was a cemetery containing many burials and several huge mounds. The Wright Square was situated between the cemetery and the Great Circle, with no entry on the far side. The only way into the Octagon Earthworks without climbing over earthen walls was through the open corners of the Octagon, where barrier mounds provided two choices for entering the space. Walkways linked the Great Circle to the Square, the Square to the Ellipse, and the Ellipse to the Octagon. No walkway was built between the Octagon and the Great Circle. A small circle built on the southeast side of the Octagon with an entryway opening toward the east is one of the eight points of alignment for observing the long lunar cycle. There were many earthen circles of varying sizes along the walkways within the Newark Earthworks complex, each with low walls and an entryway that designated a space apart that invited social activities: dance, prayer and preparation, meetings, socializing with family and friends, feasting, and ceremony.

The Newark Earthworks were remarkably intact when surveyed by settlers in the mid-1800s. For much longer than two thousand years, generations of indigenous groups of different cultures lived in the area around present-day Newark, Ohio. After European contact, new American Indian groups moved into the area and lived among the Newark Earthworks for generations until the late 1700s, including historic tribes such as the Shawnee. These historic, multiethnic tribal groups were the last American Indian stewards of the Newark Earthworks.

Today when visitors approach the Newark Earthworks, they cannot experience the place as the builders intended. Streets and highways obliterated the small circular enclosures and most of the walkways connecting the large geometric earthen enclosures. Only two of the largest earthworks were protected from outright destruction: the Great Circle and the Octagon Earthworks. The Octagon has been leased to a private country club that has modified the ancient site to accommodate a golf course encompassing an underground watering system, golf greens, bronze memorial markers, sand traps, benches, ball washers, restrooms, paved golf cart paths cutting through the earthen walls, outbuildings inside the Octagon, parking lots, a clubhouse with tennis courts, and a large outdoor pool. Across the complex, a plowed-over corner of the huge square and some of the walled walkways remain. The built water features in the complex are gone. Housing developments, factories, and roads crowd the remaining earthen structures. A visitors' center built at the Great Circle's entryway blocks the view of the eastern horizon. Large trees, some of them nearly a hundred years old, grow inside the enclosures and through ancient walls. The northern section of the Octagon has not been maintained, attracting animal and human activity. To appreciate fully the significance and beauty of the place requires patience, careful study of the nineteenth-century survey maps, the Ancient Ohio Trail website,[2] an excellent tour guide, and imagination.

Most of the stories told about the purpose of the Newark Earthworks do not balance archaeological insights with informed American Indian perspectives. Descriptions and tour narratives sometimes appear to be influenced by stereotypical ideas about precontact American Indian culture. Much of what is understood about the Newark Earthworks has been deduced from the earthen architecture, analysis of the artifacts from grave mounds and from early historic settler accounts. Archaeological research has not yet definitively linked the earthworks builders to any contemporary tribal nations, culture, or language group, and there are no tribal governments in Ohio. As a consequence, there are no tribal stewards currently watching over the Newark Earthworks in the twenty-first century.

Maintenance of Native Knowledge in the Settlement Era

The Great Lakes region was home to dynamic American Indian cultures. Before 1492, the ancestors of contemporary American Indians lived throughout the Ohio valley for thousands of years. Estimates of precontact North American indigenous populations in North America are between 5 and 20 million, with a thousand unique cultures and linguistic groups.[3] For at least five thousand years, earthworks have been built by many cultures throughout the eastern third of North America. While the precise geometry of the Hopewell culture earthworks are unique to Ohio, conical mounds contained burials and were a ubiquitous part of the landscape. Ancient conical mounds alongside those more recently constructed indicate a shared knowledge maintained over time, geographical place, and across different cultures. It was likely the case that everyone living in the precontact woodlands cultures understood that conical mounds were the graves of not only their relatives but of ancestors who lived long before them. How else does one explain the existence of thousands of burial mounds dating back thousands of years, most of which were destroyed only during the settler era?

The construction of earthworks architecture continued well into the historic era. The three major earthworks-building cultures in the Ohio region are

Adena (800 BCE to 100 CE), Hopewell (100 BCE to 400 CE), and Fort Ancient (1000 CE to 1650 CE). People of the Fort Ancient culture were present in Ohio 150 years *after* European contact in 1492. There were hundreds of American Indian groups in the Ohio valley during the ancient and historic eras, and visitors often ask whether knowledge about the purpose of the Newark Earthworks might have persisted, handed down in oral histories, through families, documented on birch bark, perhaps woven into textiles or inked onto skin or cloth. It is a reasonable query; tribal groups had been in the region for thousands of years, accumulating specialized systems of knowledge. However, after European contact, the retention of cultural knowledge was jeopardized by centuries-long catastrophic factors: pandemics, military and settler violence, abrupt and recurring cultural disruptions, multiple forced migrations, and war. Indigenous people from the east, south, and north began moving into the Ohio valley. The first permanent American settlement in Marietta, Ohio, in 1788, was accompanied by a militia, because previous attempts to establish an outpost were unsuccessful. Indian communities in the Ohio valley experienced the consequences of population pressures, settler demands for land, and violence. The loss of individuals and families with specialized cultural knowledge must have had an enormous impact on the quality of life and on the natural environment. While American Indian people adapted to the new social and political order of the settler economy into the mid-1800s, they were survivors of traumatic, devastating events. Many had journeyed from other parts of the country, suffering tragic losses of family and community.

Even before tribal groups were forced out of the region, widespread excavations and grave robbing were commonplace. At occasions of gathering for signing treaties or at the time of imminent departure from their homes in forced removals, tribal members often expressed concerns about leaving their homeland. A public statement from 1842 describes the anguish and confusion of the Wyandotte people, which had sought to live peaceably within a settler community until several of their members were murdered by white settlers:

> My people, the time for our departure is at hand. A few words remain only to be said. Our entire Nation has gathered here for farewell. We have this morning met together for the last time. . . .
>
> . . . It remains only for me to say farewell. . . .Here our dead are buried. We have placed fresh flowers upon their graves for the last time. No longer shall we visit them. Soon they shall be forgotten, for the onward march of the strong White Man will not turn aside for the Indian graves. Farewell—

> Farewell Sandusky River. Farewell—Farewell our hunting grounds and homes. Farewell to the stately trees and forests.[4]

As tribal groups were being forced from the area, the first surveys of the Newark Earthworks were garnering international excitement in the mid-1800s.[5] Farms, homes, businesses, a canal, railroad tracks, and roads were encroaching on the Newark Earthworks complex as well as abandoned Indian villages and areas. As a result, huge numbers of stone tools and artifacts of indeterminate age were found during farming and construction that disturbed or destroyed graves and burial mounds. The people who used the Newark Earthworks and other Hopewell culture sites buried their relatives with skillfully crafted items including copper cutouts, beautifully carved stone pipes in animal shapes, carved and etched gorgets and game balls, tablet carvings, jewelry, and regalia. The publication of Squier and Davis's *Ancient Monuments of the Mississippi Valley* in 1848 marked the point when archaeology became the dominant frame of reference through which the public learned about the Newark Earthworks and the cultural items of interest to the public. Although some of the larger earthen enclosures and conical burial mounds were preserved in Newark and around Ohio, most were destroyed by construction, as in the case of the ancient cemetery at Newark. Throughout the state, inadvertent discovery of cultural items spurred on excavations, the stated intent being to document what was left of ancient American Indian culture before it disappeared under plow and town.[6] This rationale was easily justified once the items found in burial mounds were defined as art and artifact instead of being part of a human grave carefully prepared by relatives and buried with finality.

By 1850, all formally recognized groups and tribes were forced out of Ohio into Oklahoma, Missouri, and Kansas or to the north into the lands of Michigan and Wisconsin.[7] Tribes faced cultural upheavals and intense challenges in their new homes in the southern plains; others defended their existing homelands in the Upper Great Lakes region as they fought to stay in their ancestral lands. Once tribes were forced to leave and no longer had physical, social, or political presence in the region, their claims to the land they occupied for generations was lost, and they ceased to be active participants in Ohio affairs. Consequently, Ohio's turbulent and complex American Indian history became simplified and romanticized as reflected in news stories, school curricula, popular fiction, political proclamations, and the narratives of local historical societies. For a long while, American Indians of all eras were effectively "erased" from both Ohio history and the land on which they had lived and from the collective memories of most Ohio citizens.

The absence of tribal presence in Ohio encouraged a proprietary and obsessive focus on the earthworks, funerary items, and the ancestors interred inside enormous burial mounds in Newark and elsewhere. At the turn of the twentieth century new knowledge rapidly accumulated about the Hopewell culture through excavations and examinations of cultural items, and for decades the archaeological models were considered successful at generating theories and methodologies. By comparison, living American Indians did not have explanations about the Newark Earthworks, which served to discourage further investigation of associations between ancient and historic American Indian cultures.

The keen archaeological interest in buried artifacts and graves did not escape the notice of collectors and looters. The availability of artifacts, combined with the absence of strong legal protections for ancient and historic cultural items, created incentives for a market in ancient and historic cultural items, a situation that continues unchanged and unregulated to the present day.[8] Earthworks and burials on private land in Ohio are considered property of the landowners. Because so many artifacts continue to be traded and found through farming, urban development, and artifact hunting, large collections of precontact cultural items are held privately by individuals.[9]

By and large, indigenous precontact cultural items and places are not considered American legacies to be protected for appreciation by future generations. When the enigmatic "Wray Figurine" was found in 1881 under the largest burial mound at the Newark Earthworks cemetery, the stone figurine became the property of the landowner. In 1984, a news story featured a couple who had searched for and located an ancient burial on their farm. The newspaper published a photograph of the human remains with the story. Since the burial had been excavated on private property, the couple was legally entitled to own the remains and to place the burial items inside a historic covered bridge on their farm.[10] In the summer of 2012, it was reported that remains of a precontact-era ancestor were found among items in a Newark pickup truck and repatriated to a local group claiming to represent Native American interests.[11] Overall, site management policies and cemetery laws have not discouraged looting or set guidelines for the trade in American Indian artifacts in Ohio modeled on federal policies.[12]

American Indians and the Newark Earthworks

In 1992 the Ohio Historical Society, owner of the two remaining enclosures of the Newark Earthworks, began an archaeological study to determine the age and structure of the Great Circle in Heath, Ohio. Shortly after the excavations began, American Indian communities launched a protest that lasted more than a week and attracted local and regional media attention. Although the protest drew more than a hundred people, it did not interrupt the archaeological research project.[13] And while the protest may have been defused, it was significant; Ohio citizens who were also American Indian had taken action to declare that the Great Circle was a sacred site built by their ancestors. The protestors claimed that archaeological research was a violation of that status, and further, they were interested in the cultural management of the Newark Earthworks.

Since the protest was resolved after a brief negotiation, the event soon fell away from public view. The Ohio Historical Society administrators did not take the opportunity to initiate a dialogue with Ohio's American Indian communities about the Newark Earthworks. Instead, in 1997, the society quietly granted a fifty-year extension to the existing thirty-year lease of the Octagon Earthworks held by members-only Moundbuilders Country Club. The details of the lease agreement were not publicly acknowledged until the country club proposed an expansion of its clubhouse in 1999. The plan was met with strong public opposition generated by an interest group, Friends of the Mounds, which included representatives of the American Indian community who had participated in the protest five years earlier. As a result, the Ohio Historical Society suspended the proposed expansion and belatedly initiated a management plan and accepted public comments.[14] The management plan stated that the Octagon Earthworks would be open to the public four days each year and continue to be leased to the private country club.

Over the past two decades, hundreds of American Indian people have learned about the Newark Earthworks. Since the mid-2000s the Newark Earthworks Center and the American Indian Studies Program at The Ohio State University have hosted tours of the Newark Earthworks for American Indian scholars, tribal representatives, artists, and writers. Participants of several conferences have toured the Great Circle and the Octagon Earthworks. A unique meeting took place at Ohio State University in 2011: the Society of American Indians Centennial Symposium.[15] The symposium brought together American Indian studies scholars to reflect upon the legacy of American Indian intellectuals who organized the Society of American Indians and met at Ohio State in 1911 to discuss ways to promote the survival of American Indians during massive social changes at the turn of the century. In 1911, the group took time from their meetings to travel to the Octagon Earthworks. There, they stood inside the Octagon's walls and sang in unison "America the Beautiful." One hundred years later, American Indian professors went to the Octagon on a frosty morning to honor their 1911 colleagues by singly stepping forward to offer traditional or cherished songs in their nations' indigenous languages.[16]

Since 2004, the Newark Earthworks Center and the American Indian Studies program at Ohio State University have received suggestions about new research directions from scholars, artists, and tribal citizens. Derived from a deep, personal understanding of indigenous cultures and traditions, these ideas have reinvigorated the thinking about the context and purpose of the Newark Earthworks with the promise of making connections to contemporary American Indian people and traditions. Perhaps two thousand years ago, gatherings at the Newark Earthworks foreshadowed today's intertribal powwows: nexus of song and dance; a place to seal alliances and resolve conflicts; a place for ceremony, marriages, reunions, memorials, births, and funerals. When LeAnne Howe, renowned Choctaw author, visited the Newark Earthworks, she was the first to suggest that the enormous enclosures at the Newark Earthworks complex, among other sites, could have been the venue for stickball, a dangerous and formal precontact game. Used as a strategy to settle disputes in the place of armed conflict, stickball was played on enormous ball fields, engaging hundreds of players from different villages or groups. The "Adena Pipe," a Hopewell culture pipe excavated from a burial mound in Ross County, was carved in local stone in the shape of a man who has been interpreted to be a "shaman." Howe noted that the figure could represent a ball player calling to his fellow players. Recently, images of a figurine from Spiro Mound and one from Mexico's Mayan culture have surfaced on the Internet and appear strikingly similar to the Adena Pipe. This comparison awaits further study.[17]

Federal Indian Policy and Ohio

Ohio policy makers, especially state officials and legislators, need to become more informed about the historic and ongoing relationship between the federal government and contemporary tribal governments. Tribes evicted from

Ohio continue to have cultural interests in their homelands. Preservation policies and the enactment of legislative protections for historic and ancient American Indian sites would improve with increased participation by federally recognized tribal governments in Ohio. Achievement of this goal is impeded by the fact that all federally recognized tribes have been absent from the state since the mid-1800s and by the demographic trends of the American Indian population in Ohio. In the 1950s, urban migration and the federal policy of relocation brought tribal citizens from reservations to Ohio. Over the last several decades, private organizations in Ohio have used the term "tribe" as well as specific American Indian governments' names in carrying out a variety of activities. In 2010, more than twenty-five thousand people in Ohio self-identified as American Indian, and more than sixty-five thousand people reported they were American Indian in combination with one or more other races. This represents a 25 percent increase from the 2000 census.[18]

In 2009, the Ohio World Heritage committee charged with preparing the nomination dossier turned to federal policies for guidance in seeking the inclusion of American Indian governments and individuals in the nomination process. This strategy proved useful in conveying the distinctions between federally recognized tribes and American Indian individuals: individuals may hold American Indian tribal citizenship, or individuals may be direct descendants of tribal citizens, or individuals may have distant family connections.[19] Tribes are sovereign governments akin to states; individuals can represent their experience and cultural knowledge, but they are not tribal representatives.

These distinctions are useful to ensure that tribal governments are included in Ohio's historic preservation initiatives, because American Indian involvement with the Newark Earthworks is dynamic and diverse, involving both individuals and organizations. Many Native individuals have lived near the earthworks for significant periods of time. Native individuals have participated in much of the recent history of the Newark Earthworks, educating the public about the indigenous achievements represented by earthworks, engaging in vigorous political efforts to obtain greater access to the Octagon Earthworks, and participating in the 2005–6 lunar standstill events. Local and regional organizations include American Indian centers and American Indian studies programs. American Indian centers have organized political activities and organized events at the Newark Earthworks. American Indian studies programs provide access to the global network of scholars, writers, and scientists who have expertise relating to American Indian culture, history, and traditions that inform understanding of the earthworks landscape in Newark and Ohio. Organizations such as the National Congress of American Indians and the Native American Rights Fund address issues of national interest to tribal nations. They report on the status of federal legislation and lobby on behalf of tribes. The Newark Earthworks have been featured in the Morning Star Institute's National Prayer Day for Sacred Places' annual listing, which reaches a national audience.[20]

Contemporary American Indian tribal nations are recognized at the federal level by the president, Congress, and federal agencies. Only federally recognized tribes have the authority to represent tribal governments:

> Indian Nations are sovereign nations with formal governments, and have been for thousands of years. That Tribal governments inherently have sovereignty is recognized in Article I of the Constitution: "The Congress shall have power to . . . regulate commerce with foreign Nations, and among the several States, and with the Indian Tribes." Just like every other government, Indian tribes have the authority to make their own laws and enforce them for the safety and welfare of their communities. In addition, Tribal governments are dedicated to protecting our unique Indian cultures and ways of life. Just as states and foreign countries have a unique relationship with Washington, so do Tribal Nations. The fundamental bond is the same: Tribal governments represent, and are accountable to, the citizens who elect them.[21]

Tribes that are federally recognized have a government-to-government relationship with the US government. They interact with federal agencies in developing services for tribal citizens and economic development initiatives for the continued support of tribal activities. Tribes develop cultural preservation policies relating to their tribes' historic and ancestral lands, and they have experience in managing their historic and cultural assets and seek to develop mutually beneficial relationships with local and state governments and private organizations and businesses. Tribes determine citizenship criteria.[22]

Tribal Stewardship of the Newark Earthworks

At the time of the lunar standstill in 2005 at the Newark Earthworks and over two and half years, there were no American Indian government representatives participating as experts or advisors. The archaeological and astronomical knowledge of the earthworks was illuminating, and the local community appreciated each moonrise, but it soon became apparent that the meaning of the earthworks and the reasons why tracking the long moon cycle was so important continued to elude the research and local community. Over time, the partner organizations collaborating on the effort to attain nomination of the Hopewell earthworks as World Heritage Sites learned that most American Indian visitors had no prior knowledge about the Newark Earthworks. In fact, most of the scholars, scientists, writers, leaders, and tribal representatives who visit Newark are learning about the Newark Earthworks for the first time.

The recognition of the need to contact tribes with histories in the Great Lakes and Ohio valley regions emerged gradually: World Heritage and preservation projects needed to be informed by tribal histories and cultural knowledge with potential to be relevant to the meaning and purpose of the Newark Earthworks. The Ohio State University American Indian Studies Program collaborated with the Newark Earthworks Center to seek out scholars with close connections to their tribal governments and American Indian government leaders in order to invite them to tour the Newark site and begin a dialog. The overall goal was to bring balance to the understanding of the earthworks from the time they were built into the present-day through an exchange of information with American Indian people knowledgeable about their histories and traditions.

In 2007, Chief Glenna Wallace and a delegation of the Eastern Shawnee Tribe of Oklahoma traveled to Ohio and accepted an invitation to visit the Newark Earthworks. Shortly afterward, formal tribal outreach programs were initiated, and the author traveled to tribal government offices in Oklahoma. Subsequently the Newark Earthworks Center was invited to present information about the Newark Earthworks and World Heritage at the annual "To Bridge a Gap" conference cosponsored by the US Forest Service and Oklahoma tribes. Discussions with tribal officials with heritage interests in Ohio have resulted in strong support for legislative reform and a better understanding of American Indian governments' interests and concerns. In time, these dialogs should contribute to American Indian leadership for reassessing historic and precontact site management plans and making significant contributions to cultural heritage in Ohio. In 2011, the Eastern Shawnee Tribe of Oklahoma formally endorsed a resolution accepted by the National Congress of American Indians to support World Heritage recognition for the Newark Earthworks as included in the Hopewell Ceremonial Earthworks.[23]

The World Heritage nomination process for cultural sites encourages research about historic and contemporary indigenous connections to earthworks and related places, providing opportunities for American Indian governments and their cultural leaders to reconnect with the land they were forced to leave in the nineteenth century. In turn, they can contribute a wealth of cultural knowledge about the Ohio landscape and acknowledge their role as the most recent indigenous stewards of the earthworks before tribes were forced to leave. Some contemporary tribes describe concentrations of earthworks in their historic lands and mound-building traditions in their pasts. Tribal languages may include words relating to the architecture, construction, shapes, and purpose of earthworks. At the Chickasaw Cultural Center, an exhibit shares the word for "mound builders of the Mississippi and Ohio River valleys: *onchaba ikbi'*." Through the exhibit, curators explain that the Chickasaw territory included New England, the Midwest, and part the Southeast, where mounds were built from five hundred to two thousand years ago.[24]

The Muscogee (Creek) Nation website states, "The Muscogee (Creek) people are descendants of a remarkable culture that, before 1500 AD, spanned all the region known today as the Southeastern United States. Early ancestors of the Muscogee constructed magnificent earthen pyramids along the rivers of this region as part of their elaborate ceremonial complexes."[25]

American Indian people brought a vision into the physical world to build a place where the sky and earth meet to transform the lives of their relatives and their future generations, and people could have traveled the rivers and tributaries from great distances to experience them. The Octagon Earthworks provided a place of quiet respite for the courageous scholars attending the Society of American Indian meetings in 1911. The protest at the Great Circle in 1992 energized an entire community to take action so the earthworks would be preserved into the future. In 2011, scholars of the Society of American Indians Centennial Symposium visited the Octagon Earthworks to sing in their native languages to honor the ancients and the sacred place they built.[26] American Indian tribes and their citizens and descendants share a common history; they are related to the people who lived in the Americas for thousands of years and who imagined and built the earthworks in Newark. The current stewards of the Newark Earthworks need the cooperation and good will of American Indian governments and leaders. The time has come for the return to American Indian stewardship of the earthworks.

Notes

1. See, most notably, Squier and Davis, *Ancient Monuments of the Mississippi Valley.*

2. Regarding the Ancient Ohio Trail website (http://www.ancientohiotrail.org), see John Hancock's contribution to this volume.

3. See Thornton, *American Indian Holocaust and Survival;* Stannard, *American Holocaust;* and Mann, *1493.*

4. Farewell speech by Chief Squier Grey Eyes, June 12, 1843, http://www.wyandot.org/farewell.htm.

5. Squier and Davis, *Ancient Monuments of the Mississippi Valley.*

6. See D. Thomas, *Skull Wars,* 140.

7. I acknowledge Gail Zion, a community historian and lifelong resident of Newark, for her 2010 application of the concept of a "history of erasure" to Newark.

8. For instance, H. Johnson, Auction Block, reports sales of Midwest precontact artifacts for more than $458,000 at an Ohio auction house. That magazine's Auction Block section provides current auction reports for Indian items.

9. Mary Yost, "Basement Houses Indian Museum: Whitehall Couple Has Thousands of Artifacts," *Columbus Dispatch,* Oct. 22, 1978, B5.

10. See Dan Baird, "Indian's Remains Found," *Columbus Dispatch,* Dec. 9, 1984, B3. Information about the covered bridge can be found at http://www.dalejtravis.com/bridge/ohio/htm/3502320.htm.

11. See Kimberly Gasuras, "Bones Found in Crestline Junkyard Are Native American, Old," *News Journal* (Mansfield, OH), July 25, 2012.

12. See, for example, "Cultural Protection and NAGPRA," National Congress of American Indians website, http://www.ncai.org/policy-issues/community-and-culture/cultural-protection-and-nagpra.

13. See Jim Woods, "State Jumps into Indian Mound Dig Controversy," *Columbus Dispatch,* July 15, 1992, B5.

14. Ohio Historical Society, *Newark Earthworks Historic Site Management Plan.*

15. Chadwick Allen, "Introduction."

16. See Margaret Noodin, "Bundling the Day and Unraveling the Night."

17. See the image comparing stone figurines of ballplayers captioned "Ball Player Composite" accompanying the article "The Mesoamerican Origin of North American Stickball," http://peopleofonefire.com/?s=Mesoamerican+Origin+of+North+American+Stickball.

18. See Tina Norris, Paula Vines, and Elizabeth Hoeffel, "The American Indian and Alaska Native Population: 2010," 2010 Census Briefs, 7, 10, http://www.census.gov/prod/cen2010/briefs/c2010br-10.pdf.

19. See Toensing, "Updated Federally Recognized Tribes List Published."

20. "June 16–24 Set for 2012 National Sacred Places Prayer Days," Native American Rights Fund website, http://www.narf.org/nill/resources/Prayer_Sacred_Sites.pdf.

21. Jefferson Keel, "Sovereignty and the Future of Indian Nations."

22. See, for example, "Frequently Asked Questions: What is the legal status of

American Indian and Alaska Native tribes?," Bureau of Indian Affairs, US Department of the Interior website, http://www.bia.gov/FAQs/.

23. National Congress of American Indians Resolution PDX-11-060, "Support the Nomination of Ohio Earthworks to Become World Heritage Sites" was passed by the Subcommittee on Human, Religious and Cultural Concerns, the Litigation and Governance Committee, and the general membership at the Portland, OR, conference in October 2011. http://www.ncai.org/resources/resolutions/support-the-nomination-of-ohio-earthworks-to-become-world-heritage-sites.

24. Great Civilization Exhibit, Chickasaw Cultural Center, Sulphur, Oklahoma, http://www.chickasawculturalcenter.com/explore-exhibits-artifacts.html.

25. Muscogee (Creek) Nation website, http://www.muscogeeNation-nsn.gov/Pages/History/history.html.

26. Noodin, "Bundling the Day and Unraveling the Night."

MARY N. MacDONALD

Whose Earthworks?

Newark and Indigenous Peoples

Monuments erected long ago in Ohio, including earthworks located in Newark and Heath in Licking County, claim the attention of a worldwide audience and at the same time have particular significance for indigenous communities. While we all recognize the earthworks as monuments of a culture that flourished prior to European incursion into the Americas, they may well have greater significance for contemporary Native American cultures than for settlers and the descendants of settlers who now make their homes in Ohio. Since there is considerable interest in having the earthworks at Newark (along with a couple others in Ohio), designated as UNESCO World Heritage sites, I want to bring the concerns of UNESCO for the world importance of the site into conversation with the US federal Native American Graves Protection and Repatriation Act (NAGPRA) and with the United Nations Declaration on the Rights of Indigenous Peoples (UNDRIP). As we think about the surviving earthen architecture in Licking County and the people who created it, I suggest that we consider the stake that Native Americans—and other indigenous peoples—have in the past and future of these constructions and sites.

Several scholars contributing to this volume have devoted many years to the study of the Newark Earthworks and have extensive knowledge of the structures and the artifacts recovered from them. While immensely grateful to them for sharing their data and their interpretations, I am myself a latecomer to the earthworks. An Australian-born historian of religions who for many years has studied religions of Oceania, since the late 1980s, living in Syracuse, New York, I have taken a special interest in Native American cultures and in particular those of central New York that, indeed, have links to the ancient cultures of Ohio. I bring a religious studies perspective and a concern for the ethics of scholarship to our shared study of the earthworks.

The Newark Earthworks and other mounds and earthen edifices around the world are starting points for thinking about the making of cosmologies and cultures. They encourage us to study how communities construct places and ways of life that flourish for a time and then, in the face of ecological and historical circumstances, yield to change. It seems that, in what is now the Newark region, people gathered to make connections to both the seen and the unseen worlds through material and symbolic mechanisms of exchange. If we think of religion as a process of making powerful and life-giving connections, as many practitioners and scholars do, then these places are sites of religious significance. Even if we prefer not to use the term "religion," we are left with the fact that both practical and symbolic activities took place at Newark connecting people to a world they knew and inhabited.

As we learn elsewhere in this volume, the ancient mounds at Newark include the largely intact Great Circle and Octagon Earthworks, as well as the more thoroughly effaced Wright Earthworks. Scholars have concluded that these features were built between 100 BCE and 500 CE by people referred to as Hopewellians who lived along rivers of the Northeast and Midwest of what is now the United States.[1] These people, they suggest, were linked in an exchange system that extended from the southeastern United States to the Canadian shores of Lake Ontario using both waterways and land routes. Some believe that Hopewell populations originated in what is now western New York and moved south into Ohio, while others think that Hopewell groups from what is now western Illinois spread along the Illinois River and into southwestern Michigan.[2] Somewhere there are descendants of those people, but they are many generations away from the Hopewellians. The Newark Earthworks have often been described as a "ceremonial center," a place where numerous sorts of rites, probably including rituals for the dead, were carried out. There is, however, also evidence that they were places of social gathering, trade, and astronomical observations. No doubt modern frames like "religion," "economics," and "astronomy" fall short of capturing what the Mound Builders thought they were about, and we don't know what they called the place or themselves.

On a drizzly spring afternoon in 2011, following a morning of lectures by experts on the Newark site, I stood with them in the Great Circle, the largest circular earthwork in the Americas, marveling at the science and the physical labor that had gone into the alignment and construction of the earthen walls, which are 14 feet high on the eastern side, 6 feet high on the western side, and 1,200 feet across. Activities within the walls in ancient times may well have included astronomical observations, trade, social/political gatherings, storytelling, and entertainments. Ritual practitioners probably carried out ceremonies that drew pilgrims to the place, while merchants gathered to exchange minerals, textiles, medicines, and foods. Probably the roles of merchant and pilgrim overlapped. Meanwhile, farmers from outside the walls may have engaged in agriculture to support themselves and those who gathered at the earthworks. We surmise that the trade and rituals—material and religious exchange activities—carried out by the ancient residents and visitors were thought to keep the physical, social, and spiritual aspects of the world in balance and to pass on their knowledge to the next generation.

Long after the Hopewell-era residents, astronomers, ritual specialists, storytellers, pilgrims, and traders were gone from Licking County, the Great Circle served, as we learn in Richard Shiels's essay in this volume, a wide variety of uses in advance of its present status as a park managed by the Ohio Historical Society. Current visitors—tourists, students, members of Native American groups, and others—come today to learn of the past, to imagine this place in another age, and even to make connections in their own lives. Engagement with the earthworks may be part of the ongoing human quest for meaning. In 2012, for the first time the annual powwow sponsored by the Native American Indian Center of Central Ohio (NAICCO) was held at the earthworks. Reporting on it in *Indian Country Today*, Stephanie Woodard wrote, "This year's location was extraordinary, emphasizing the deep history and powerful spiritual connections of Native people on this continent and in Ohio."[3]

Later on that rainy afternoon in 2011, we visited the Octagon Earthworks, now home to Moundbuilders Country Club. As explained in Ray Hively and Robert Horn's essay, there is solid evidence to support the notion that the Octagon complex was a lunar observatory in which astronomers tracked the movement of the moon. Although golfing and its associated social activities might seem to trivialize a sacred place, the golf club has forestalled the urban and agricultural development that has been so destructive of other Ohio mounds. Even avowedly secular human beings have awe and respect for places that human communities have shaped and sustained in the past, and it is that sensibility that prompts state and national agencies and UNESCO to preserve them as part of the national and universal human heritage.[4]

Prior to coming to Newark I had visited other prehistoric sites in North America such as Mesa Verde and Chaco Canyon and numerous Mesoamerican sites such as Caracol in Belize, Tikal in Guatemala, and Teotihuacan in Mexico. I had also been to Stonehenge and Avebury in the United Kingdom. Each site tells part of a story, a story of people who inhabited a land and interacted with it. With physical labor and ritual actions they shaped the earth to their purposes and sought to play their part in facilitating the fecundity of land, community, and cosmos. Archaeologists, astronomers, historians, and social scientists are at work around the world trying to fill in gaps in these stories, trying to establish the purposes to which mounds and buildings and surrounding land were put, while philosophers, psychologists, and religious studies scholars ponder the values and meanings embraced in the cultures and their edifices.

Informed, then, by this historical background and hypothesizing about the original conceptions of these ancient sites, my special interest is in the more recent—and future—management of those constructions and landscapes, especially as management decisions influence and involve indigenous peoples, including those from parts of the world *other than* North America. The Newark Earthworks were designated a National Historic Landmark in 1964 and in 2006 named the "official prehistoric monument of the State of Ohio." Thus, on federal and state levels, postcolonial America seeks to forge connections with landscapes that were shaped by the country's indigenous peoples.

Additionally, we learn in this volume about ongoing efforts to have the Hopewell Ceremonial Earthworks, a broader complex of Ohio sites that includes Newark, awarded UNESCO World Heritage designation. In 2011, I was able to visit another component of that broader complex, Fort Ancient in Warren County, site of a culture that flourished around 1000–1750 CE that is believed to be descended from the Hopewell; and I likewise visited the well-known Great Serpent Mound in Adams County, which is the largest serpent effigy mound in the world and is thought to have been built by the Fort Ancient culture. I came to see, then, that beyond Newark's interconnected Great Circle, Octagon, and Wright Earthworks, there is a more extensive picture. Newark is part of a much larger region with links reaching to the north, south, east, and west in the Americas and with affinities to other regions of the world.

Together these numerous Ohio mounds pose challenges not only for interdisciplinary study but also for the vexing matters of preservation and management. Examining the material structures and artifacts their builders have left us, scholars work to reconstruct their social structures, economies, knowledge of various kinds, ideologies, and cosmologies. Contemporary study of these earthworks, and of indigenous cultures and religions more generally, is being carried out in the context of an increased awareness of the consequences of the encounter of indigenous peoples and settler peoples from the Age of Discovery to recent times. Thus, the ethics of scholarship today is more attentive to the concerns of indigenous stakeholders than was the case in earlier times. As Native Americans have made common cause with other aboriginal peoples worldwide, "a discourse of indigeneity" has emerged. Part of that discourse concerns the connections of indigenous peoples today with ancestors who left their marks on landscapes around the world. The proposal to have the Hopewell Ceremonial Earthworks awarded UNESCO World Heritage site designation, therefore, provides an opportunity to consider not only the possibility that these sites have significance for everyone and that, as Marti Chaatsmith alerts us, they hold particular meaning for Native Americans, but also that the Ohio earthworks may have a very special significance for indigenous peoples in other parts of the world.

In recent decades indigenous peoples have insisted on the right to participate in decision making with regard to sites in their own territories and sites that have significance for their local and extended communities.[5] Peoples' lives are shaped by places—by natural landscapes and constructed habitats, by homes and business centers and ceremonial sites. Religions are concerned with how we orient ourselves in the cosmos—with how we place ourselves geographically, socially, historically, and spiritually. The earthworks provide opportunities for scholars of religion, in collaboration with colleagues in other fields, to reflect on religion in the making in the Americas as people simultaneously shaped the earth and shaped their worlds of significance.

In time, genetic genealogy may be able to link the ancient Hopewellian peoples who are believed to have constructed the Newark Earthworks with contemporary groups whose ancestors, for various reasons, left what is now Ohio for other places. There are vestiges of migration histories in the oral accounts of many Native American peoples, and there are written references to the movements of peoples after the arrival of Europeans.[6] We know that both before and after colonial intrusion, many Native American groups entered and eventually left Ohio. Europeans in search of land and resources disrupted indigenous ways of life and subsistence and religion. They caused movements of groups into and out of Ohio. For instance, the Seneca of western New York, allied with European fur traders, came west into Ohio as the supply of beaver diminished in Haudenosaunee territories. It could be that fragments of stories of the Earthworks at Newark or elsewhere in Ohio have been preserved among peoples who lived in the region after the Hopewellians left, such as among the Shawnee, whose ancestors, some believe, came from the Fort Ancient area, or among the Seneca, who came to the region much later.

While it makes sense, then, to say, in the spirit of UNESCO, that the Newark Earthworks belong to all of us, we could also make a case that they belong *in a special way* not only to Native Americans but also to the indigenous peoples of the world who are increasingly entering into solidarity with each other and forming alliances to ensure that their heritage and contemporary ways of life are recognized and respected. A consequence of the Age of Discovery and the era of colonialism was that indigenous peoples were marginalized in their own ancestral lands. Their achievements in architecture, astronomy, and agriculture were underestimated. Their languages and literatures (both oral and written) were ignored or even derided. Their cultures were regarded as inferior to Western civilization. Their religions were denigrated and at times outlawed. Thus, as we bring our various competencies to bear on the earthworks, I recommend that we look to the US federal law known as NAGPRA (the Native American Graves Protection and Repatriation Act) and the United Nations Declaration on the Rights of Indigenous Peoples (UNDRIP) for guidance on how to construe the relationship of Native Americans and other indigenous peoples to the earthworks. These documents don't give us easy answers, but they do suggest an ethical stance that acknowledges the humanity and dignity of the Mound Builders and of indigenous communities that today recognize a relationship to them.

As Duane Champagne and Carole Goldberg's essay explains, NAGPRA (1990) is a US federal law that provides a process for museums and federal agencies to return certain Native American cultural items—human remains, funerary objects, sacred objects, and objects of cultural patrimony—to lineal descendants, to Indian tribes, and to Native Hawaiian organizations.[7] In the 1960s, thefts from museums and archaeological sites around the world led to a 1970 UNESCO Convention on the Means of Prohibiting and Preventing the Illicit Import, Export and Transfer of Ownership of Cultural Property. Legislation with similar intent to NAGPRA has been enacted in several parts of the world, and the United States has entered into agreements with other countries regarding Native American remains and artifacts that have been housed in overseas collections. While contemporary Native American groups may not be able to make a case under NAGPRA for assuming ownership of the earthworks, it seems to be in the spirit of NAGPRA that they be involved in the care and custodianship of the sites for, more than later arriving settler peoples who live in the United States today, they are connected, as descendants of the original inhabitants of the land, to the ancestors who built the Newark structures and other mounds in the region. Some, of course, are more closely connected than others. NAGPRA, as Greg Johnson along with Champagne and Goldberg has noted, is legislation that is evolving in its application. The passage of the legislation and the efforts over the past two decades to put it into practice have made the point that Native Americans are the rightful custodians of their sacred patrimony. For indigenous communities nationwide, the implementation of NAGPRA has been a spiritual and emotional journey. It has meant the return of human remains that can now be properly interred according to traditional practice and the restoration of sacred objects to their communities. The process has included careful recording of inventories of human remains and cultural objects held in museums and extensive conversations between Native American elders and museum curators. Varying points of view and sensibilities have made for rough moments as well as points of shared appreciation.

A 1995 BBC documentary film, *Bones of Contention: Native American Archaeology,* for instance, focused on the tensions between museum curators and Native Americans over human remains held in museums. On the one hand, the documentary suggests, scientists would like to examine the remains in order to better understand matters of migration and health, while, on the other, Native Americans would like to see them disposed of with appropriate ceremonies. Some of the speakers in the documentary think that both may be possible. At one point in the film there is a discussion of so-called Kennewick Man, the name for the skeletal remains of a prehistoric man found on a bank of the Columbia River in Kennewick, Washington, in 1996. Scientific examination dated this skeleton at between 5,650 and 9,510 years old. Several Native American tribes claimed the skeleton, but in 2004, subsequent to the film and after a prolonged battle in the courts, the US Court of Appeals for the Ninth Circuit ruled that a cultural link between the tribes and Kennewick Man could not be genetically established and, therefore, permitted the scientific study of the remains to continue. In the film, a young Native American man makes the case that it does not matter to which tribe Kennewick Man might belong. He is Native American, and his remains should be given burial by Native Americans. If we argue in this way—for "Native American" identity in addition to a local tribal or national identity such as Wyandotte,

Delaware, or Miami—we could say that the Newark Earthworks are Native American constructions and should be in the care of Native Americans. We might go even further and say that they are indigenous constructions that the indigenous peoples of the world should celebrate in solidarity. Perhaps, though, more simply, the lesson to take from NAGPRA is that Native Americans should have a voice in the discussion of the earthworks, occasions to study the earthworks, and opportunities to work in educational centers and visitor centers associated with them. The Ohio State University Newark Earthworks Center, the Ohio Historical Society visitor center at the Great Circle, and other groups promoting study and tourism of the earthworks can facilitate dialogue and cooperation between indigenous residents and visitors as well as between "settler residents" and visitors, while institutions and scholars who study indigenous cultures can help ensure that indigenous personnel are included in their teams.

While NAGPRA is a significant stage in an ongoing process of acknowledging the rights of indigenous peoples, UNDRIP, a declaration that resulted from decades of work by indigenous leaders from the United States and abroad, is also instructive. Today there are people who identify as Hopi and Navajo and Shawnee and Haudenosaunee, and there are many who acknowledge both Native American and European heritage. I have noted that in addition to claiming a local identity, many indigenous peoples also identify as Native American or Indian, asserting a social and strategic alliance with a larger group. In this sense they take on shared responsibility with the larger group for the care of ancestral sites and the continuation of ancestral traditions. The web expands farther when the native peoples of the Americas make common cause with the Native peoples of Africa and Asia and Oceania and speak of the ways and the rights of indigenous peoples.

UNDRIP has come out of this discourse of indigeneity. The declaration is not a legally binding document, but nonetheless it is significant for setting out the individual and collective rights of indigenous peoples, including their rights to culture, identity, language, employment, health, and education. The declaration, which takes a human rights approach, is an important step in the conversations among indigenous communities worldwide—part of their process of coming to a common understanding—and it is also a step in the conversation between indigenous communities and the global community. The declaration emphasizes the rights of indigenous peoples "to maintain and strengthen their own institutions, cultures and traditions, and to pursue their development in keeping with their own needs and aspirations." UNDRIP prohibits discrimination against indigenous peoples and "promotes their full and effective participation in all matters that concern them and their right to remain distinct and to pursue their own visions of economic and social development." Like NAGPRA and similar laws in other countries, but on a global scale, the declaration affirms respect for the heritage of indigenous peoples. It is not a perfect document and, given that the member states of the UN are not perfect either, it could not be expected that they would all find it completely satisfactory. The declaration can be seen as part of an ongoing conversation—a conversation about how the world community and indigenous communities within that larger community should work together and separately in caring for places and traditions. The declaration was adopted by the UN General Assembly during its sixty-second session at UN headquarters in New York on September 13, 2007. At the time of its release the UN Permanent Forum on Indigenous Issues described it as setting "an important standard for the treatment of indigenous peoples that will undoubtedly be a significant tool towards eliminating human rights violations against the planet's 370 million indigenous people and assisting them in combating discrimination and marginalization."[8]

The idea for the declaration emerged in 1982 when the UN Economic and Social Council set up a Working Group on Indigenous Populations to develop human rights standards that would protect indigenous peoples. However, its history goes back further to the encounter of cultures in the Age of Discovery, to the arrogance of colonizers, to the resistance of indigenous peoples, to attempts at mutual understanding. In 1985 the Working Group on Indigenous Populations began drafting a Declaration on the Rights of Indigenous Peoples, and this was completed in 1993. The following year it was approved by the Sub-Commission on the Prevention of Discrimination and Protection of Minorities. The draft went to the Commission on Human Rights, which established a working group to examine its terms. After several revisions the final version was adopted on June 29, 2006, by the Human Rights Council. It was then referred to the General Assembly, where it went to a vote on September 13, 2007. The vote was 143 countries in favor, 4 against, and 11 abstaining. Those voting against were Australia, Canada, New Zealand, and the United States. These are all settler nations having their origins as British colonies. They have large nonindigenous settler populations and smaller indigenous populations, and they voiced concerns about how their current land tenure systems would be affected by implementation of the provisions in the declaration. It is one thing to acknowledge that the land was never *terra nullius* but another to agree on how it should be shared by indigenous and settler peoples today. However, despite their misgivings, since 2007 Australia, Canada, New Zealand, and the United States have endorsed the declaration.

UNDRIP consists of a preamble and forty-six articles. Articles 11 and 12 have particular implications for sites such as the Newark Earthworks even though the whole document asserts the respect due to indigenous peoples and their cultural heritage. Each of those articles has two parts. The first part of Article 11 says: "Indigenous peoples have the right to practice and revitalize their cultural traditions and customs. This includes the right to maintain, protect and develop the past, present and future manifestations of their cultures, such as archaeological and historical sites, artifacts, designs, ceremonies, technologies and visual and performing arts and literature." Part 2 of Article 11 says: "States shall provide redress through effective mechanisms, which may include restitution, developed in conjunction with indigenous peoples, with respect to their cultural, intellectual, religious and spiritual property taken without their free, prior and informed consent or in violation of their laws, traditions, and customs."

In the spirit of Article 11 we could make a case for assistance from the US government and the State of Ohio for the preservation and study of the Newark Earthworks and other sites.

The first part of Article 12 says: "Indigenous peoples have the right to manifest, practice, develop, and teach their spiritual and religious traditions, customs and ceremonies; the right to maintain, protect, and have access in privacy to their religious and cultural sites; the right to the use and control of their ceremonial objects; and the right to the repatriation of their human remains." The second part of Article 12 reads: "States shall seek to enable the access and/or repatriation of ceremonial objects and human remains in their possession through fair, transparent and effective mechanisms developed in conjunction with indigenous peoples concerned."

Article 12, then, overlaps with NAGPRA provisions and gives direction for care to be taken with human remains and objects found at historic sites. Moreover, it asserts the right to privacy in carrying out ceremonies. Although this might not be an issue for the Newark Earthworks, since there has not been a continuity of ceremonial practice, it would be significant for some sites. In 2010, when visiting Chaco Canyon, I was told that Navajo and Hopi came regularly to carry out ceremonies at the canyon because they believe that their ancestors came from this place. At Newark, there is the possibility of the introduction of contemporary rituals, such as the powwow sponsored by the Native American Center of Central Ohio.

In sum, the Newark Earthworks are simultaneously part of the heritage of the indigenous peoples of North America, part of the heritage of all who live in the United Sates today, part of the heritage of the indigenous peoples of the world, and part of world heritage. How can we respectfully negotiate these strands of heritage so that astronomers, mathematicians, philosophers, religious practitioners, scholars, and citizens can all draw on the experiences and practices of the past as they connect through material remains with the Hopewellians who once lived in Ohio? The Ohio Historical Society, the Newark Earthworks Center, and our various academic associations all have roles to play. In the spirit of NAGPRA and UNDRIP we are challenged to include indigenous scholars in our research on the earthworks and to cooperate with the people of Licking County and beyond in making the earthworks more known and accessible to the indigenous peoples of the world.

Notes

1. "Hopewellian" is an archaeological term derived from the Hopewell site in the Scioto River drainage of south central Ohio. After exploring a group of mounds located on the property of the Hopewell family in 1891–92, Warren K. Moorehead applied the designation "Hopewellian" to the various cultures that participated in a widespread exchange system in midwestern and eastern North America. It is not known what the various groups now described as Hopewellian called themselves.

2. See, for example Dancey, *The First Discovery of America*.

3. Woodard, "A Pow Wow amongst the Largest Geometric Earthworks Complex in the World."

4. The Newark Earthworks are part of the Hopewell Ceremonial Earthworks, sites nominated in January 2008 by the US Department of the Interior for potential submission to the UNESCO World Heritage List. In addition to being of "outstanding universal value," sites proposed for World Heritage status must meet at least one of the criteria listed on the UNESCO home page, at http://whc.unesco.org/en/criteria/.

5. In May 2012, a group of indigenous leaders from around the world, assembled in New York for the Eleventh Session of the UN Permanent Forum on Indigenous Issues, sent to UNESCO a Joint Submission on the Lack of Implementation of the UN Declaration on the Rights of Indigenous Peoples in the Context of UNESCO'S World Heritage Convention. The submission reiterated concerns voiced a year earlier about three World Heritage nominations that were under consideration by the World Heritage Committee at its thirty-fifth session and that the signatories of the 2011 submission

advised should be deferred. These were Kenya Lake System in the Great Rift valley (Kenya), Western Ghats (India), and Trinational de la Sangha (Congo / Cameroon / Central African Republic). All three nominations had been made, it was claimed, "without due consideration of Indigenous cultural values." The 2012 submission claims that the World Heritage Committee largely ignored the earlier submission, inscribing Kenya Lake System on the World Heritage List without obtaining "the free, prior, and informed consent of the Endorois." The other two nominations had been referred back to the states submitting them for, among other things, further consultation with the indigenous communities. At the time of the United Nations Permanent Forum on Indigenous Issues Eleventh Session, the revised nominations were not available for public scrutiny.

6. For a discussion of Native American ways of remembering in relation to places, see, for example, Nabokov, *A Forest of Time.*

7. Native Hawaiians are more closely identified with other Polynesian peoples than with Native Americans. However, because of the annexation of Hawai'i by the United States, NAGPRA is attempting to deal with the patrimony of both Native Hawaiians and Native Americans.

8. United Nations Forum on Indigenous Issues, "Frequently Asked Questions: Declaration on the Rights of Indigenous Peoples," http://www.un.org/esa/socdev/unpfii/documents/FAQsindigenousdeclaration.pdf.

PART VI The Newark Earthworks in the Context of Law and Jurisprudence

Ancient and Ongoing Possibilities

DUANE CHAMPAGNE & CAROLE GOLDBERG

The Peoples Belong to the Land

Contemporary Stewards for the Newark Earthworks

How do indigenous peoples understand, relate to, and interpret ancient places such as the Earthworks at Newark? While it is difficult to survey the views of hundreds of tribal communities, many tribal nations have taken actions and expressed their viewpoints on the physical remains of ancestors, funerary materials, and specific places that are significant in their tribal teachings and traditions. Tribal efforts to protect sacred places are found in the movements to gain legislative and federal administrative protections in the American Indian Religious Freedom Act, the Native American Graves Protection and Repatriation Act, the National Historic Preservation Act, presidential executive orders, legal cases, and state-level legislation to protect sacred sites and burials, as well as tribal initiatives to provide protection. Tribal communities are seeking new and better ways to protect their cultural, religious, and historical legacies.

This essay addresses two themes concerning contemporary American Indian views and possible actions toward efforts to renew and preserve ancient sites such as the Earthworks at Newark. The first characterizes indigenous motivations, values, and practices about ancient holy places. What are the goals and methods of action for indigenous peoples for preserving and renewing ancient religious places, especially when they have been removed from them for centuries? And what are the forms of organization and coalitions of indigenous nations when working on national and international issues of preserving cultural heritage?

The second theme works within the framework of US law and explores the ways in which tribal communities have expressed their views in the law and the constraints and opportunities afforded by treaties, case law, congressional acts, and federal implementation. What rights and possibilities are afforded to federally recognized tribal nations for recovering access to places, ancestral remains, and objects of cultural patrimony?

Understanding the motivations, forms of organization, and legal context within which indigenous nations seek to preserve their cultural and historical heritages will help inform contemporary approaches to preserving and renewing the Earthworks at Newark. Indigenous peoples with legal rights or cultural attachments to the earthworks heritage may want to participate. The tribes, to the extent possible, will want to participate on their own terms, with their own goals, and within their own forms of social and cultural organization. Greater understanding of indigenous perspectives and organizations will help create more effective multicultural coalitions in support of renewing and preserving the Earthworks at Newark.

Indigenous Views on Land and Place

Land and spiritual places are of central importance to indigenous nations. Despite having been moved around the continent, American Indians uphold traditions and community memory of living in specific territories for thousands of years. Many traditions suggest tribal communities occupied their present lands from time immemorial.[1] While many American Indian communities were removed from their traditional territories during colonial periods, most communities continue to value their histories and their presence on the land. Along with land come sacred places, creation teachings, and the landmarks of the tradition and storytellers.[2] Place has economic, political, and spiritual meaning for indigenous peoples. At specific places, the creator formed the world, instructed the people how to live, gave them laws, values, ceremonies, and institutions.[3]

Indigenous peoples did not own land in the Western sense of fee-simple holding. Rather the people belong to the land, like the plants, animals, places, and sacred bundles. "We do not own the land, we are of the land, we belong to it," according to Lenape teachings.[4] Usually the creator allocated territory, cultural meaning, and rules of order.[5] Indigenous peoples' territories are holy lands given by the creator for a purpose and are full of events and meaning that are little known to outsiders and sometimes lost to tribal peoples themselves because of forced separations. Resistance to the removal policy, initiated in 1830, was mustered by the more traditional tribal members, who did not want to leave their sacred and holy places.[6] The Black Hills, despite physical separation from tribal reservations, remain sacred and part of the creation instructions of the Lakota, while Bear Butte is a sacred place for both the Cheyenne and Lakota peoples.[7]

The Indigenous Cultural Preservation Movement

In the contemporary world, tribal peoples have pursued legal, cultural, and legislative avenues to protect and recover access to land and sacred places.[8] These efforts have yielded some success through US law, such as through passage of the American Indian Religious Freedom Act (AIRFA), the Native American Graves Protection and Repatriation Act (NAGPRA), and amendments to the National Historic Preservation Act (NHPA).

AIRFA, passed by Congress in 1978, addresses indigenous interests in preserving sacred places. Together with a 1994 presidential executive order, AIRFA directs federal officials and agencies to respect Indian sacred sites and consult with relevant tribal communities about the disposition of traditional tribal territories currently within federal public land administration.[9] The 1992 amendments to the NHPA[10] require federal agencies to consult with tribes and Native Hawaiian organizations whenever the agencies issue permits or take action related to development on state or private lands if that development affects properties of religious and cultural significance to those Native groups. Though NAGPRA is primarily concerned with protecting Native American cultural objects and human remains, it establishes a framework for documenting and valuing tribal connections to particular places, a precondition for successful claims by indigenous groups to participate in the management and protection of those places. These measures underscore that sacred places should be accessible to Indian communities for ceremonies, respected, and preserved. Multitribal organizations such as the National Congress of American Indians and the Native American Rights Fund were the primary advocates for their passage.[11]

Some tribal communities have organized nonprofit organizations under state law and purchased land to preserve locations and plants significant in their traditional histories and sacred teachings. At the state level, tribal communities have sought to protect burials and sacred places through recognition and protection under state laws. While the Indian communities have been active in protecting their heritages whenever possible, their record of success has been limited for lack of land ownership, lack of implementation of the law, little understanding by the general public, and often too little political influence.[12]

Native American communities remain deeply interested in their sacred and ceremonial places, including sites outside the boundaries of their present reservations. Most American Indian communities left their homelands under economic, military, or political pressures. Under their own interpretations, American Indians believe they retain cultural rights to land and sacred places.

American Indian Heritage and the Earthworks at Newark

It is difficult to tie the Earthworks at Newark to any particular tribal community. The historical record is ambiguous, and present-day tribal communities do not routinely perform rituals there. Perhaps rituals as originally performed there have not taken place since at least 500 CE. Some argue that the Adena, Hopewell, and Mississippian cultures were created by a succession of different migrating communities. However, it can be just as well argued that descendants of the same tribal peoples created and participated in all three cultural patterns. Current archaeological thought suggests that Adena cultural communities were joined by newcomers and, with some exchange of ideas and culture, formed the Hopewell culture.[13]

Ancient American Indians built geometric earthen enclosures upon more than six hundred sites within present-day Ohio.[14] The diverse and decentralized pattern of the earthworks suggests many local clusters of lineage-based communities sharing economic, political, religious, or ceremonial practices to varying degrees.[15] The earthworks and mounds may have been organized by a spiritual network or hierarchy that tied locally autonomous lineage groups together in ceremonial relations, if not political and economic relations.[16] For instance, the emergence of shared practices among American Indian communities, such as smoking pipes for ceremonial, social, and political purposes, became widespread.[17] While tribal communities share many distinct general patterns, such as strong emphasis on this-worldly orientations rather than otherworldliness as in Christianity, tribal forms of worship are diverse, distinct, and numerous.[18]

The people who practiced ceremonies at the Earthworks at Newark most likely practiced local and distinct ceremonial rituals. Specific concepts of death, upper world, lower world, and elements of fire, earth, water, and air are inscribed in the earthworks' construction.[19] Similar understandings are found in the worldviews of the historical tribal nations in the region. Many tribal communities may have attended ceremonies at the Newark Earthworks and could understand its purpose and relations to the cosmic order. Nonlocal tribal communities would show great respect for the ceremonies and the religion of that earthworks, even though they had their own distinct ceremonies, mounds, or earthworks with their own specific meanings. While tribal communities may not share the same ceremonial rituals, they respect that other tribal nations carry out their ceremonial practices according to their own creation teachings and understandings.

Colonial Contact and the Ohio Indian Diaspora

The many historical tribal communities that lived in Ohio have not left commentary about the thousands of earthworks and mounds found there. The historical tribal communities would have known the land very well, but the historical record is silent about what the Indians thought about the sites. During the 1500s, diseases decimated the Mississippian culture peoples living in the Ohio region. Infectious diseases introduced by Europeans caused losses in population in the range of 70 to 90 percent. Often old and young people died first. Much cultural knowledge, even the social and economic ability to conduct ceremonies, was lost during this period.[20] The surviving communities were more decentralized and egalitarian than previously.

The western colonial expansion, especially in the eastern United States, including Ohio, resulted in further displacement of numerous Indian nations. Many of the eastern indigenous communities lived for hundreds if not thousands of years in their homelands. Peoples such as the Delaware were forced economically and politically to sell land and retire further westward, passing through Ohio only to be forced to remove again and again, eventually ending up in Ontario and Oklahoma.[21] The retreating eastern nations often met resistance from nations farther west, and the westward retreat resulted in wars and battles with other Indian nations.

During the 1640s, the Iroquois sought hunting and fur-trading agreements with the tribes in the Ohio region but were rebuffed because many nations, both Algonquian and Iroquoian speakers living there, were allied to the French in New France. The Iroquois, supported by Dutch and British weapons, prevailed among many tribes in the Ohio area, and many retreated farther west.

By 1700, however, the British began to occupy the Mohawk valley, and the Iroquois decided that the British were equally as threatening to their national interests as were the French in New France. Thereafter the Iroquois helped organize a confederacy of western Indian nations for trade, military, and diplomatic purposes. The Western Indian Confederacy, long occupying and defending the Ohio region, signed the Treaty of Greenville on August 3, 1795.[22]

In the Treaty of Greenville, the United States negotiated with the Western Confederacy as collective owner of southern Ohio, including present-day Licking County, and the Earthworks at Newark.[23] The confederacy adopted the position that the Indian nations "held their lands in common and that unanimous consent of all would be necessary to cede any part of it."[24] The Americans honored the Western Confederacy's demands by recording collective ownership and cession of lands in southern Ohio. For the United States, the Treaty of Greenville identified the owners of the land in Ohio. These tribes were the Wyandot, Shawnee, Chippewa, Wea, Kickapoo, and Kaskaskia, as well as bands among the Delaware, Ottawa, Potawatomi, and Miami. The Wea signed for the Piankeshaw.[25]

At the treaty ceremonies, Little Turtle, the famous Miami chief, recounted Miami Nation responsibilities for land occupied in western Ohio from time immemorial. Little Turtle said: "I have now informed you of the boundaries of the Miami nation where the Great Spirit placed my forefathers a long time ago and charged him not to sell or part with his lands but to preserve them to his posterity. This charge has now been handed down to me."[26] Other nations within the Western Confederacy most likely held similar views.[27]

The confederated nations ceded much of southern Ohio to the United States but retained hunting rights within the ceded territories.[28] Consequently, the confederated Indian nations retained access to all six hundred or more sacred ceremonial sites in Ohio. Only after the removal treaties, starting in 1830, when hunting and gathering rights were extinguished, did the Ohio Indians lose access to the earthworks and mounds. Certainly the confederated Indians noticed and understood that the earthworks and mounds were constructed for spiritual and ceremonial purposes.

After 1830, federal removal policy pressured many eastern tribal nations to migrate farther west. Between 1830 and 1842 many of the Ohio Indian nations signed removal treaties. For many tribes, removing west meant leaving spiritual places and the land and places given in the tribe's creation teachings. The religious seriousness of Shawnee removal and the organization of travel was recorded by David Robb, one of the agents assigned to manage the removal of Indians from present-day Auglaize County. Robb recounted:

> After we had rendezvoused, preparatory to moving, we were detained several weeks waiting until they had got over their tedious round of religious

> ceremonies, some of which were public and others kept private from us. One of their first acts was to take away the fencing from the graves of their fathers, level them to the surrounding surface, and cover them so neatly with green sod, that not a trace of the graves could be seen.
>
> When their ceremonies were over, they informed us they were now ready to leave. They then mounted their horses, and such as went in wagons seated themselves, and set out with their "high priest" in front, bearing on his shoulders "the ark of the covenant," which consisted of a large gourd and the bones of a deer's leg tied to its neck. Just previous to starting, the priest gave a blast of his trumpet, then moved slowly and solemnly while the others followed in like manner, until they were ordered to halt in the evening and cook supper. The same course was observed through the whole of the journey.[29]

The tribal nations, when removing west, left the burial grounds of their ancestors, carried away their sacred bundles and sacred fires, and, in the new land, reestablished religious organization and ceremonies and anointed sacred places.[30] The Indians renewed their ceremonial life on new land farther west and continued their cultural lives.

The Sacredness and Renewal of the Earthworks at Newark

The sacred bundles and ceremonies were carried with the people on their involuntary journeys. If the Indians nations were to return to Ohio, could they renew their ceremonies and spiritually recharge their sacred places? The answer to this question is probably yes. It may depend on the specific beliefs of each tribe. As Robb noted, some ceremonies were for public consumption, but others were not. Some ceremonies could be performed for the community, other Indians, and non-Indians. Other ceremonies were performed secretly and were probably the knowledge and responsibility of a small group of elders.

Any ceremonies practiced by the indigenous group(s) that originated the Earthworks at Newark are not in the living memory of contemporary tribal communities. Nevertheless, the earthworks were and are a sacred place. Contemporary tribes respect the earthworks and may wish to perform ceremonies there to show respect and honor their spirits. These contemporary ceremonies would not be the same ceremonies that were performed more than two thousand years ago. Their intent, however, may be very similar.

Were foreign or public ceremonies prohibited at the Earthworks at Newark? This would depend on the specific features of the cultural and ceremonial complex of the indigenous community or communities that ministered to the earthworks. Some cultures have very exclusive ceremonies and religious membership, while other communities are willing to accept and use religious ceremonies and ideas from other tribal nations. For some traditions, there are multiple paths to participating in the sacred. For example, contemporary Cheyenne elders practice Catholicism, belong to the Native American Church, and perform traditional Cheyenne sun dance ceremonies.[31] The main rule is that the different religious rituals and doctrines should not be mixed.

There is some reason to believe that those who ministered to the Earthworks at Newark welcomed other nations and ceremonial practices. First, there are no large permanent settlements at the earthworks. The nearby settlements were small, seasonal, or perhaps primarily temporary camps for taking part in ceremonies.[32] In their full sixty-mile extension, the earthworks were not defended. Many local communities participated ceremonially and helped preserve what has been called the "Great Hopewell Road" and the ceremonial centers at both ends sixty miles apart. The roadway was a short distance to numerous earthworks and mounds, indicating the participation of many communities and a supporting network of ceremonial centers.[33] Remarkably, there appears to be little disruption or degradation of the earthworks complex by human violence during the period before Europeans arrive in Ohio. The earthworks were treated with great respect for at least two thousand years. Pilgrimages, trade, gatherings, and ceremonies ministered by coalitions of communities in the region could have taken place.

Second, the Earthworks at Newark appear to have been spiritually decommissioned at some point. The spiritual decommissioning may have happened when the associated communities were forced or decided to migrate or when the community members turned to different ritual or religious beliefs. As in the Shawnee case cited above, ceremonial places can be decommissioned and also renewed. A contemporary renewal ceremony may not be the same as the original ceremonial rituals but nevertheless can show continuity, respect, and honor for the place and its past sacred history.

Third, even though some ceremonies performed at the Newark Earthworks were esoteric, most likely public ceremonies with invited guests from other nations were also performed. The pattern of public social ceremonies is common among many tribal communities.

If the Earthworks at Newark were selected as a World Heritage site, then people of many nations around the world would be invited to visit. Many indigenous peoples from around the world would probably want to attend and perform ceremonies. All nations and religions would be invited to pay their spiritual respects to such an ancient and public sacred place, as long as the site is treated with reverence and no harm is done to the earthworks.[34] The more respect and honoring of spiritual forces that embody or protect the earthworks, the better. Damage and disrespect will bring harm to the participating communities. People of all nations and religions with positive spiritual intent would be welcome and consistent with both indigenous religious traditions and the World Heritage designation.

If an American Indian nation claims to have direct connections to the Earthworks at Newark and wants to practice nonpublic ceremonies, this too should be honored. The oral histories and religious understandings of indigenous peoples should be respected and accommodated, as long as general public access is maintained at other times and no physical or spiritual harm comes to the site. Perhaps the most powerful ritual might be restoration of the full sixty miles of the Great Hopewell Road between Newark and Chillicothe.[35] Recovering the entire complex would be the ultimate show of respect and spirituality, returning wholeness to the earthworks ceremonial complex.

Linguistic Ties and Identities

The nations of the Western Confederacy are most likely the remnant peoples of ancient Ohio and are composed of Algonquian and Iroquoian language family speakers. Other Algonquian and Iroquoian linguistic speakers may claim connections, ancestry, or cultural affiliation to the nations of the Western Confederacy or to the ancient peoples of Ohio. If a tribal community has oral history connections and traditions to peoples of the Ohio region, then these traditions should be recognized. For example, some Sioux nations have traditions that locate their ancestors in present-day Wisconsin and the Ohio region, as well as ties to Iroquoian-related-speaking peoples in the east and southeast.[36] Siouan peoples with such traditions should be invited to participate and have their traditional histories respected. They should be invited to participate in any alliance, confederacy, or organization dedicated to preserving and renewing the Earthworks at Newark.

Indigenous Ways of Organizing Preservation and Renewal

The movement of Indian nations caused by European colonial expansion creates problems for contemporary NAGPRA proceedings, in which tribal groups seek repatriation of cultural objects and human remains. The legal standard requires that a tribe making such a claim prove, by a preponderance of evidence, that these materials are "culturally affiliated" with the tribe. Cultural affiliation is defined as a "relationship of shared group identity,"[37] a relationship that may be demonstrated by many types of evidence, including geographic, biological, oral historical, and archaeological.[38] Many burial objects and ancestral remains are hard to identify clearly with a particular community. There may be multiple claimants and multiple supporting narratives.

The National Museum of the American Indian (NMAI) came up with a culturally informed method for managing such situations.[39] The museum staff contacted and researched all tribal nations that might have lived in a specific region within historical memory or documentation. Tribes could also make oral history claims. The NMAI repatriation committee and staff did not try to decide who was right or which nation most likely constituted the descendants of a particular disposition. Rather, in Indian style, they consulted with all groups that had a possible claim. All claims were respected and noted. This method often took years of patient diplomatic work among as many as seventy tribal nations in some instances. The goal was not to determine a "winning" nation. Rather, the aim was to identify all claimants and then work out an agreement about how the disposition of specific ancestral remains or funerary objects could be effected with the approval of all nations. If more than one nation decided to engage in ceremonies at the reburial of ancestors or funerary objects, then all nations that felt strongly enough to conduct ceremonies were allowed to do so. The key here was to give respect to the historical interpretations of all nations, working out a consensual process for action.[40]

In addition, the NMAI works with coalitions or organizations of tribal nations that collectively represent their interests and actions on repatriation petitions. Such contemporary tribal organizations and NMAI procedures may be a model for including tribal nations to support preservation and renewal of the Earthworks at Newark. The tribal repatriation coalitions often represent the interests of nations with shared cultural backgrounds. In the case of the Earthworks at Newark, nations from varying national, historical, and cultural traditions have interests in participating and consequently forming a coalition of supporting tribal nations.[41] Each nation participating in the preservation and renewal of the earthworks should enter as an equal partner with the other indigenous nations. From an indigenous point of view, such an organization would be an international alliance, not too different from the old Western Confederacy in cultural organization and process. Such an alliance or confederation could play a significant role in the effort to preserve, restore, and renew the Earthworks at Newark and in supporting a proposal to the UNESCO World Heritage Committee.

The US Legal Framework for Reclaiming Access to Tribal Territory

The earthworks are situated on land that, as noted earlier, was ceded by the Western Confederacy through the Treaty of Greenville in 1795. Several of the tribes included within that confederacy, including the Wyandot, Ottawa, and Shawnee, were still living in Ohio in the 1830s. The Wyandot and Ottawa occupied some of their old villages, north of the treaty line. But with passage of the Removal Act in 1830,[42] these tribes were under extreme pressure to cede their remaining lands and move west.[43] The Shawnee entered into two treaties in 1831,[44] and the Wyandot finally succumbed to a treaty in March 1842.[45] Subsequently, the territory comprising the earthworks became public land of the State of Ohio and later was transferred to more local control, eventually winding up in the ownership of the Ohio Historical Society in 1933.

Any confederation of indigenous nations that may seek a role in preserving, restoring, and renewing the earthworks is likely to encounter resistance from the current occupant—a country club, calling itself Moundbuilders Country Club, that leases the land from the Ohio Historical Society. The lease was first made with the society's predecessor in 1910 and has been renewed to run until 2078.[46] Because private rather than federal land is involved, most of the federal measures offering protection to indigenous sacred and culturally significant sites will not apply. Only the National Historic Preservation Act (NHPA) is directed at private land. But so long as the country club does not seek a permit or any other action from the federal government in connection with its use of the land, the requirements of NHPA will not be triggered, no matter how historically and culturally significant the earthworks may be.

One alternate strategy the confederation might consider is securing return of the land. US law largely rules out such a strategy, however.[47] Federal Indian law makes no provision for restoration of land that was ceded through a federal treaty, such as the Treaty of Greenville.[48] Indeed, in 1946, Congress passed the Indian Claims Commission Act,[49] designed to extinguish all tribal claims to lands that had been ceded in unfair treaties. The act offered compensation for lost lands but required tribes to forgo any future claims to those lands. Among the proceedings that led to monetary settlements under the Indian Claims Commission Act were a variety of claims by tribes that were parties to the Treaty of Greenville, including a claim by the Shawnee, whose village was probably the closest Native settlement to Newark.[50]

Recent legal scholarship suggests yet another strategy that the confederation of tribes might pursue. This scholarship emphasizes that exclusive ownership was never the foremost mode of indigenous relationship to land and that a more appropriate conception of relationship to the land within Native cultures would focus on stewardship, rather than ownership, in partnership with other interested and mutually respectful users.[51] A stewardship approach aims "to secure Indian entitlements to property without transferring title from the current (non-Indian) owner."[52] These entitlements would include access rights so that indigenous peoples can prevent desecration of the sites and continue to fulfill their custodial responsibilities to the land through ceremonies and stewardship. Such rights would be exercised in cooperation with other users, who might include natural-resource development corporations, recreationalists, and environmental constituencies, all of whom might desire access and usage rights as well.

In order to invoke stewardship rights, however, an indigenous group must establish some threshold entitlement to participate in the decision-making group responsible for the land. The Native understanding of connection to, as well as respect and responsibility for, the land should be sufficient to support such entitlement. It is not difficult to envision, however, that non-Native claimants to the earthworks would argue against the participation of the confederation of tribes, based on assertions that the earthworks long predate the presence of the historical tribes of Ohio. Similar arguments have been made to prevent repatriation of human remains and cultural objects under NAGPRA, as in the case of the so-called Kennewick Man, a set of nine-thousand-year-old human remains found in Washington State.[53] Archaeologists argued in that case that the tribes with historical ties to the land where the remains were found should not be entitled to repatriation because the remains were so old that "cultural affiliation" could not be established.

Recent rule making under NAGPRA, however, has suggested a way of conceptualizing entitlements for the confederation tribes that does not depend on proving, for legal purposes, cultural continuity between the Hopewell culture peoples who built the earthworks and the historical Ohio tribes. Under NAGPRA, when tribes are unable to demonstrate that cultural objects and human remains are "culturally affiliated" with any current federally recognized group, the items are declared "culturally unidentifiable" and ineligible for repatriation. NAGPRA provides, nonetheless, that an administrative body, the NAGPRA Review Board, shall devise procedures for the "disposition" of culturally unidentifiable items.[54] In 2010, the Review Board finally issued such regulations.[55] Significantly, these regulations specify that culturally unaffiliated human remains found on any land—public or private—shall be disposed of to the tribe "from whose tribal land, at the time of the excavation or removal, the human remains were removed." If the land is not tribal land at the time of excavation, the regulations provide that the culturally unaffiliated human remains shall be disposed of to the tribe or tribes "recognized as aboriginal to the area from which the human remains were removed." The regulations further state, "Aboriginal occupation may be recognized by a final judgment of the Indian Claims Commission or the United States Court of Claims, or a treaty, Act of Congress, or Executive Order."[56] The Treaty of Greenville land cessions, affirming the position of confederation tribes, would thus entitle the successors to these treaty tribes to lay claim to all human remains found in the ceded territory, regardless of their age and demonstrated cultural affiliation. Furthermore, and ironically, the very Indian Claims Commission process that produced the extinguishment of tribal ownership claims to land in Ohio also now serves to establish that these are aboriginal lands of the claiming tribes for purposes of disposition of "culturally unidentifiable" human remains under NAGPRA.

In effect, the new NAGPRA regulations affirm the vision of tribal relationship to land that is advanced in this essay. These regulations presuppose that even if a tribe cannot demonstrate specific cultural continuity with the human remains according to NAGPRA, its connection to the land upon which these remains were found puts it in a special position to care for them. Either because of characteristics believed to be common to most indigenous peoples or because of a belief in the special sensitivity of one indigenous group to the ceremonies and practices of others (or both), the tribe with ties to the land is treated, under NAGPRA, as entitled to stewardship over the remains that are found there. Conveniently, the national NAGPRA database lists all the tribes associated with the Treaty of Greenville land cession.[57]

Models of Indigenous Stewardship

In recent decades, indigenous nations have provided examples of cooperative initiatives with one another and with nonindigenous groups to achieve effective stewardship of lands to which they all may have ties. In order to be recognized as legitimate participants in such efforts, indigenous nations have argued not only that they have special cultural connections to the land but also that they have knowledge that may help preserve and sustain the land for the benefit of all future users.[58] These indigenous partnerships with nonindigenous groups demonstrate that a concept of stewardship focusing on responsibilities and nonexclusive use or access rights can achieve a more satisfactory outcome than a concept of exclusive property rights that typically results in denying any indigenous participation.

A positive illustration of Native stewardship is the Chumash Village in Malibu, California, where Chumash people care for the land and share their ethic of caring with the wider public. Around 2000, allied with local environmentalists, they formed a nonprofit called Wishtoyo ("rainbow" in Chumash)[59] and approached the County of Los Angeles, which owned a strip of beach near the site of eight Chumash villages dating back at least eight thousand years. For $100 a year, the county leased the land to Wishtoyo, which agreed to re-create a village where both tribal ceremonials and public education programs could take place.[60] Now there are half a dozen *aps*, or traditional Chumash dwellings, several ceremonial sites, a sweat lodge, and plank canoes built in the traditional way. Invasive plants have been removed and indigenous plants restored. The watershed of adjacent Nicholas Canyon has been revived, after decades of illegal dumping of cement and other harmful matter. Indigenous groups from around the world have joined the Chumash there in performing ceremonies. Simultaneously, Native stewardship of the site has made it possible for groups of students, teachers, local community members, and others to learn about Chumash culture and what it means to treat the land with respect. As Chumash elder Charlie Cooke noted at an event marking the opening of the village, "This is a special area and a learning center for all people, not just Native American people."[61]

Other examples of successful Native stewardship, accommodating broad uses while respecting Native cultures, include the Sinkyone Intertribal Wilderness Area in northern California[62] and the Native American Land Conservancy in southeastern California.[63] Both demonstrate the value of multitribal coalitions, alliances with environmentalists and non-Native governments, and selective use of legal tools such as conservation easements.

Conclusion

The strategies described above have been pursued nationwide, restoring indigenous groups to stewardship roles over land from which they may have been separated for more than a century.[64] For instance, the Indian Country Conservancy, an offshoot of the much larger Trust for Public Land, has protected more than 200,000 acres for about seventy tribes.[65] What these initiatives share is a recognition by non-Native interests of the special relationship between indigenous peoples and their territories, a relationship that attaches to all those who came before them. These initiatives also share an expectation that the tribal relationship will coexist with larger public interests in access to the land and that a partnership of Native and non-Native institutions is necessary to sustain the sites.

There are models for protecting the Earthworks at Newark as a sacred indigenous site while acknowledging the broader cultural interest. This essay has argued for a deeper understanding of the connection of the Shawnee, Wyandot, and other Ohio tribes to that place and has shown how NAGPRA supports that understanding. In partnership with federal, state, and private entities, these tribes could take a leading role in protecting and interpreting the earthworks as a World Heritage site.

Notes

1. Wilkinson, *The People Are Dancing Again*, 12.

2. McCoy, "The Land Must Hold the People," 421.

3. Champagne, *Social Order and Political Change*, 13–49.

4. Hitakonanu'laxk,' *The Grandfathers Speak*, 2.

5. Zolbrod, *Diné bahane'*.

6. Wilkinson, *The People Are Dancing Again*, 220; and Champagne, *Social Order and Political Change*, 143–71.

7. Champagne, *Social Change and Cultural Continuity among Native Nations*, 285–90.

8. B. Brown, *Religion, Law, and the Land*, 6–7, 125–29, 172.

9. See Yablon, "Property Rights and Sacred Sites," 1623.

10. 16 USC § 470a(d)(6).

11. See, for example, Trope and Echo-Hawk, "The Native American Graves Protection and Repatriation Act."

12. Deloria, "Secularism, Civil Religion, and the Religious Freedom of American Indians," 9–20; Trafzer, "Serra's Legacy."

13. Silverberg, *The Mound Builders of Ancient America*, 226.

14. W. Mills, *The Archaeological Atlas of Ohio.*

15. Abrams, "Hopewell Archaeology," 169–204.

16. Bernardini, "Hopewell Geometric Earthworks," 331–56.

17. Steinmetz, "The Sacred Pipe in American Indian Religions."

18. Champagne, *Social Change and Cultural Continuity among Native Nations*, 25–44.

19. A summary of the worldview of late woodlands cultures is given in Joseph Campbell, "Part 2: The Northern Americas," in *Way of the Seeded Earth*, 116–221.

20. Vickers, *A Companion to Colonial America*, 48–49.

21. Clinton A. Weslager, *The Delaware Indian Westward Migration.*

22. Treaty of Greenville, 7 Stat. 49 (1795), http://avalon.law.yale.edu/18th_century/greenvil.asp.

23. Ohio—Present-Day Tribes Associated with Indian Land Cessions 1784–1894, National NAGPRA Online Databases, http://www.nps.gov/history/nagpra/ONLINEDB/INDEX.HTM.

24. Hine and Faragher, *The American West*, 119–20.

25. Treaty of Greenville.

26. Little Turtle at the Treaty of Greenville, http://www.accessgenealogy.com/native/turtle/turtle-treaty-greenville.htm.

27. See McCoy, "The Land Must Hold the People."

28. Treaty of Greenville, article 7.

29. Robb, "Indian Characteristics and Customs," 299–300. See also D. Lucas, "Our Grandmother of the Shawnee: Messages of a Female Deity."

30. Champagne, *Social Order and Political Change*, 145, 164–71, 178–83, 214–18, 228–37; and Wilkinson, *The People Are Dancing Again*, 11–12, 364–74.

31. Fieldwork by Duane Champagne, personal communication from member of Dog Soldier Society, Northern Cheyenne Reservation, 1983.

32. Dancey and Pacheco, *Ohio Hopewell Community Organization*; and Shiels, "On the Pilgrim Road."

33. Price, "Hopewell Road."

34. Harjo, "Sacred Places and Visitor Protocols," 81–82.

35. Lepper, "Tracking Ohio's Great Hopewell Road," 52–56.

36. Saponi Nation of Ohio, http://saponi-ohio.org/8001.html.

37. 25 USC § 3001(2).

38. 25 USC § 3005(a)(4).

39. A statute similar to NAGPRA applies to the Smithsonian Museum, to which NMAI belongs. See 20 USC § 80q et seq.

40. For a description of repatriation procedures at the NMAI, see National Museum of the American Indian, *Programs and Services Guide*, 55–59; and Henry, "Challenges in Managing Culturally Sensitive Collections at the National Museum of the American Indian."

41. See also Kerber, *Cross-Cultural Collaboration.*

42. Act of May 28, 1830, 4 Stat. 411.

43. O'Donnell, *Ohio's First Peoples,* 110–26.

44. Treaty with the Seneka and Shawnee, July 20, 1831, 7 Stat. 351; Treaty with the Shawnee, Aug. 8, 1831, 7 Stat. 355.

45. Treaty with the Wyandot, Mar. 17, 1842, 11 Stat. 581.

46. Vanishing History, *Columbus Monthly,* www.columbusmonthly.com/content /stories/2009/11/vanishing-history.htm.

47. A separate question is whether international law, especially the 2007 United Nations General Assembly Declaration on the Rights of Indigenous Peoples, would provide for indigenous rights to own or manage the land ceded in the 1795 treaty.

48. The only federal prohibition on tribal transfers applies to transfers that take place without federal permission. See 25 USC § 177. A few tribes, such as the Taos Pueblo in New Mexico and the Tule River Indian Tribe of California, have had lands restored by Congress when the lands were in federal ownership.

49. Pub. L. No. 79-726, 60 Stat. 1049 (1946).

50. See, for example, *Shawnee Tribe of Indians of Oklahoma et al. v. United States,* 40 Ind. Cl. Comm. 161 (1977). For a map showing the old Shawnee village, 1794, see Royce, *Indian Land Cessions in the United States,* plate 156.

51. Carpenter, Katyal, and Riley, "In Defense of Property," 1022.

52. Ibid., 1114.

53. See *Bonnichsen v. United States,* 367 F.3d 864 (9th Cir. 2004).

54. 25 USC § 3006(c)(5).

55. 43 CFR § 10.11.

56. Ibid. at § 10.11(c).

57. See Ohio—Present-Day Tribes Associated with Indian Land Cessions, 1784–1894, which lists forty contemporary tribes associated with tribes included in the Treaty of Greenville.

58. For a notable instance of a successful argument about special knowledge justifying stewardship rights, see Haberfeld, "Government-to-Government Negotiations," 127.

59. One of the coauthors of this essay serves on the board of directors of the Wishtoyo Foundation.

60. Fred Alvarez, "A Homecoming of Sorts for Malibu's First Residents," *Los Angeles Times,* Nov. 17, 2005, B3.

61. Quoted in ibid.

62. Kathie Durbin, "Rediscovering the Lost Coast," 18.

63. Native American Land Conservancy, "About the NALC,"http://nalc4all.org /about.htm.

64. See Middleton, *Trust in the Land.*

65. Pattie Logan, "Tribes Use Land Conservancies to Reclaim Ancestral Grounds," *High Country News,* Sept. 24, 2011, http://www.hcn.org/issues/43.15/tribes-use-land-conser vancies-to-reclaim-ancestral-grounds, accessed May 2, 2014.

GREG JOHNSON

Caring for Depressed Cultural Sites, Hawaiian Style

My first clue that the Newark sites are depressed was the amount of care being shown them by a handful of dedicated locals from several walks of life and professions. A healthy site might occasion admiration, celebration, and even adulation. Newark is worthy of all of these, but what I perceived most directly was concern. These local individuals, including Richard Shiels, Marti Chaatsmith, and Brad Lepper, and their efforts, including the work required to host the symposium out of which this volume springs, express and manifest what Native Hawaiians would call *mālama* (care). Mālama is usually regarded as a manifestation of a *kuleana* (responsibility), whether to a family line, a religious sensibility, or something more vague but no less powerful—a tugging on one's *na'au* (gut). As I began to see Hawaiian-like patterns of care at the Newark sites, I started to think about the ways recent Hawaiian practices concerning archeological sites, historic memorials, and ancestral burials might inform an understanding of dynamics unfolding in Ohio. In what follows I outline some of my thoughts with reference to fieldwork I have conducted over the past several years. My aim is to explore this relationship of depression and care by way of Hawaiian examples to shed some comparative light on possible futures of the Newark sites.

Despite their grandeur and significance, there is no denying that the Newark sites are depressed. They are urban and suburban; nature and (Native) culture stand at a rather profound remove. The sites have been repurposed several times over, including as an amusement park complete with a horse-racing track, a fairground, a golf course, and a city park. Parts of the surrounding area are economically depressed, and this low-level gloom seems to seep over to the sites. Nor does the physicality of the sites bespeak much loving attention over recent decades. Additionally, the sites have an image problem with respect to their more famous mound neighbors and with respect to much flashier World Heritage sites. This image problem is perhaps best attested in the negative: aside from the care of some devoted locals, the sites seem nearly lost on the local imaginary. In my limited experience with the local population, some people in Newark are not even aware of the sites, let alone of their magnitude and deep histories.

The Newark sites are not alone in their condition, however. In Colorado, for example, one can find remarkably neglected Ancestral Puebloan ruins within a few miles of Mesa Verde National Park and World Heritage site. This pattern can be found globally, of course. This malaise extends even to paradise. Hawaiʻi has two World Heritage sites, Volcanoes National Park and Papahānaumokuakea, both of which are known more for their natural than their cultural components. Apparently, these are cared for well, if by radically different means: volcanoes elicit tourism, whereas Papahānaumokuakea is a marine reserve with strictly limited access. Other sites in Hawaiʻi also receive considerable and visible mālama, including various national monuments like Puʻu Kohalā, smaller state-protected sites like Moʻokini Heiau, and privately managed sites like the magnificent Piʻilani Heiau. Numerous locally cared-for sites exist as well, some of which I describe below. But disregard, neglect, and outright violence characterize the fate of numerous depressed sites throughout the islands. The most egregious example of a mistreated site is Kahoʻolawe, known as the Target Island because the US military used it for bombing practice for several decades, notwithstanding the presence of many ritual sites and burial grounds on the island.[1] Sites in downtown Honolulu and Waikiki have been assaulted by a less obviously aggressive but equally destructive modality of violence—unmitigated development. This continues to the present, even, at times, at the hands of a church and the state government. Depressed sites are not limited, however, to those in urban areas or damaged by military activities. In contemporary Hawaiʻi they can be found in agricultural areas and remote villages and on sandy beaches.

Mālama can be found in all of these places, too. What has inspired me about the Hawaiian capacity to care for challenged sites is the same thing that has pushed me to sharpen my understanding of living cultural processes. Mālama today is diverse in its contours and manifold in its expressions, and its practitioners are far from unanimous in their stated positions. Precisely for these reasons, mālama is a cultural engine that repays analysis with potentially instructive comparative insights. My aim here is to develop an account along these lines by way of three examples of contemporary mālama. The first two take us to Maui, one *ma uka* (upland) and the other *ma kai* (seaward), the third to Hawai'i Island. The first example roughly parallels Newark insofar as it is a magnificent ancient site that has been repurposed and is now being cared for by devoted locals. The second case is presented as an example of ways to care for the spirit of a place in the context of legal compromise. The third focuses upon the labor and pains behind memory work, considering an example that is not archeological but the lessons of which speak to reviving depressed sites, especially in contexts that depend upon state and federal funding.

At the outset, I would like to say a little about social memory and practices of mālama in broad strokes. Social memory is negotiated, ropelike (many strands make up a coil), frequently discontinuous, reconstituted by various means according to various ends, and almost always has multiple stakeholders.[2] All of this is as true in Hawai'i as anywhere else. And as with all colonized places, disruptions and dispossession have exacerbated the variability and contentious aspect of articulating social memory. Further, as in most American Indian contexts, social memories were outright ruptured at times by many-tentacled colonial institutions.[3] Increasingly, however, Hawaiian scholars and the vibrancy of contemporary Hawaiian cultural life are making plain that far more continuities of culture have survived than are generally recognized by current histories.[4] That said, by the mid-twentieth century Hawaiian culture was undeniably suppressed, if not depressed. Mālama, for example, was not a word on people's lips; indeed little Hawaiian was spoken at all.

Hawai'i has experienced a profound cultural resurgence since the 1960s. This has been a broad cultural renaissance, which has included language, dance, art, various textile practices, and attention to Hawaiian natural and cultural resources. In terms of recovered mālama, three manifestations of culture in action stand out: protection of Kaho'olawe, rejuvenation of open-ocean canoeing, particularly with Hokule'a,[5] and the burial protection movement, particularly as catalyzed by the events at Honokahua on Maui in the late 1980s.[6] From the vantage point of 1959, the year of Hawai'i's contested statehood, few could have foretold Hawaiian mālama today. Might Newark be on the cusp of similar processes of reawakened mālama today? At a minimum, we should not rule this out in our collective assumptions about the site and its possible futures.

Repurposed, Again: Kula Ridge

Widespread fascination with the unusual features of the Newark sites is wholly understandable. The various walls, mounds, openings, and their unusual alignments stand as a grand puzzle, an incitement to curiosity for all but the most bereft of imagination. Visiting the sites and hearing various experts discuss possible explanations of the anomalous features called to mind a Hawaiian site I had come to know over the past few years, Kula Ridge on Maui. While not nearly as well documented as the Newark sites, Kula Ridge is roughly analogous in several respects. It too has hard-to-explain but immediately intriguing standout features, including 10–15-foot-tall tower structures and very unusual wall formations. Another similarity to Newark is the history of repurposing of the site, in this instance from ancient agricultural land to ranch land to a proposed low-income subdivision. Finally, Kula Ridge shares in the depressed quality of Newark insofar as the local community has been slow to recognize its past significance and potential future integrity if properly cared for. This is beginning to change.

The story I wish to tell about Kula Ridge is about the difference a devoted caregiver can make, even when concrete successes are elusive. Mālama can yield results beyond, beside, and in addition to site preservation. It can promote reawakened sensibilities concerning place, history, community, and belonging. Marti Chaatsmith's efforts, among those of others, strike me as being headed in this direction at Newark. Dana Naone Hall, a well-known poet and burial rights activist, is at the center of the story on Maui. Kula Ridge is but the latest of sites receiving her mālama in a decades-long career of caregiving —in fact, Naone Hall is one of the figures who restored mālama to its central place in Hawaiian discourse and practice. Emerging from the intersection of cultural sensibilities and environmental immediacies, Native Hawaiians have emphasized mālama ʻaina, mālama kai, and mālama wai (care for land, sea, and fresh water) as central pillars of the restored Hawaiian tradition. Naone Hall's edited volume *Mālama: Hawaiian Land and Water* is one product of this set of concerns.[7] In the late 1980s, as a result of the crisis, struggle, and cultural innovations at Honokahua, where 1,110 *iwi kūpuna* (ancestral remains) were disinterred for a hotel, Naone Hall began to assert and extend the practice of mālama to the ancestors, particularly their burial sites.[8] She has been active on the Maui and Lanaʻi Island Burial Council for more than seventeen years, and her influence on burial protections has been profound statewide over this time. Characteristic of her mālama practice, Hall attends equally to burial disturbances, no matter their magnitude or visibility. Currently, she is fighting for the integrity of more than six hundred burials at a church in downtown Honolulu while simultaneously taking up the cause of a single individual disturbed at Kula Ridge.[9]

Much of Naone Hall's mālama labor at Kula Ridge has been imagination work. Her aim has been to nurture imaginations of Kula Ridge; she wants the state, the developer, and even locals to regard it as something more than old ranch land, something more than an economic opportunity. Mālama in this mode is about ministering not only to sites and graves but also to people's imaginations. Naone Hall is keenly aware that reimagined futures necessitate poetic visions of the past as present. This has long been her gift: an uncanny capacity to restore imagination through vision. She possesses an equally remarkable gift in communicating her insights to others. At Kula Ridge, however, the task has been tough. As with Newark, surface-level history—thin but stark—seems to prevent alternative visions for most. Ranching detritus at Kula, like fairways and greens at Newark, stand as obdurate facts, saying: "This is that state of the land; to imagine otherwise is fantasy." This situation is made more difficult because of the disarray and ineffectiveness of the Hawaiʻi State Historic Preservation Division. The state burial law makes room for acts of traditional imagination—oral histories, for example—but the current broken state of affairs enables little space for anything beyond narrowly focused fiscal thinking on the part of the government. Absent state pressure, developers have little motivation to reenvision place and history, particularly when to do so might delay their projects or cost them money. This context has prevented Hall from having legal success so far in her efforts to stop development at the site.

Nonetheless, her efforts stubbornly continue. She regularly makes the state and county aware of the threat to the site, and she has expanded her reach to planning boards, land-use commissions, and other bureaucratic entities. For our purposes, we should attend to another realm of her success amid challenges. Modestly but persistently, Naone Hall is making headway with locals on Maui, planting the seeds of a reimagined Kula Ridge—one that buds from its cultural past: an agriculture land of intense productivity and life, a ritual sphere of continuous activity, and a sacred burial area remembered by the land itself. The name of the gulch that cuts through the land is Keahuaiwi, altar of the bones. Over the past few years the narrative community of Keahuaiwi has expanded exponentially. Naone Hall was one of its only voices as recently as 2009. Now one hears the sounds of mālama—of community concerns, of rekindled connections, of emplacement—on an increasing number of lips around the island.

My point in telling the story of Kula Ridge is to emphasize something I have detected in a number of burial protection and repatriation contexts that is relevant to understanding the dynamics in play at Newark. Many times repatriation and burial activists have met with real legal successes, especially since the passage of the Native American Graves Protection and Repatriation Act of 1990 (NAGPRA) and related state laws.[10] Increasing awareness about and influence of the United Nations Declaration on the Rights of Indigenous Peoples holds out the promise of similar successes on the global stage.[11] But there have been and will be failures, too. Native claims are not always strong enough to prevail, laws are not always capacious enough to make room for certain forms of mālama, and administrative bodies are not always competent enough or adequately funded to discharge their responsibilities. In the face of manifest failures of institutions and policies, I want to draw attention to the secondary and tertiary effects of community-level mālama. Even when access is denied to sites, claims to affiliation rejected or ignored, and so forth, another form of mālama success can often be detected: revitalized community awareness of sites and narrative engagement with them. This is manifest in a simple but contagious way: when stakeholder communities, however near or far, tell the story of the place themselves. Heritage sites can be "owned" and cared for in a range of ways, including through story. I do not mean story for story's sake. I mean stories that grow communities in their telling. Similar to what is happening on Maui, back on the mainland, communities from Ohio to Oklahoma are telling the story of Newark today thanks to the work of Marti Chaatsmith and others, and that is a mālama victory worth celebrating in its own right.

Practice in Compromised Places: Moʻoloa

Having just described Dana Naone Hall's workweek mālama practice, now let me tell you how she spends many of her weekends, which are likewise filled with mālama. My point in recounting the following is to suggest ways in which caring for sites can have a reverberating effect whereby adjacent areas, sometimes entire regions, come to enjoy nurturing attention. Waves of love, as it were, can move outward from sites to enfold often neglected contiguous areas. This is a story, then, about how to broaden the reach of mālama imagination and practice. It is also a story about crafting long-range victories out of legal compromises. This is the story of Moʻoloa (Long Lizard), an oasis of several acres of native plants and Native Hawaiian learning and relaxation. In view of real estate prices on Maui, Moʻoloa occupies a rather improbable site just inland from one of the most beautiful stretches of beach in all of Hawaiʻi, Oneloa (Big Beach). How did a Hawaiian *hui* (organization) come to possess this land, and what do they do there?

Moʻoloa has a deep history of human occupation. The larger region is known as Honuaʻula, a place well remembered in oral traditions and in contemporary poetry.[12] It is a place legendary for its fishing, beauty, and views. Honuaʻula is also known for its *alanui,* its ancient pathway. Like the Great Hopewell Road that connects the Newark sites to their neighbors and to larger trading and travel channels, many prominent ancient sites in Hawaiʻi are connected by "roads." These of course served quotidian purposes, but they also figure in the oral tradition for their ritual functions. Such alanui were traveled by various priests of the *akua* (deities), who made circuits around the islands to perform rituals and receive offerings. Lono, the god of love, play, and agriculture, among other related spheres, was celebrated by rituals along the portion of the trail-cum-road that traverses the shoreline lands of Honuaʻula. Fruits of the harvest, the rewards of planting, were offered up to Lono's priestly representatives during the Makahiki season. The cadence and rhythm of qualities associated with Lono are what Naone Hall and others aspire to restore to Maui at Moʻoloa: care for the land, peaceful repose, relaxation, and general fruitfulness of body and soul.

This emphasis upon Lono qualities is long overdue in the development of Honuaʻula and in Hawaiʻi generally. For too long, says Naone Hall, attention has been focused upon Kū and his warlike qualities.[13] Historically, going back hundreds of years, Lono and Kū have been regarded by Hawaiians as dialectically governing the world and its cycles through a complementary but antagonistic relationship.[14] Kū orchestrated the necessary but violent elements of aggression and struggle; Lono's time ushered in peace, sexuality, and celebration. Naone Hall wants this Lono constellation to prevail at Moʻoloa. Restorative *mana* (power, energy) is sought for the land itself and for its children. But Naone Hall is no stranger to Kū energy. Indeed, in a most traditionally dialectical fashion, there would be no Lono season at Moʻoloa today if there had not been a Kū season yesterday.

Yesterday in this case was the mid-1980s, a time of rapid hotel development on Maui. At that time a Japanese corporation, Seibu, was in the process of building a hotel and golf course complex in Makena (a region of Honuaʻula) that would put modernity on the doorstep of Oneloa. More alarming still, from a local point of view, the development plans included closure of the old Makena Road in order to privatize the area between the hotel and the sandy beach for exclusive use by hotel guests. Erasure of the road in this manner would not only scrub out the tracks of the ancestors; it would inhibit contemporary Native Hawaiians' access to the shore in that area. Fighting this double offense, Naone Hall, her husband Isaac Hall, and several other concerned locals, including Leslie Kuloloio, formed Hui Alanui o Makena (Group Caring for the Makena Road). After protracted struggle, the hui reached a settlement with the developer. It was not an outright victory, for the hotel was built and remains there today. But as a result of the settlement the hotel was pushed inland, off the Alanui o Makena. The alanui was repurposed, and its current state bespeaks the uneasy compromises that make for limited protection of Hawaiian pasts today: it is a groomed path through hotel grounds, complete with a "comfort station" for visitors. As Naone Hall asks, "A comfort to whom?"[15]

The settlement had another offshoot, one that gave life to Moʻoloa. The Seibu Corporation deeded three acres of prime land to a nonprofit organization to be established by Hui Alanui o Makena in the settlement agreement. Members of the group were too busy to do much with the land itself for years. It was still Kū's season for Naone Hall. She would soon fight the battle over Honokahua mentioned above and then go on to years of burial council work. Other members of the hui were likewise consumed with struggles to mālama the land, water, and culture of Maui. While these struggles have not subsided, some real victories have been won along the way. In a limited sense, one might say that by end of the first decade of the twenty-first century, the season of Lono had returned for some of these veteran activists, whose attention now turned to Moʻoloa.

Immediately adjacent to legendary Oneloa, Moʻoloa was hardly pristine in the 1980s. The land deeded to the hui had received little attention for decades. More recently, it had become a landfill for the developer. Lono faced a challenge. But Naone Hall and others summoned a vision of tranquility and education for Moʻoloa. With little more than will power and volunteer sweat, they nurtured the site, removing buried asphalt and concrete, invasive species, and other detritus, revealing the volcanic cinder layer just below the surface. Then they began the work of *kanu* (planting), which continues today. Moʻoloa is now becoming a veritable garden. Kukui trees, various palms, *ki*, banana, *lilikoi*, and a range of indigenous shrubs and ground cover have taken the place of *kiawe* trees and thorny, nonnative plants. The hui has made Moʻoloa available to local Hawaiian groups with various missions (language immersion and cultural studies, for example) and even recreational groups—very much in the spirit of Lono—so long as the visitors lend a hand, planting here, watering there, pulling a weed on the way out. The results of such mālama are palpable. For my own part, I have been privileged to spend a number of days at Moʻoloa over the past several years. Anxious by nature, I am relieved of this burden when tending the ki cuttings and watering trees at Moʻoloa. My son Hayden feels the same way, as do most of the *malahini* (guests) I have spoken with. Most of all, I see that the spirit of Lono has found a contemporary home when I see the faces of Dana Naone Hall, her husband Isaac, and their dog Moʻo when they arrive at their beloved Moʻoloa.

I am not suggesting that there is a direct correlation between Moʻoloa and Newark, of course. My point is to suggest how some patterns of mālama found in Hawaiʻi might have some loose relevance for Newark and, more broadly, for thinking about care of challenged sites in general. Beyond a feel-good story about recovered dignity at a depressed site, one point of my version of the story of Moʻoloa has been to suggest one way that indigenous and activist communities can promote rejuvenation at specific locations as a means to mālama broader regions. Mālama at Moʻoloa shares in, harnesses, and feeds the broader mana of Honuaʻula. Born out of a compromise settlement, it stands as a clear victory of mālama spirit today.

Memory Takes Work: Miloliʻi

My hands grew tired fairly quickly as we worked our way through the big pile of *lauhala* (pandanus leaves) in Halealoha Ayau's driveway outside Hilo. He and I were flattening and rolling the leaves into strips that would then be used for weaving burial baskets. *Poʻo* (head) of Hui Mālama I Na Kupuna o Hawaiʻi Nei, a tremendously active and successful repatriation organization, Ayau was preparing for a major event that would garner considerable media attention.[16] After twenty years of negotiations, he had secured repatriation agreements from several prominent British museums, including the London Museum of Natural Science. In the scope of British resistance to repatriation globally, this represents quite a victory. With aid from various agencies, including the Office of Hawaiian Affairs, Ayau would be traveling to England in 2011 and 2012 in order to reclaim hundreds of *iwi kūpuna* (ancestral bones) and bring them home. Placement in proper burial containers would be a sign to the bones that they were back in the care of their own people. Thus it was that our mundane task was to bring about a hopeful future. I told Ayau that I was glad to lend a hand and happy that he let me. "No problem," he said, "and, besides, this isn't a kapu activity." In other words, preparing the lauhala was not properly ritualistic, so an uninitiated person could participate. Then he said something that clued me in to the difficulties of unmarked cultural labor. "Kainani [his wife] and I have held a number of lauhala workshops to help get this done, but few people show up. But they'll come for the reburial ceremony when the kūpuna come home."[17]

This comment put the proverbial finger on something I had been thinking about quite a lot: how to give adequate theoretical attention to the uncelebrated labor that goes into memory work. I have written about Ayau elsewhere, with special attention to the ways he articulates and embodies "living tradition."[18] Now he was pushing me to think afresh about the hard work of making tradition live. Like others, I admire Ayau for his charisma and success, and my written work had focused on just this aspect of his persona. Now I wish to draw attention to moments like lauhala preparation and one of its modern parallels, grant writing. Before the ancestors can be buried, they require a basket, which requires flattened lauhala. Before Ayau can repatriate or protect burials, he writes grant applications tirelessly. The work he does is expensive and time consuming and thus requires resources beyond what he can subsidize personally. The more I have come to appreciate this and the fact that grant writing consumes so much of his time, the more I have become frustrated with media, popular, and academic accounts (even my own) of Ayau and activists like him.

Ironically, Ayau's detractors occasionally note his grant-writing success, implying that he uses Hawaiian causes to fill his own bank account.[19] This imbalance of reporting and appreciation strikes me as unfair and, for our purposes, analytically impoverished. Ayau's work has helped to restore ancestral fishponds, rebuild *heiaus* (temples), and protect threatened rock art sites and numerous graves and has led to the repatriation of hundreds of *iwi kūpuna*. All of this and more—including the following story about a community memorial—would not have been possible if Ayau had not spent hours upon hours writing grant applications. For more than twenty years he has performed this task, writing state and federal agencies for support. Beyond writing grant proposals, he has participated in grant review processes and grant administration. Many years he spends two weeks in Washington, DC, evaluating grants for the federal Administration for Native Americans. This has connected Ayau to native peoples and projects far and wide and has kept his grant-writing skills sharp. What follows is a brief account of one of his recent grant-based projects.

On July 25, 2010, Ayau and I drove south from his home in Hilo, through Hawai'i Volcanoes National Park and the Ka'u district, toward the southern edge of the Kona side of Hawai'i Island. Our destination was "the last traditional fishing village in Hawai'i." A small and impoverished community, Miloli'i was made famous by a video recorded there in the mid-1990s by the famous Hawaiian singer IZ (Israel Kamakawiwo'ole).[20] It featured shots of the village and of locals singing along with his songs and his renditions of Hawaiian folk songs dating back to ballads written by Queen Lili'uokalani. Those shots have taken on a metonymic function, standing as a symbol of a traditional community not yet wholly lost to the forces of capitalism. This identity as the "last traditional fishing village" has been embraced locally; at least a sign at the town park declares as much. But how does the "last" village last? How does it sustain its identity in the face of forces that have made it the last of its kind? These forces are bearing down hard. Kids in the village seem uninterested in learning traditional *opelu* (mackerel) fishing methods, which are time consuming and depend upon delayed gratification because the fish must be trained to come to a certain area. In place of delayed gratification, alarming numbers of village youth have turned to drugs, and of these the harshest kind, especially "ice" (crystal meth). In response to this crisis, a serious question is being taken up by parents and the elders: How do we survive the loss of our children? It is a classic tragic question. One answer has been to look back to the past.

Specifically, the community has rallied around commemorating five village fishermen whose boat sunk in a storm in 1949. It was a huge blow to the small community to lose six able-bodied adults in one moment and has left its mark since. In terms of social memory, one prevalent narrative about the tragedy concerns the costs of modernity. The community had pooled resources to buy its first modern fishing boat, and this group of men was bringing it to Miloli'i on its maiden voyage when it sank. The accident pulled the promise of modernity out from under them in several respects, not least of which was the confidence to attempt similar endeavors in the future. By most accounts, the community has been a shadow of its former self ever since. Some years ago a respected and successful fisherman from the village, Walter Paulo, decided to address the weight of social memory directly. He proposed bringing ritual conclusion to the event by properly marking the occasion through a community-sponsored memorial. His idea inspired others in the community but was slow to take hold at the level of concrete action.

This is when Halealoha Ayau became involved. Upon hearing about Paulo's vision, he made contact with him, proposed his idea for writing the project into a grant proposal to the Administration for Native Americans, and did so. Ayau's efforts were rewarded, and soon the community had the resources to implement Walter Paulo's vision.[21] This grant was the difference between and an idea and its realization. That is why we were there. Ayau was coordinating a community meeting about the memorial, which he and his crew would build the following week. The design was for two stone pillars, a large slab of stone for names and commemorative comments, and a thatch roof. One aspect of the grant project was to teach local youth traditional skills for building with *pōhaku* (stone). In the course of preparing for construction, it had come to Ayau's attention that some members of the community felt that other deceased members of the community should be likewise honored. Therefore, Ayau proposed that the stone bear two sets of names: those of the fishermen on one side, and those of other deceased members of the community on the other, with special attention to those who have no proper memorialization in the cemetery. The purpose of the meeting we attended was to get a list of names for this second component. It was fascinating to behold. Slow to speak at first, various community members began to perform memory work together, thinking aloud about former residents, lost connections, grappling here and there about dates and name spellings. Two hours later Ayau had an agreed-upon list. A blank slate took on history. It is as substantial as stone, thanks to Ayau's hard days quarrying grant awards.

My reason for including the story of Miloliʻi here is to draw attention to frequently underappreciated forms of mālama, especially starkly bureaucratic ones like grant writing, petitioning state and federal agencies for action, or working with UNESCO for heritage site designation. Visiting Newark and listening to stories there, I heard plenty to indicate that many people on the ground are involved in "invisible" or "offstage" mālama, forms of care that are given out of public view, with little audience, and frequently no appreciation. These kinds of tasks are the lifeblood of serious, abiding mālama. Whether by way of Dana Naone Hall attending meeting after meeting to make sure the State of Hawaiʻi administers its burial law with integrity or Halealoha Ayau sitting down to write another grant proposal, mālama at this level is keeping Hawaiian culture vital. At Newark it is clear that many communities share in similar forms of unseen mālama labor, including indigenous activists, historians, archeologists, and others. My hope is that this labor as a whole gets due recognition and, moreover, that the diverse interests groups performing this labor see and respect the efforts of others, even when sometimes they appear to be at cross-purposes with their own agendas.

Conclusion: A Polynesian Newark?

Care and neglect. These are fundamental to the human condition, whether at the level of the individual or at the level of culture. Neglect, in particular, seems to thrive in spaces between cultures, in gaps between claims to histories. Heritage site protection is manifestly about assuaging neglect, but according to whose terms and values? Care for the physicality of the past is differently imagined, diversely practiced, and often contentious across lines of solidarity, where these are ethnic, regional, or otherwise construed. This is not least because different groups bring forward divergent claims to ownership (legal or moral) of sites, differing interpretations of history, and highly variable comfort levels with relationships of narrative to place, especially with claims to continuity and relationships over time. Absence of care, then, can set in even when—or precisely because—multiple groups lay claim to a site. Different expressions of care may become mutually canceling or so muddled in practice as to leave the appearance of neglect. A cacophony of agendas is not a recipe for care. And yet seldom does a monolithic or unified vision of care emerge in settings fraught with the politics of the past. The question then becomes, What are some actionable pathways to care? In Hawaiian terms, How and under what conditions can mālama prevail?

I am of the opinion that no general answer will be found to this question, though we can aspire collectively to catalog and model success stories when we see them, in hopes that some roughly translatable principles and practices emerge over time. The three cases of mālama in action I have sketched may offer some inspiration in this direction. None of them mirror conditions at Newark, and I am quite aware that Hawaiʻi and Ohio are rather different places. Beyond the obvious difference in physical contexts—one archipelagic and the other midcontinental—among the discrepancies one might note is the stark difference in the visibility of Native people in the respective places. As a scholar of comparative religion, I am not put off by these differences. I find them "good to think," to invoke Levi-Strauss's classic quip. Such manifest incongruities relieve me of the burden of forcing direct corollaries in my analyses. My approach has been oblique, aiming for some modest insight about Newark framed in terms of several Hawaiian places and the people who care for them. The spirit of this volume, as I understand it, aims in just this direction: to shed light in unexpected ways, to unsettle thoughts about Newark with a little travel elsewhere.

At the very broadest level, I wish to emphasize the following point in closing. For scientists, land managers, administrators, and politicians, I would offer encouragement not to be afraid of living indigenous traditions as their practitioners continue to emerge. In Hawai'i, the most vibrant sites are those that receive local mālama. This has meant such radical things as Hawaiians performing rituals at sites, engaging them as they would have been utilized centuries ago.[22] Sometimes this process involves modest alteration of sites, if only through human presence upon them. Occasionally some Hawaiians have engaged or made claims upon sites in ways that appear "invented" by scholars and other observers.[23] This dynamic has been in play at Newark and seems likely to remain relevant. Thorny territory this, but one response is simply to ask what it means in terms of mālama. If such groups and their practices bring attention and care to a site, then what is the real cost? More radically, another response is to say: Show me uninvented traditions, and we can talk. In any case, my point to various stakeholders is simply to forestall fear and criticism of ways of engaging the sites that do not map directly onto archeological knowledge or "preservationist" agendas. The benefits of lived engagement with sites will redound to all—a pulse will be felt, interest will be generated, and care will follow.

Notes

1. Blackford, "Environmental Justice, Native Rights, Tourism, and Opposition to Military Control."
2. On social memory theory, see, for example, Huyssen, *Present Pasts.*
3. Concerning the colonization of Hawai'i, see, for example, Merry, *Colonizing Hawai'i;* Kauanui, *Hawaiian Blood.*
4. See, for example, Silva, *Aloha Betrayed.*
5. See, for example, Finney, *Sailing in the Wake of the Ancestors.*
6. Naone Hall, "Sovereign Ground."
7. Naone Hall, *Mālama.*
8. For more on Naone Hall's role at Honokahua, see G. Johnson, *Sacred Claims.*
9. G. Johnson, "Varieties of Native Hawaiian Establishment."
10. 25 USC 3001, 1990. On NAGPRA generally, see, for example, Fine-Dare, *Grave Injustice.*

11. United Nations Declaration on the Rights of Indigenous Peoples, http://www.un.org/esa/socdev/unpfii/documents/DRIPS_en.pdf.

12. See Naone Hall's poem "Signs," in *Mālama.*

13. For a discussion of Kū sensibilities in contemporary settings, see Kame'eleihiwa, *Native Land and Foreign Desires.*

14. On *akua,* Malo, *Hawaiian Antiquities,* and Valeri, *Kingship and Sacrifice.*

15. Naone Hall, *Mālama,* 148.

16. On Hui Mālama generally, see Ayau and Tengan, "Ka Huaka'i O Na 'Oiwi"; Kunani, "Stone by Stone, Bone by Bone."

17. Ayau, personal communication, July, 24, 2010.

18. Johnson, *Sacred Claims.*

19. Burl Burlingame, "Group Picked to Bring Remains Instead Gives 205 Sets to State," *Honolulu Star-Bulletin,* Dec. 30, 2000, http://archives.starbulletin.com/2000/12/30/news/story2.html.

20. Israel Kamakawiwo'ole, *The Man and His Music* (DVD) (Honolulu: Mountain Apple, 1995).

21. The three-year, $400,000 grant was awarded to an organization headed by Ayau, Hui Ho'oniho.

22. See Tengan, *Native Men Remade.*

23. For a discussion of "invention" theories and their relevance to contemporary Hawai'i, see G. Johnson, "Authenticity, Invention, Articulation."

WINNIFRED FALLERS SULLIVAN

Imagining "Law-Stuff" at the Newark Earthworks

For those who primarily study the earthbound works of people living today or in the relatively recent past through their written words, the Newark Earthworks initially presents an undeniably awesome but frustratingly silent landscape. Who were these workers of the earth who looked to the skies? How might we conjure their lives? How might we, like our nineteenth-century spiritualist ancestors, make the dead speak and tell us their secrets? What does it mean for our work when all of the words are supplied by us, not by them? Who can speak for them?

As one reads the work of archaeologists and astronomers, views efforts to reconstruct an understanding of the earthworks in their own time, and listens to current would-be representatives of those builders, a possible interlocutor emerges out of the mists of time—and the miracle of contemporary video technology—to populate our imagination. We are invited by the astronomers to stand on Observatory Mound in the avatar of a native inhabitant and imagine ourselves in another world—some two millennia ago—a world that is visible to us today principally in the traces of massive earthen constructions with their seductively precise astronomical alignments.[1] We are led to see this iconic figure, the shaman, as the natural denizen of the space and the natural leader of his people.[2]

A local Ohio reporter describes a recent visit to the Great Circle: "On this mound, a Native American shaman might have stood above drums drumming and rattles rattling and hundreds of people standing in the earthen circle below, all waiting for the moon to rise beyond the circle, beyond the adjoining octagonal mound, ascending above the horizon."[3] We were introduced to the shaman first at the Octagon, but now he has moved to the Great Circle—as though he is the only individuated human who can be imagined being in this place.[4] Newark. Hopewell. How little any of these names tell us.

The seemingly neutral geometric and local contemporary designations are at once simply descriptive and confoundingly opaque.

The opacity seems to lead irresistibly to the sacred—another word that stands in for the limits of our knowledge and our experience. We are invited to think of these spaces as mysterious and sacred, places that were once, and perhaps still are or should be, dedicated to burial or worship or pilgrimage or heavenly contemplation or communication with the divine. The plea is that they now deserve protection because of their enduring sacred value, the natural wonder of these spaces, their connection to historic tribes or local and national history, or, perhaps, their link to a worldwide cultural heritage—now threatened. But often it seems it is simply because we moderns are so desperately in need of the sacred.

There is by now a rich literature analyzing the seemingly endless appropriation of Indian things for majority purposes.[5] One strand of this history of exploitation is the figuring of the Indian as religious. While sometimes that figuration has been for the purposes of rejecting Indian religious ways as idolatrous, it has also been for the purpose of parasitically supplying a missing piece. The notion is that a sense of the sacred is something precious that we lack, something that Indians have and that we can get from them. Next to what Mesoamerican scholar Kay Read calls the noble and the nasty Indian in the imagination of non-Indians is the shaman—the holy Indian.

Peering through the mist of time and mystification, who else might we see? Are these places best thought of as religious places, as "ceremonial centers," as places of pilgrimage? If we stand with the members of the imagined assembled crowd, how can we imagine the work that they did? How did they live and govern their lives? What role did law play in this place?

Law Then

Native American legal studies have flourished in the last few decades. The rich resources of oral history, ethnography, and archival research, along with an increasing number of legally trained Native scholars, are yielding an ever more complex understanding of the legal lives of preconquest peoples and the legal borrowings, impositions, and transformations that have occurred in the subsequent half millennium. The Newark Earthworks are a part of this story.

It is not a simple task. How do we move past the figure of the shaman to the kinds of events and transactions that would be of interest to a legal scholar? Lacking archives and with a highly attenuated oral history, the political and legal ways of those who lived in much earlier times seem still out of reach —just beyond our view.

I begin from the assumption that where there are people, there is law and legal jurisdiction, not just because I am a lawyer and lawyers are naturally hegemonic in their inclinations, although that is certainly the case, or because we live in a peculiarly law-obsessed time and place, although that is also the case, but because noticing law tells us something important about human societies. Human life is always everywhere subject to rules and governed by those who know how to make and break them. All human societies we know about have had rules about marriage, burial, trade, governance, food, and crime, among many other matters—including religion—and they have all had ways of settling disputes that inevitably arise. As Robert Cover says, "We inhabit a *nomos*—a normative universe. We constantly create and maintain a world of right and wrong, of lawful and unlawful, of valid and void."[6] And by "we," he means all humans. Cover also believed that "no set of legal institutions or prescriptions exists apart from the narratives that locate it and give it meaning."[7] Law is universal, and law is universally socially and culturally embedded. Everywhere there is what Karl Llewellyn and Adamson Hoebel in *Cheyenne Way* called "law-stuff."[8]

So how do we conjure "law-stuff" in this particular place? Who might there be there to do the law jobs? And what roles might we imagine for the assembly?

These earthworks are believed to have been constructed during the Middle Woodland period by a culture we call the Hopewell, a cultural efflorescence that centered in the Ohio River valley but whose catchment extended north into present-day Canada, west into what we call the Dakotas, east to the Delaware, and south to the Gulf of Mexico. The material record suggests that those who lived here were agriculturalists and that they had trade relationships that extended all over North America and into Central America. We don't know why they built these earthworks. We know that they are not burial grounds and that they are not defensive works. We are not sure where the people who built them lived. There is evidence of settlement nearby, but not enough dwellings have been discovered for the number of people imagined to have gathered in the Great Circle. Were these places only used at certain times of the year? Are the dwellings that have been found dwellings for caretakers or religious guardians? What do the roads that connected the Newark Earthworks to others in the vicinity imply about their activities? How do we put together the picture being developed by those archaeologists who study agriculture and daily life with the picture being developed by those who study the earthworks and the archaeoastronomers?

Assembling the fragments of what one thinks one knows about any society into a picture of a whole society is always itself fragmentary and provisional. Furthermore, those who have tried to imagine the lives of Hopewell peoples, like other scholars, bring their own theories of what makes societies tick. Is life driven by economic considerations, or by power, or by reproduction? There are so many questions. But using the rich resources of legal history and legal anthropology, we can begin to make some educated guesses about how law might have worked in a society that apparently did not have an aristocratic class (we surmise that from the fact that burial mounds did not develop in Ohio until much later) but that may have had trade and other relations with peoples all over North America.

Let us go back to the Octagon and the figure of the shaman. Do we imagine him or her to have any relationship to law? If we move around the globe to Iceland at roughly the same time or just a little later, we find another such solitary figure standing before an annual assembly in a magnificent outdoor space. He is the lawspeaker of ninth-century Iceland.

Medieval Iceland was a society without "kings, counts, or monks" as William Ian Miller puts it.[9] It was a society without a state, but it was not a society without law or one without inequality and violence. Norwegian refugees settled Iceland in the mid-ninth century. Icelanders lived in dispersed households, the basic unit of social organization and legal personality, but were organized into a hierarchy of local chieftainships. They met once a year at the Thing, a national assembly that functioned as both the legislative and the judicial branch of government. (The Icelandic parliament is still called the Thing.) Provision was made in the law for care of the poor and compensation of loss from accident. There was no executive branch. Enforcement of decisions by the courts was by the litigants themselves through the elaborate protocol of the feud.

At the annual Thing, the Allthing, the lawspeaker, the only paid officer of Icelandic government at the time, recited the oral law, standing on a prominent outcropping, the Law Rock.[10] He was the repository of the law. It was his responsibility to know the law and consult with other elders if he needed to fill in bits of his knowledge or there was some ambiguity. It is said that when Iceland converted to Christianity around 1000 CE, the lawspeaker of the time, in response to the arrival of Danish missionaries, lay under his cloak for a day and a night and then rose and declared that henceforth Iceland would have one law and one religion. Until that time there had been both pagans and Christians in Iceland.

Miller describes the intensive event that was the annual Thing:

> Even though Thing meetings accounted for but three weeks of the year, those weeks were more intensely lived and more anxiously anticipated than any other random three weeks. These gatherings were a time for making and renewing acquaintances, contracting marriages, and exchanging invitations to feasts. Above all, the Things were the arenas for intensive legal action; they were the locations where successes and failures were unambiguously on display, where prestige and honor was competed for, and won and lost. Reputations, if not exactly made there, were on display there. The Thing was the one place where a large audience was assured and hence the stakes in social interactions occurring there were higher.[11]

Miller is not so interested in religion as law, so he does not mention the religious event that the Thing must have been as well.

We know about medieval Icelandic law because of the sagas and law codes that were written down four or five centuries later. The law of the Thing was a largely oral culture. The Icelanders' law was a law without writing. What we have—maybe only what we always have about times and places before we live—is echoes of their words. Theirs was a society without a state and without writing but not, importantly, a society without law. Using mostly only the sagas, Miller reconstructs how Icelanders managed their society using law, without any king or sheriffs. The community saw that the law was enforced and justice done. It is a very remarkable story.

How might these few details about Icelandic law help us to think about Hopewell law? Like the Hopewell in the Middle Woodland period, the Icelanders had no real lordship. It was not until later that burial mounds, the signature of aristocratic culture, began to be built. Like the Hopewell, the Icelanders were a pastoral and agricultural people who lived in dispersed settlements. Like the Hopewell, the Icelanders had dealings with their neighbors. They bought and sold things, exchanged gifts, settled disputes over injuries and insults. They punished miscreants. The penalty was fines or outlawry, although sometimes settlements were arranged. They were concerned with honor and prestige. And they were subtle experts at law and legal thinking. Like that of the Hopewell, perhaps, the time of this society was limited, giving way to the big men who began to get rich and build monuments to themselves. In the thirteenth century, Iceland became subject to the King of

Norway and became an increasingly hierarchical society with a larger division between rich and poor.

Since Miller's book was written two decades ago, much more has been written about the conditions and possibilities of government in stateless societies.[12] Thinking of law at the earthworks provides an opportunity to imagine how that might have been done in this place. If the earthworks were a place of periodic assembly of Hopewell peoples, it is not unlikely that it was an occasion for the settling of legal matters. We might imagine marriages arranged, contracts formed, and disputes arbitrated. And we might imagine that, like that of the Icelanders, the law that governed these events drew on understandings of the human person and society that were learned from narratives of the people's history and purposes. But the lawspeaker was not in charge of everyone. He was a resource for the law. It was his job to tell anyone who asked what the law was. There were juries and arbitrators to decide cases. Perhaps it worked something like this at the earthworks. For a time.

Does this imagined picture have any place today at the Earthworks?

Law Today

In June 2002, Barbara Crandell was arrested for criminal trespass when she entered the Moundbuilders Country Club to pray on Observatory Mound. Julie Shaw tells the story in the *Newark Advocate:*

> Considered by some Native Americans as an elder or grandmother, the 73-year-old woman of Cherokee descent regularly prays at the various mounds and earthworks in Ohio. "It's something made by native people, my people. It's a connection to my heritage, my race of people," Crandell said from her Thornville home. She has said that she went to pray at the earthworks on June 26 for the people affected by the fires in the West. She got tired and went to rest on the observatory mound, she has said. The observatory mound where she was sitting lies near the fairway to the No. 10 hole. Golfers on the 10th tee inside the earthen circle sometimes first shoot their balls toward the observatory mound, then on their next shots, try to drive the balls toward the 10th hole, which lies on the other side of the circle. According to Crandell's account, she was taunted by two golfers, then asked to leave by the country club's president, Skip Salome. Salome left and returned with two Newark police officers. After arguing with Salome and the officers, Crandell threw her cane at Salome, then was arrested and taken off the course.[13]

Crandell was convicted by a jury of criminal trespass. On appeal the court held that because the Ohio Historical Society is a private entity, her civil rights claim was not applicable, as there was no state action.[14]

Again we see the lone figure, the romantic hero of our imagination—and maybe of hers. But law is not a story of lone figures. Law is a social story. What law can we see at Observatory Mound today? Both public and private law governs her activities. There is the law of her religious tradition that compels her. Prayer is a rule-governed activity; rules determine what you say and where you say it, to whom it is addressed, whether it is silent or spoken, how your body should be arranged. There are also the rules of the golf course. And, like all places in the United States, the area encompassed by the Newark Earthworks is subject to a mind-boggling amount of government legal regulation, by the City of Newark, by Licking County, by the State of Ohio, and by the federal government, to start with. But there are underlying Indian land claims and international law haunting this place as well.[15]

Take the Octagon, where Barbara Crandell made her small act of protest, her claim to prioritize a law not of golf or of trespass but of obligation to the gods. Who owns this land? The area in which the earthwork we call the Octagon resides was given to the Ohio Historical Society in 1933 and leased by the Moundbuilders Country Club (a lease that has subsequently been extended through 2087). Modern legal title to land in this part of Ohio begins with the Northwest Ordinance and the Treaty of Greeneville that settled the Indian wars that followed the Revolution.[16] Ohio was the first of the western states carved out of the Northwest Ordinance to be surveyed and divided into homesteads. Various parcels were conveyed for various purposes. Present-day Licking County, where the Newark Earthworks are located, was part of the United States Military District, a portion of Ohio that was divided and given to continental soldiers after the war.[17] The City of Newark was founded in 1802. On what does all of this ownership rest? A murky and bloody claim to title founded in discovery and conquest, followed by treaties and purchases of dubious legality, subject to ongoing litigation.[18]

Widening the lens, we will see national and international legal regimes laying claim to jurisdiction. As others discuss in this volume, the land on which the Newark Earthworks stand is subject to federal legislation protecting Indian objects and remains, including the Native American Grave Repatriation Act. The United Nations Declaration on Rights of Indigenous Peoples announces the intentions of the world toward subject peoples. And finally the 1972 World Convention on the Protection of the World Cultural and Natural Heritage is enlisted to recognize and cultivate these sites.

How does one begin to imagine a law of the earthworks, a law particular to the earthworks that honors their history, in such a thicket of "law-stuff"? Is there a legal line to be drawn from the people of the Octagon to the people of Newark? Can we make the kind of connection that those who claim its sacred status imply? It is significant to remember that the Icelanders did not invent and maintain their law in isolation. Their law drew on the Nordic and Germanic traditions they brought with them, interacted with Christian laws, and was in ongoing and evolving negotiation with Norwegian legal demands. And yet the Icelandic parliament is still called the Thing and still traces its history to the lawspeaker and the Law Rock.

What does it mean to notice the multiplicity of law, to notice that whatever story it might tell itself, or that political theologians might tell, the state does not have a monopoly on violence or law. One version of legal pluralism, like religious pluralism, is the simple observation that there is a lot of "law-stuff," just like there is a lot of religion stuff—all jostling up against each other. Global legal history is being retold with these ideas in mind. The best new work on legal multiplicity shows how law is continually made.[19]

Indigenous legal studies are flourishing today around the world. Some of the most interesting of this work argues that sovereignty is not a zero-sum game. Rejecting the notion in political theory that sovereignty means absolute control over a particular territory, these writers are making the case that you cannot understand law with these concepts alone.[20] Building on critiques of sovereignty by scholars in both indigenous legal studies and others, Justin Richland, like Miller, shows us how the Hopi, without a state, make room for their law in a context of overlapping jurisdiction.[21]

Richland recounts the efforts of the Bureau of Indian Affairs in the 1930s to persuade the Hopi that they should reorganize as a "tribe," a designation that made no sense in the Hopi sociopolitical order because Hopi villages were self-governing and autonomous, and even within villages power was significantly dispersed. In Richland's telling, Oliver La Farge, representing the federal government, was eventually successful at overcoming Hopi resistance but only after significantly adapting his account to Hopi ways of government. Richland explains:

> Thus it was that La Farge faced not only the decentralization of Hopi communities into several villages, but the thoroughly ambiguous dispersal and diffusion of social power within each Hopi village into various clans, ritual sodalities, and their overlapping authorities when he first came to the

> reservation to proffer the BIA's idea of a single Hopi tribe under a unified tribal government. And the representatives of the various Hopi villages made sure to remind La Farge of this every time he spoke to them. Sometimes the questions the Hopi villagers put to La Farge were direct, as during one of his first meetings in the village of Shungopavi, where he reports that he was asked, "Was I sure this wouldn't hurt the clan organization?" and flatly told "We are already organized."[22]

Organization can take many forms.

Richland suggests that, rather than focusing on the limits to Hopi sovereignty, in the past or today, by focusing on "jurisdiction"—the reach of the law—one can see how a decentralized society like the Hopi maintains autochthonous legal and political ways within the limits of its sovereignty. Detailing several intravillage property disputes from recent cases before the Hopi tribal courts, Richland shows how disagreements over Hopi tradition provided the occasion for the tribal court to reinforce traditional Hopi practices with respect to the protection of traditional knowledge and how "tradition operates as a mode of jurisdiction, pointing up the limits of an emergent Hopi sovereignty and the potentiality for exceeding those limits."[23]

Drawing from these two brief legal histories—that of the Icelanders and that of the Hopis—one might begin to suggest a framework for thinking of law at the Newark Earthworks outside the constraining political ideology of the nation-state and empire. I am not suggesting that such horizontal and pluralistic ideas of law provide perfect justice but that understanding law this way allows us to see the people assembled at the Circle or the Octagon more fully. Not all societies are as law inclined as the Icelandic or the Hopi, but all the ones we know about use law to make a space for regularizing human relations, resolving disputes, and performing justice.

Conclusion

If we assume that all people have "law-stuff"—that is, that they have governments, rules that they live by, and a need for dispute resolution—we can begin to imagine some of the possibilities at the earthworks. Lacking writing, perhaps they had "lawspeakers" as the Icelandic peoples had, persons who were responsible for the oral transmission of the law. Perhaps they used the occasion of periodic pilgrimages or festivals or ceremonies to convene courts, to settle inheritance and descent matters, and to arrange marriages. Disputes might have been settled by chiefs or law specialists of some kind. Oracles may have been consulted to resolve difficult evidentiary questions. In difficult cases, ordeals may have been used. Certainly there were institutions relating to property use, trade agreements, and practices of exchange. Perhaps taxes were collected. Perhaps executions were performed. Certainly all of these practices would have depended on and reinforced what we would call religious cosmologies and anthropologies. Perhaps the calendars they saw in the sky organized legal work as well as religious work.

Perhaps over time these religious cosmologies and anthropologies changed or were refitted as political and legal practices changed. If we imagine shifts to more centrally organized political formations from more dispersed settlements, maybe the new chiefs or kings produced legal codes and began to insist that legal decisions conform to royal purposes. And if there was a hierarchical order, for example, as Steve Lekson suggests was the case in Chaco Canyon,[24] perhaps there were sumptuary laws as in the Aztec capitals, laws that regulated what commoners could wear. But everything we know about such changes suggests that there was ongoing pushback and the continuing presence of legal pluralism even in these kingdoms and empires.

The modern effort to separate law and religion has turned out not to work very well. As Bruno Latour might say, we have never been modern.[25] Separation has only produced hybrid forms.

We know that there is a lot of law that does not fit the statist definition of law: customary law of various kinds, transnational law, religious law. But we find it hard to imagine law without the apparatus of the modern state. Why would people obey law that is not backed up by the violence of the modern state? Where does law come from if not from the modern legislature?

Newark invites us to think in the gaps of the American legal imagination. And if one of the purposes of preserving these places is to educate our children about other ways of living, perhaps we can see a lawspeaker beckoning to them as well as the shaman. Imagining the people who built these remarkable places as immune from law does not serve them or us. It reproduces a postcolonial politics in which native peoples are idealized and infantilized. Why should we not imagine them as having a sophisticated law to match their sophisticated astronomy and building capacity? Why should we not imagine a law of Newark that was not produced by the UN or the US government?

Notes

I would like to thank Kay Read and Kathleen Self for teaching me some things about how to think about oral societies.

1. See Hively and Horn, "Hopewell Cosmography at Newark and Chillicothe, Ohio."

2. Center for the Electronic Reconstruction of Historical and Archaeological Sites, *EarthWorks*. Thanks to John Hancock for providing me with a copy of this CD-ROM. See also Hancock, "The Earthworks Hermeneutically Considered."

3. Julie Shaw, "Mounds Home to Varying Opinions: Earthworks the Site of a Curious Dilemma," *Newark Advocate Reporter*, Aug. 11, 2002, http://www.newarkadvocate.com/news/stories/20020811/topstories/409718.html.

4. For a recent exploration of the problematic figure of the shaman, see Kehoe, *Shamans and Religion*.

5. M. Brown, *Who Owns Native Culture?*

6. Cover, "The Supreme Court, 1982 Term—Foreword," 4.

7. Ibid.

8. Llewellyn and Hoebel, *The Cheyenne Way*, 42.

9. William Ian Miller, *Bloodtaking and Peacemaking: Feud, Law, and Society in Saga Iceland* (Chicago: University of Chicago Press, 1990).

10. William R. Short, "Viking-Age Laws Legal Procedures," http://www.hurstwic.org/history/articles/society/text/laws.htm.

11. Miller, *Bloodtaking and Peacemaking*, 21.

12. See, for example, James C. Scott, *The Art of Not Being Governed: An Anarchist History of Upland Southeast Asia* (New Haven, CT: Yale University Press, 2009).

13. Shaw, "Mounds Home to Varying Opinions."

14. *State v. Crandell*, 2003-Ohio-2512.

15. For a marvelous introduction to the backstories of Indian law and federal Indian law, see Goldberg, Washburn, and Frickey, *Indian Law Stories*.

16. Knepper, *The Official Ohio Lands Book*, is a thorough account of the history of the appropriation, surveying, and granting of land in Ohio.

17. Ibid.

18. R. Williams, *The American Indian in Western Legal Thought*.

19. See, for example, Benton and Ross, *Legal Pluralism and Empires, 1500–1850*.

20. See, for example, Bruyneel, *The Third Space of Sovereignty*. Frank and Goldberg, *Defying the Odds*, uses a dual concept of sovereignty, political sovereignty, and cultural sovereignty to describe the Tule River Tribe's autonomy.

21. Justin B. Richland, "Hopi Tradition as Jurisdiction: On the Potentializing Limits of Hopi Sovereignty," *Law & Social Inquiry* 36 (2011): 201–34. Richland cites Bradin Cormack's important book *A Power to Do Justice*.

22. Richland, "Hopi Tradition as Jurisdiction," 14.

23. Ibid., 220.

24. Lekson, *A History of the Ancient Southwest*.

25. Latour, *We Have Never Been Modern*.

APPENDIX

What Is So Special about the Newark Earthworks?

Fifteen Viable Replies

RICHARD D. SHIELS (historian of American religion): The two-thousand-year-old Newark Earthworks, despite a convoluted history of neglect and re-purposing during the past two hundred years, provide (among other things) a uniquely evocative educational resource with which to reconceptualize the history of Ohio and America, which has roots far deeper than Euro-American occupation.

BRADLEY T. LEPPER (archaeologist, Hopewell specialist): The Newark Earthworks were, during the Hopewell era, a pilgrimage destination with enormous drawing power, evoking a vast ritual context, a "ceremonial machine" or "monumental engine of world renewal" in which to orchestrate highly choreographed Hopewellian mortuary rituals, the goal of which "may have been nothing less than the regeneration of the Earth."

RAY HIVELY (astrophysicist) and ROBERT HORN (philosopher): The Newark Earthworks provide a uniquely elaborate instance of archaeoastronomical planning wherein enormous and stunningly sophisticated geometrical configurations track the movements of celestial bodies, most notably an 18.6-year lunar cycle, as well as integrating the built features with the local topography of streams, valleys, and "hilltop observing stations."

HELAINE SILVERMAN (archaeologist, Andean specialist): The Newark Earthworks, like the Nazca Lines in Peru, were, during the Hopewell era, a highly venerated pilgrimage destination, a heterotopic sacred site, designed in ways that reflect decidedly religious priorities, a place that would have impressed ancient visitors as exceptional in the extreme, an awe-inspiring contrast to the broader landscapes within which they lived their ordinary lives.

STEPHEN H. LEKSON (archaeologist, Southwest specialist): The Newark Earthworks, like Chaco Canyon, is a site much revered for its supposed conception as a "sacred place" and a center for "rituality," which may, however (again like Chaco), actually owe its impressive scale and configuration less to ostensibly "spiritual" priorities than to more to prosaic political and economic factors (i.e., the exceptional religiosity of both sites may have been much overestimated).

TIMOTHY DARVILL (British archaeologist): The Newark Earthworks are an underappreciated site that deserves credit as a top-tier ceremonial center insofar as they provide outstanding exemplification of common themes such as geometrical precision, linkages of the living with the dead, and "cosmological structuring" that one observes at much more high-profile ancient centers such as Stonehenge, China's Temple of Heaven, and Çatalhöyük in Turkey.

JOHN E. HANCOCK (architect and architectural historian): The Newark Earthworks persist as continuously meaningful "works of architecture" (rather than simply "archaeological *sites* that hold knowledge about distant cultures"), which contemporary audiences can best appreciate through the sorts of digital technologies and "augmented reality" utilized on the dynamic web-based guide to the earthworks known as the "Ancient Ohio Trail."

THOMAS BARRIE (architect and architectural historian): The Newark Earthworks were—and remain—a "liminal place" or "place of mediation" at which contemporary visitors, not unlike their ancient Hopewellian counterparts, are afforded a special opportunity to reflect upon such profound matters as "the ontological significance of home," the ongoing precarities of the natural environment and the inescapable "mystery of death."

MARGARET WICKENS PEARCE (geographer, indigenous cartography specialist): The Newark Earthworks exemplify characteristically Indigenous mapping practices, which (unlike Western cartography) are "process-oriented," "dispersed" and "embodied" so that visitors—either in the ancient past or in the future—are themselves "mapmakers" who may contribute to a continually regenerative mapping process that was begun by earlier generations of Hopewell surveyors.

THOMAS S. BREMER (American religionist, tourism and travel specialist): The Newark Earthworks endure as both a site of "modern religiosity" and a tourist attraction insofar as the site remains an appealing travel destination to which highly reverent visitors (i.e., antitourists of sorts) undertake meaning-making journeys, which nonetheless include the characteristically touristic practices of "aestheticizing," "commodifying" and "ritualizing."

MARTI L. CHAATSMITH (sociologist and Indian scholar): The Newark Earthworks enable a fortuitous sort of two-way exchange: on the one hand, increasing appreciation of the Ohio mounds serves as a resource for contemporary native peoples, especially those with some historical connection to the region, to retrieve a sense of their own history; on the other hand, Indian stakeholders increasingly emerge as resources for the preservation and thoughtful management of the Ohio mounds.

MARY N. MACDONALD (historian of religions, Australian specialist): The Newark Earthworks—the contested status of which reflects the broader ongoing processes of colonialism—not only "belong *in a special way*" to all American Indians, an affiliation that ought to be recognized and respected, but, moreover, to the indigenous peoples of the entire world, and thus management of the site ought to be informed also by the United Nations Declaration on the Rights of Indigenous Peoples.

DUANE CHAMPAGNE (sociologist and American Indian studies specialist) and CAROLE GOLDBERG (lawyer and scholar of law): The Newark Earthworks were, in the Hopewell era, a major pilgrimage destination, which exemplified a native land ethic of "stewardship" wherein many groups felt both special investments and special responsibilities in connection to the place, but none claimed exclusive ownership, a nonhegemonic and cooperative model that should inform the future management of the site.

GREG JOHNSON (religionist, North America and Hawai'i specialist): The Newark Earthworks presently stand as a "depressed cultural site" insofar as the mounds occasion admiration, celebration, and even adulation, but most of all "concern"—that is, a sense of distress, which evokes a corrective sensibility like that which Native Hawai'ians call *mālama* (care), which entails both legal and extralegal strategies of site reclamation and management.

WINNIFRED FALLERS SULLIVAN (historian of religions and law): The Newark Earthworks provide a context in which to reconsider our often-romanticized imaginings of native peoples by giving serious consideration to the possibility that this place was designed less to facilitate ritual engagements with the divine or perhaps the honored dead than as a forum in which to undertake a distinctive sort of "law-stuff," that is, "a space for regularizing human relations, resolving disputes, and performing justice."

BIBLIOGRAPHY

Abrams, Elliot M. "Hopewell Archaeology: A View from the Northern Woodlands." *Journal of Archaeological Research* 17, no. 2 (Jan. 2009): 169–204.

Adler, Judith. "Origins of Sightseeing." In *Travel Culture: Essays on What Makes Us Go,* edited by Carol Traynor Williams. Westport, CT: Praeger, 1998.

Agarwal, Sabrina C., and Bonnie A. Glencross. *Social Bioarchaeology.* Chichester, West Sussex, UK: Wiley-Blackwell, 2011.

Albarella, Umberto, and Sebastian Payne. "Neolithic Pigs from Durrington Walls, Wiltshire, England: A Biometrical Database." *Journal of Archaeological Science* 32 (2005): 589–99.

Allen, Chadwick. "Introduction: Locating the Society of American Indians." *Studies in American Indian Literatures* 25, no. 2 (Summer 2013): 3–22.

Asad, Talal. *Genealogies of Religion: Discipline and Reasons of Power in Christianity and Islam.* Baltimore: Johns Hopkins University Press, 1993.

Ashbee, Paul. *The Bronze Age Round Barrow in Britain: An Introduction to the Study of the Funerary Practice and Culture of the British and Irish Single-Grave People of the Second Millennium B.C.* London: Phoenix House, 1960.

Atwater, Caleb. "Description of the Antiquities Discovered in the State of Ohio and Other Western States." *Archaeologia Americana* 1 (1820): 105–267.

Aveni, Anthony F., ed. *Foundations of New World Cultural Astronomy: A Reader with Commentary.* Boulder: University Press of Colorado, 2008.

———. "The Nazca Lines: Patterns in the Desert." *Archaeology* 39, no. 4 (1986): 32–39.

———. "Order in the Nazca Lines." In *The Lines of Nazca,* edited by Anthony F. Aveni, 41–113. Philadelphia: American Philosophical Society, 1990.

Ayau, Edward Halealoha, and Ty Kāwika Tengan. "Ka Huakaʻi O Na ʻOiwi: The Journey Home." In *The Dead and Their Possessions: Repatriation, in Principle, Policy, and Practice,* edited by Cressida Fforde, Jane Hubert, and Paul Turnbull, 171–89. New York: Routledge, 2002.

Bachelard, Gaston. *The Poetics of Space.* Boston: Beacon, 1994.

Barrett, John C. "Chronologies of Landscape." In *The Archaeology and Anthropology of Landscape,* edited by Peter J. Ucko and Robert Layton. London: Routledge, 1999, 21–30.

Barrett, John C., and Kathryn J. Fewster. "Stonehenge: *Is* the Medium the Message?" *Antiquity* 72 (1998): 847–52.

Barrie, Thomas. *The Sacred In-Between: The Mediating Roles of Architecture.* Abingdon, Oxon, UK: Routledge, 2010.

Basso, Keith H. *Wisdom Sits in Places: Landscape and Language among the Western Apache.* Albuquerque: University of New Mexico Press, 1996.

Bell, Catherine. *Ritual Theory, Ritual Practice.* New York: Oxford University Press, 1992.

Bender, Herman E. "Medicine Wheels or 'Calendar Sites.'" *Time and Mind* 1, no. 2 (2008): 195–206.

Benton, Lauren, and Richard Ross, eds. *Legal Pluralism and Empires, 1500–1850.* New York: New York University Press, 2013.

Bernardini, Wesley. "Hopewell Geometric Earthworks: A Case Study in the Referential and Experiential Meaning of Monuments." *Journal of Anthropological Archaeology* 23 (2004): 331–56.

Bernhardt, Jack E. "A Preliminary Survey of Middle Woodland Prehistory in Licking County, Ohio." *Pennsylvania Archaeologist* 46, no. 1–2 (1976): 39–54.

Bertemes, François, Peter F. Biehl, Andreas Northe, and Olaf Schröder. "Die neolithische Kreisgrabenanlage von Goseck, Ldkr. Weißenfels." *Archäologie in Sachsen-Anhalt,* 2 (2004): 137–45.

Blackford, Mansel. "Environmental Justice, Native Rights, Tourism, and Opposition to Military Control: The Case of Kaho'olawe." *Journal of American History* 91, no. 2 (2004): 544–71.

Bolnick, Deborah A. "The Genetic Prehistory of Eastern North America: Evidence from Ancient and Modern DNA." PhD diss., University of California, Davis, 2005.

Bradley, Richard. *Altering the Earth: The Origins of Monuments in Britain and Continental Europe.* Edinburgh: Society of Antiquaries of Scotland, 1993.

———. *The Past in Prehistoric Societies.* London: Routledge, 2002 .

———, ed. "Sacred Geography." Special issue, *World Archaeology* 28, no. 2 (1996).

Bremer, Thomas S. "A Touristic Spirit in Places of Religion." In *Faith in America,* vol. 2, *Religious Issues Today,* edited by Charles H. Lippy, 37–57. Westport, CT: Praeger, 2006.

Brine, Lindesay. *Travels amongst American Indians: Their Ancient Earthworks and Temples: Including a Journey in Guatemala, Mexico and Yucatan, and a Visit to the Ruins of Patinamit, Utatlan, Palenque and Uxmal.* London: S. Low, Marston, 1894.

Brophy, Kenny. "Water Coincidence? Cursus Monuments and Rivers." In *Neolithic Orkney in Its European Context,* edited by Anna Ritchie. Cambridge: McDonald Institute for Archaeological Research, 2000.

Brose, David S., and N'omi Greber, eds. *Hopewell Archaeology: The Chillicothe Conference.* MCJA Special Paper Number 3. Kent, OH: Kent State University Press, 1979.

Brown, Brian Edward. *Religion, Law, and the Land: Native Americans and Judicial Interpretations of the Land.* Westport, CT: Greenwood, 1999.

Brown, Michael F. *Who Owns Native Culture?* Cambridge, MA: Harvard University Press, 2004.

Bruyneel, Kevin. *The Third Space of Sovereignty: The Postcolonial Politics of U.S.-Indigenous Relations.* Minneapolis: University of Minnesota Press, 2007.

Buikstra, Jane E., and Lane A. Beck. *Bioarchaeology: The Contextual Analysis of Human Remains.* Amsterdam: Academic, 2006.

Buikstra, Jane E., and Douglas Charles. "Centering the Ancestors: Cemeteries, Mounds and Sacred Landscapes of the Ancient North American Midcontinent." In *Archaeologies of Landscape: Contemporary Perspectives,* edited by Wendy Ashmore and Bernard Knapp, 201–28. Oxford: Blackwell, 1999.

Burks, Jarrod. "Geophysical Survey at Ohio Earthworks: Updating Nineteenth Century Maps and Filling the 'Empty' Spaces." *Archaeological Prospection* 21 (2012): 5–13.

———. "Recording Earthworks in Ohio, Historic Aerial Photography, Old Maps and Magnetic Survey." In *Landscapes through the Lens: Aerial Photographs and the Historic Environment,* edited by David C. Cowley, Robin A. Standring, and Matthew J. Abicht, 77–87. Oxford: Oxbow Books, 2010.

Burks, Jarrod, and Robert A. Cook. "Beyond Squier and Davis: Rediscovering Ohio's Earthworks Using Geophysical Remote Sensing." *American Antiquity* 76, no. 4 (2011): 667–89.

Burl, Aubrey. *Prehistoric Avebury.* New Haven, CT: Yale University Press, 2002.

Byers, A. Martin. "The Earthwork Enclosures of the Central Ohio Valley: A Temporal and Structural Analysis of Woodland Society and Culture." PhD diss., State University of New York at Albany, 1987.

———. "The 'Heartland' Woodland Settlement System: Cultural Traditions and Resolving Key Puzzles." In Byers and Wymer, *Hopewell Settlement Patterns, Subsistence, and Symbolic Landscapes,* 276–96.

———. "Is the Newark Circle-Octagon the Ohio Hopewell 'Rosetta Stone'? A Question of Archaeological Interpretation." In *Ancient Earthen Enclosures of the Eastern Woodlands,* edited by Robert C. Mainfort and Lynne P. Sullivan, 135–53. Gainesville: University Press of Florida, 1998.

———. *The Ohio Hopewell Episode: Paradigm Lost and Paradigm Gained.* Akron, OH: University of Akron Press, 2004.

———. *Sacred Games, Death, and Renewal in the Ancient Eastern Woodlands: The Ohio Hopewell System of Cult Sodality Heterarchies.* Lanham, MD: AltaMira, 2011.

Byers, A. Martin, and DeeAnne Wymer, eds. *Hopewell Settlement Patterns, Subsistence, and Symbolic Landscapes.* Gainesville: University Press of Florida, 2010.

Cajete, Gregory. *Native Science: Natural Laws of Interdependence.* Santa Fe, NM: Clear Light, 2000.

Campbell, Joseph. *Way of the Seeded Earth: Mythologies of the Primitive Planters.* New York: HarperCollins, 1994.

Carneiro, Robert. "A Theory of the Origin of the State." *Science* 169 (1970): 733–38.

Carpenter, Kristen A., Sonia K. Katyal, and Angela R. Riley. "In Defense of Property." *Yale Law Journal* 118 (2009): 1022–125.

Carr, Christopher. "Historical Insights into the Directions and Limitations of Recent Research on Hopewell." In Carr and Case, *Gathering Hopewell,* 51–70.

———. "Scioto Hopewell Ritual Gatherings: A Review and Discussion of Previous Interpretations and Data." In Carr and Case, *Gathering Hopewell,* 463–79.

———. "Social and Ritual Organization." In *The Scioto Hopewell and Their Neighbors:*

Bioarchaeological Documentation and Cultural Understanding, edited by D. Troy Case and Christopher Carr, 151–288. New York: Springer, 2008.
Carr, Christopher, and D. Troy Case, eds. *Gathering Hopewell: Society, Ritual, and Ritual Interaction*. New York: Kluwer Academic/Plenum, 2005.
———. "The Gathering of Hopewell." In Carr and Case, *Gathering Hopewell*, 19–50.
Carrasco, Davíd, ed. *To Change Place: Aztec Ceremonial Landscapes*. Niwot, CO: University Press of Colorado, 1991.
Castleden, Rodney. *The Stonehenge People: An Exploration of Life in Neolithic Britain, 4700–2000 BC*. London: Routledge, 1990.
CERHAS (Center for the Electronic Reconstruction of Historical and Archaeological Sites). *EarthWorks: Virtual Explorations of the Ancient Ohio Valley*, exhibit and CD-ROM. Cincinnati: University of Cincinnati CERHAS, 2007.
Champagne, Duane. *Social Change and Cultural Continuity among Native Nations*. Lanham, MD: Altamira, 2006.
———. *Social Order and Political Change: Constitutional Governments among the Cherokee, the Choctaw, the Chickasaw, and the Creek*. Stanford, CA: Stanford University Press, 1992.
Chaudhuri, Jean, and Joyotpaul Chaudhuri. *A Sacred Path: The Way of the Muscogee Creeks*. Los Angeles: UCLA American Indian Studies Center, 2001.
Chidester, David, and Edward T. Linenthal. Introduction to *American Sacred Space*, edited by David Chidester and Edward T. Linenthal, 1–42. Bloomington: Indiana University Press, 1995.
Cobb, Charles R., Jeffrey Maymon, and Randall H. McGuire. "Feathered, Horned, and Antlered Serpents: Mesoamerican Connections with the Southwest and Southeast," in *Great Towns and Regional Polities in the Prehistoric American Southwest and Southeast*, edited by Jill E. Neitzel, 165–81. Albuquerque: University of New Mexico Press, 1999.
Coe, Michael D. "Religion and the Rise of Mesoamerican States." In *Transition to Statehood in the New World*, edited by Grant D. Jones and Robert R. Kautz. Cambridge: Cambridge University Press, 1981.
Coleman, Simon, and John Elsner. *Pilgrimage: Past and Present in the World Religions*. Cambridge, MA: Harvard University Press, 1995.
Cormack, Bradin. *A Power to Do Justice: Jurisdiction, English Literature, and the Rise of Common Law, 1509–1625*. Chicago: University of Chicago Press, 2008.
Cover, Robert M. "The Supreme Court, 1982 Term—Foreword: *Nomos* and Narrative," *Harvard Law Review* 97 (1983): 4–68.
"Cultural Protection and NAGPRA." National Congress of American Indians website. http://www.ncai.org/policy-issues/community-and-culture/cultural-protection-and-nagpra.
Dancey, William S., ed. *The First Discovery of America: Archaeological Evidence of the Early Inhabitants of the Ohio Area*. Columbus: Ohio Archaeological Council, 1994.

Dancey, William S., and Paul J. Pacheco, eds. *Ohio Hopewell Community Organization.* Kent, OH: Kent State University Press, 1997.
Darvill, Timothy. "Research Frameworks for World Heritage Sites and the Conceptualization of Archaeological Knowledge." *World Archaeology* 39, no. 3 (2007): 436–57.
———. "*Scientia,* Society and Poydactyl Knowledge: Archaeology as a Creative Science." In *Paradigm Found: Archaeological Theory--Present, Past and Future; Essays in Honour of Evžen Neustrupný,* edited by Kristian Kristiansen, Ladislav Šmejda, and Jan Turek, 6–23. Oxford: Oxbow Books, 2015.
———. *Stonehenge: The Biography of a Landscape.* Stroud: Tempus, 2006.
———. *Stonehenge World Heritage Site: An Archaeological Research Framework.* London: English Heritage, 2005.
———. "The Stones of Stonehenge." *Current Archaeology* 21 (2011): 28–35.
———. "Tynwald Hill and the 'Things' of Power." In *Assembly Places and Practices in Medieval Europe,* edited by Aliki Pantos and Sarah Semple, 217–232. Dublin: Four Courts, 2004.
Darvill, Timothy, and Geoffrey Wainwright. "Stonehenge Excavations 2008." *Antiquaries Journal* 89, no. 1 (2009): 1–19.
———. "The Stones of Stonehenge." *Current Archaeology* 21, no. 252 (2011): 28–35.
DeBoer, Warren. "Ceremonial Centres from the Cayapas (Esmeraldas, Ecuador) to Chillicothe (Ohio, USA)." *Cambridge Archaeological Journal* 7, no. 2 (1997): 225–40.
———. "Little Bighorn on the Scioto: The Rocky Mountain Connection to Ohio Hopewell." *American Antiquity* 69, no. 1 (2004): 85–108.
———. "Strange Sightings on the Scioto." In Byers and Wymer, *Hopewell Settlement Patterns, Subsistence, and Symbolic Landscapes.*
De Certeau, Michael. "Practices of Space." In *On Signs,* edited by Marshall Blonsky. Baltimore: Johns Hopkins University Press, 1985.
Deloria, Vine, Jr. "Power and Place Equal Personality." In *Power and Place: Indian Education in America,* by Vine Deloria Jr. and Daniel R. Wildcat. Golden, CO: Fulcrum, 2001.
———. "Secularism, Civil Religion, and the Religious Freedom of American Indians." *American Indian Culture and Research Journal* 16, no. 2 (Spring 1992): 9–20.
———. "Trouble in High Places: Erosion of American Indian Rights to Religious Freedom in the United States." In *The State of Native America: Genocide, Colonization, and Resistance,* edited by M. Annette Jaimes, 267–90. Boston: South End, 1992.
Dodge, Martin, Rob Kitchin, and Chris Perkins, eds. *Rethinking Maps: New Frontiers in Cartographic Theory.* London: Routledge, 2009.
Doxtater, Dennis. "A Hypothetical Layout of Chaco Canyon Structures via Large-Scale Alignments between Significant Natural Features." *Kiva* 68, no. 1 (2002): 23–47.
———. "Parallel Universes on the Colorado Plateau: Indications of Chacoan

Integration of an Earlier Anasazi Focus at Canyon the Chelly." *Journal of the Southwest* 45, no. 1–2 (2003): 33–62.

Dragoo, Don W., and Charles F. Wray. "Hopewell Figurine Rediscovered." *American Antiquity* 30 (1964): 195–99.

Drennen, Robert D., and Christian E. Peterson. "Challenges for Comparative Study of Early Complex Societies." In *The Comparative Archaeology of Complex Societies*, edited by Michael E. Smith. Cambridge: Cambridge University Press, 2012.

Dunbar, Robin I. M. *The Human Story: A New History of Mankind's Evolution.* London: Faber and Faber, 2004.

Durbin, Kathie. "Rediscovering the Lost Coast: California's Sinkyone Indians Plan to Restore the Land of Their Ancestors." *Audubon* 98, no. 2 (1996): 18.

Earle, Timothy. "Economic Support of Chaco Canyon Society." *American Antiquity* 66, no. 1 (2001): 26–35.

Eliade, Mircea. *Shamanism: Archaic Techniques of Ecstasy.* Translated by Willard R. Trask. Princeton, NJ: Princeton University Press, 1964.

Essenpreis, Patricia, and David Duszynski. "Possible Astronomical Alignment at the Fort Ancient Monument." Unpublished paper presented to the annual meeting of the Society of American Archaeology, 1989.

Essenpreis, Patricia, and Michael Moseley. "Fort Ancient: Citadel or Coliseum?" *Field Museum of Natural History Bulletin* 55, no. 5 (1984): 5–26.

Fagan, Brian M. *Ancient North America: The Archaeology of a Continent.* 3rd ed. New York: Thames and Hudson, 2000.

Fine-Dare, Kathleen. *Grave Injustice: The American Indian Repatriation Movement.* Lincoln: University of Nebraska Press, 2002.

Finney, Ben. *Sailing in the Wake of the Ancestors: Reviving Polynesian Voyaging.* Honolulu: Bishop Museum Press, 2003.

Fowke, Gerard. *Archaeological History of Ohio: The Mound Builders and Later Indians.* Columbus, OH: F. J. Heer, 1902.

Fowles, Severin. "A People's History of the American Southwest." In *Ancient Complexities: New Perspectives in Precolumbian North America*, edited by Susan Alt. Salt Lake City: University of Utah Press, 2010.

Friedberg, Anne. *Window Shopping: Cinema and the Postmodern.* Berkeley: University of California Press, 1993.

Freidel, David A. and Jeremy A. Sabloff. *Cozumel, Late Maya Settlement Patterns.* Orlando, FL: Academic, 1984.

Gadamer, Hans-George. *The Relevance of the Beautiful and Other Essays.* Cambridge: Cambridge University Press, 1986.

———. *Truth and Method.* 2nd rev. ed. New York: Continuum, 1994.

Galloway, Patricia Kay. *Choctaw Genesis, 1500–1700.* Lincoln: University of Nebraska, 1996.

———. "Debriefing Explorers: Amerindian Information in the Delisles' Mapping of the Southeast." In *Cartographic Encounters: Perspectives on Native American*

Mapmaking and Map Use, edited by G. Malcolm Lewis, 223–40. Chicago: University of Chicago Press, 1998.

García, Maria Elena Bernal, and Angel Julían García Zambrano. "El Altepetl Colonial y sus Antecendentes Prehispanicos: Contexto Teórico-Historiográfico." In *Territorialidad y Paisaje en el Altepetl del Siglo XVI*, edited by Federico Fernández Christlieb and Angel Julían García Zambrano. México, DF: Fondo de Cultura Económica and Instituto de Geografía de la Universidad Nacional Autónoma de México, 2006.

Gatschet, Albert S. *A Migration Legend of the Creek Indians, with a Linguistic, Historic and Ethnographic Introduction*. Philadelphia: D. G. Brinton, 1884.

Gibson, Alex. "Dating Balbirnie: Recent Radiocarbon Dates from the Stone Circle and Cairn at Balbirnie, Fife, and a Review of Its Place in the Overall Balfarg/Balbirnie Site Sequence." *Proceedings of the Society of Antiquaries of Scotland* 140 (2010): 51–78.

Gibson, Jon L. *The Ancient Mounds of Poverty Point: Place of Rings*. Gainesville: University Press of Florida, 2001.

Giedion, Sigfried. "Introduction: The Three Space Conceptions in Architecture." In *Architecture and the Phenomena of Transition*. Cambridge, MA: Harvard University Press, 1971.

Gillings, Mark, and Joshua Pollard. *Avebury*. London: Duckworth, 2004.

Glowacki, Donna M., and Scott Van Keuren, eds. *Religious Transformation in the Late Pre-Hispanic Pueblo World*. Tucson: University of Arizona Press, 2012.

Goldberg, Carole E., Kevin K. Washburn, and Philip P. Frickey, eds. *Indian Law Stories*. New York: Thomas Reuters/Foundation, 2011.

González, José Humberto Medina, and Baudelina L. García Uranga, *Alta Vista: A 100 Años de su Descubrimiento*. México, DF: Instituto Nacional de Antropología e Historia, 2010.

Goodman, Ronald. *Lakȯta Star Knowledge: Studies in Lakȯta Stellar Theology*. Rosebud, SD: Sinṫe Gleṡka University, 1992.

Gosden, Chris, and Gary Lock. "Prehistoric Histories," *World Archaeology* 30, no. 1 (1998): 2–12.

Greber, N'omi. "Astronomy and the Patterns of Five Geometric Earthworks in Ross County, Ohio." Unpublished paper presented at the Second Oxford International Conference on Archaeoastronomy, Mérida, Yucatán, Mexico, Jan. 13–17, 1986.

———. "Recent Excavations at the Edwin Harness Mound, Liberty Works, Ross County, Ohio." *Midcontinental Journal of Archaeology*, special issue, 5 (1983).

Greber, N'omi, Robert Horn, Ray Hively, and Karen Royce. "Astronomy and Archaeology at High Bank Works." Unpublished paper presented at Midwest Archaeological Conference, Oct. 2010.

Greber, N'omi, and David Jargiello. "Possible Astronomical Orientations Used in Constructing Some Scioto Hopewell Earthwork Walls." Unpublished paper presented to the annual meeting of the Midwest Archaeological Conference, 1982.

Greber, N'omi, and Katharine C. Ruhl. *The Hopewell Site: A Contemporary Analysis Based on the Work of Charles C. Willoughby.* Boulder, CO: Westview, 1989.

Guirand, Felix. "Greek Mythology." In *Larousse Encyclopedia of Mythology,* edited by Felix Guirand, 119–33. London: Paul Hamlyn, 1959.

Haberfeld, Steven. "Government-to-Government Negotiations: How the Timbisha Shoshone Got Its Land Back." *American Indian Culture and Research Journal* 24, no. 4 (2000): 127–65.

Hall, Robert L. "Ghosts, Water Barriers, Corn, and Sacred Enclosures in the Eastern Woodlands." *American Antiquity* 41 (1976): 360–64.

Hancock, John E. "The Earthworks Hermeneutically Considered." In Byers and Wymer, *Hopewell Settlement Patterns, Subsistence, and Symbolic Landscapes,* 263–75.

Harjo, Suzan Shown. "Sacred Places and Visitor Protocols." In *American Indian Places: A Historical Guidebook,* edited by Frances H. Kennedy, 81–87. Boston: Houghton Mifflin, 2008.

Haven, Samuel F. "Report of the Librarian." *Proceedings of the American Antiquarian Society,* Apr. 1870, 39–41.

Hayes, Alden C. "A Survey of Chaco Canyon Archaeology." In *Archaeological Surveys of Chaco Canyon, New Mexico,* edited by Alden C. Hayes, David M. Brugge, and W. James Judge. Publications in Archaeology 18A. Washington, DC: National Park Service, 1981.

Hayden, Bryan, and Suzanne Villenueve. "Astronomy in the Upper Paleolithic." *Cambridge Archaeological Journal,* 21, no. 3 (2011): 331–55.

Heidegger, Martin. *Poetry, Language, Thought.* New York: Harper and Row, 1971.

Henry, James Pepper. "Challenges in Managing Culturally Sensitive Collections at the National Museum of the American Indian." In *American Indian Nations: Yesterday, Today, and Tomorrow,* edited by George Horse Capture, Duane Champagne, and Chandler C. Jackson, 59–65. Lanham, MD: AltaMira, 2007.

Hewett, Edgar L. *The Chaco Canyon and Its Monuments.* Albuquerque: University of New Mexico, 1936.

Hine, Robert V., and John Mack Faragher. *The American West: A New Interpretive History.* New Haven, CT: Yale University Press, 2000.

Hirth, Kenneth G. "The Altepetl and Urban Structure in Prehispanic Mesoamerica." In *Urbanism in Mesoamerica,* vol. 1, edited by William T. Sanders, Alba Guadalupe Mastache De Escobar, and Robert H. Cobean. México, DF: Instituto Nacional De Antropología e Historia, 2003.

———. "Incidental Urbanism: The Structure of the Prehispanic City in Central Mexico." in *The Ancient City: New Perspectives on Urbanism in the Old and New World,* edited by Joyce Marcus and Jeremy A. Sabloff. Santa Fe, NM: School for Advanced Research, 2008.

Hitakonanu'laxk. *The Grandfathers Speak: Native American Folk Tales of the Lenape People.* New York: Interlink Books, 2005.

Hively, Ray, and Robert Horn. "Geometry and Astronomy in Prehistoric Ohio."

Journal for the History of Astronomy, Archaeoastronomy Supplement, 13, no. 4 (1982): 1–20.
———. "Hopewell Cosmography at Newark and Chillicothe, Ohio." In Byers and Wymer, *Hopewell Settlement Patterns, Subsistence, and Symbolic Landscapes,* 128–64.
———. "Hopewellian Geometry and Astronomy at High Bank." *Journal for the History of Astronomy,* Archaeoastronomy Supplement, 15, no. 7 (1984): 85–100.
———. "A New and Extended Case for Lunar (and Solar) Astronomy at the Newark Earthworks." *Midcontinental Journal of Archaeology* 38, no. 1 (2013): 83–118.
———. "Rejoinder." In Byers and Wymer, *Hopewell Settlement Patterns, Subsistence, and Symbolic Landscapes,* 206–7.
———. "A Statistical Analysis of Lunar Alignments at the Newark Earthworks." *Midcontinental Journal of Archaeology* 31, no. 2 (2006): 281–321.
Hoare, Sir Richard Colt. *The Ancient History of Wiltshire.* London: William Miller, 1812.
Hodder, Ian. *Çatalhöyük: The Leopard's Tale.* London: Thames and Hudson, 2006.
Hodge, Mary G. "When Is a City-State? Archaeological Measure of Aztec City-States and Aztec City-State Systems." In *The Archaeology of City-States: Cross-Cultural Approaches,* edited by Deborah L. Nichols and Thomas H. Charlton, 209–27. Washington, DC: Smithsonian Institution Press, 1997.
Hodson, Frank Roy, ed. *The Place of Astronomy in the Ancient World: A Joint Symposium of the Royal Society and the British Society.* London: Oxford University Press, 1974.
Hooge, Paul E. "Preserving the Ancient Past in Licking County, Ohio: A Case Study." PhD diss., Ohio State University, Columbus, 1993.
Huyssen, Andreas. *Present Pasts: Urban Palimpsests and the Politics of Memory.* Palo Alto, CA: Stanford University Press, 2003.
Insoll, Timothy. "Shrine Franchising and the Neolithic in the British Isles: Some Observations Based upon the Tallensi, Northern Ghana." *Cambridge Archaeological Journal* 16, no. 2 (2006): 223–38.
Jennings, Jesse D. *Prehistory of North America.* 2nd ed. New York: McGraw-Hill, 1974.
Johnson, Greg. "Authenticity, Invention, Articulation: Theorizing Contemporary Hawaiian Traditions from the Outside." *Method and Theory in the Study of Religion* 20 (2008): 243–58.
———. *Sacred Claims: Repatriation and Living Tradition.* Charlottesville: University of Virginia Press, 2007.
———. "Varieties of Native Hawaiian Establishment: Recognized Voices, Routinized Charisma and Church Desecration." In *Varieties of Religious Establishment,* edited by Winnifred Fallers Sullivan and Lori G. Beaman, 55–71. Burlington, VT: Ashgate, 2013.
Johnson, Harmer. Auction Block, *American Indian Art Magazine* 32, no. 4 (Autumn 2007): 36–37.

Jokilehto, Jukka, et al. *What Is OUV? Defining the Outstanding Universal Value of Cultural World Heritage Properties, Monuments and Sites XVI.* Paris: International Council on Monuments Sites, 2008. http://international.icomos.org/publications/monuments_and_sites/16/index.htm.

Jones, Andrew. *Memory and Material Culture.* Cambridge: Cambridge University Press, 2007.

Jones, Lindsay. *The Hermeneutics of Sacred Architecture: Experience, Interpretation, Comparison.* Vol. 2, *Hermeneutical Calisthenics: A Morphology of Ritual-Architectural Priorities.* Cambridge, MA: Harvard University Press, 2000.

Judge, W. James. "Chaco Canyon–San Juan Basin." In *Dynamics of Southwest Prehistory,* edited by Linda S. Cordell and George J. Gumerman, 209–61. Washington, DC: Smithsonian Institution Press, 1989.

Judge, W. James, and Linda S. Cordell. "Society and Polity." In *The Archaeology of Chaco Canyon: An Eleventh-Century Pueblo Regional Center,* edited by Stephen H. Lekson, 189–210. Santa Fe, NM: School of American Research Press, 2006.

Kameʻeleihiwa, Lilikalā. *Native Land and Foreign Desires: Pehea Lā E Pono Ai?* Honolulu: Bishop Museum Press, 1992.

Kantner, John W., and Kevin J. Vaughn. "Pilgrimage as Costly Signal: Religiously Motivated Cooperation in Chaco and Nasca." *Journal of Anthropological Archaeology* 31 (2012): 66–82.

———. "Rethinking Chaco as a System." *Kiva* 69 (2003): 207–27.

Kauanui, Kēhaulani. *Hawaiian Blood: Colonialism and the Politics of Sovereignty and Indigeneity.* Durham, NC: Duke University Press, 2008.

Kaul, Flemming. *Ships on Bronzes: A Study in Bronze Age Religion and Iconography.* Copenhagen: National Museum of Denmark, Dept. of Danish Collections, 1998.

Keel, Jefferson. "Sovereignty and the Future of Indian Nations." Presidential address, 8th Annual State of Indian Nations Address, National Congress of American Indians, National Press Club, Washington, DC, Friday, Jan. 29, 2010. http://www.ncai.org/attachments/Testimonial_mFyKzTBbinnLZAeOHdFmYvHkGDHFzsaBEGlYJPNqbusBxJsZDur_8th_Annual_State_of_Indian_Nations_Address_2010.pdf.

Kehoe, Alice Beck. *Shamans and Religion: An Anthropological Exploration in Critical Thinking.* Long Grove, IL: Waveland, 2000.

Kelley, Klara, and Francis Harris. "Traditional Navajo Maps and Wayfinding," *American Indian Culture and Research Journal* 29, no. 2 (2005): 85–111.

Kerber, Jordan E. *Cross-Cultural Collaboration: Native Peoples and Archaeology in the Northeastern United States.* Lincoln: University of Nebraska Press, 2006.

Kirshenblatt-Gimblett, Barbara. *Destination Culture: Tourism, Museums, and Heritage.* Berkeley: University of California Press, 1998.

Kitchin, Rob, Chris Perkins, and Martin Dodge. "Thinking about Maps." In *Rethinking Maps: New Frontiers in Cartographic Theory,* edited by Martin Dodge, Rob Kitchin, and Chris Perkins, 1–25. London: Routledge, 2009.

Knepper, George W. *The Official Ohio Lands Book.* Columbus, OH: Auditor of State, 2002.

Kroeber, Alfred L., Donald Collier, and Patrick H. Carmichael. *The Archaeology and Pottery of Nazca, Peru: Alfred L. Kroeber's 1926 Expedition.* Walnut Creek, CA: AltaMira, 1998.

Latour, Bruno. *We Have Never Been Modern.* Cambridge, MA: Harvard University Press, 1993.

Lawler, Andrew. "America's Lost City." *Science* 334 (2011): 1618–23.

Lekson, Stephen H., ed. *The Archaeology of Chaco Canyon: An Eleventh-Century Pueblo Regional Center.* Santa Fe, NM: School of American Research Press, 2006.

———, ed. *The Architecture of Chaco Canyon, New Mexico.* Salt Lake City: University of Utah Press, 2007.

———. *The Chaco Meridian: Centers of Political Power in the Ancient Southwest.* Walnut Creek, CA: AltaMira, 1999.

———. *A History of the Ancient Southwest.* Santa Fe, NM: School for Advanced Research Press, 2009.

———. "Lords of the Great House: Pueblo Bonito as a Palace." In *Palaces and Power in the Americas, from Peru to the Northwest Coast,* edited by Jessica Joyce Christie and Patricia Joan Sarro, 99–114. Austin: University of Texas Press, 2006.

Lepper, Bradley T. "The Archaeology of the Newark Earthworks." In *Ancient Earthen Enclosures of the Eastern Woodlands,* edited by Robert C. Mainfort and Lynne P. Sullivan, 114–34. Gainesville: University Press of Florida, 1998.

———. "The Ceremonial Landscape of the Newark Earthworks and the Raccoon Creek Valley." In Byers and Wymer, *Hopewell Settlement Patterns, Subsistence, and Symbolic Landscapes,* 97–127.

———. "Commentary on DeeAnne Wymer's 'Where Do (Hopewell) Research Answers Come From?'" In Byers and Wymer, *Hopewell Settlement Patterns, Subsistence, and Symbolic Landscapes,* 329–32.

———. "The Great Hopewell Road and the Role of Pilgrimage in the Hopewell Interaction Sphere." In *Recreating Hopewell,* edited by Douglas K. Charles and Jane E. Buikstra, 122–33. Gainesville: University Press of Florida, 2006.

———. "An Historical Review of Archaeological Research at the Newark Earthworks." *Journal of the Steward Anthropological Society* 18, no. 1–2 (1989): 118–40.

———. "The Newark Earthworks and the Geometric Enclosures of the Scioto Valley: Connections and Conjectures." In *A View from the Core: A Synthesis of Ohio Hopewell Archaeology,* edited by Paul J. Pacheco, 224–41. Columbus: Ohio Archaeological Council, 1996.

———. "The Newark Earthworks: Monumental Geometry and Astronomy at a Hopewellian Pilgrimage Center." In *Hero, Hawk, and Open Hand: American Indian Art of the Ancient Midwest and South,* edited by Richard F. Townsend, Robert V. Sharp, and Garrick Alan Bailey, 72–81. Chicago: Art Institute of Chicago; New Haven, CT: Yale University Press, 2004.

———. *The Newark Earthworks: A Wonder of the Ancient World.* Columbus: Ohio Historical Society, 2002.

———. *Ohio Archaeology: An Illustrated Chronicle of Ohio's Ancient American Indian Cultures.* Wilmington, OH: Orange Frazer, 2005.

———. "Processions." In CERHAS, *EarthWorks.*

———. "Tracking Ohio's Great Hopewell Road." *Archaeology* 48, no. 6 (1995): 52–56.

Lepper, Bradley T., and Tod A. Frokling. "Alligator Mound: Geoarchaeological and Iconographical Interpretations of a Late Prehistory Effigy Mound in Central Ohio, USA." *Cambridge Archaeological Journal* 13, no. 2 (2003): 147–67.

Lepper, Bradley T., and Jeff Gill. "The Newark Holy Stones." *Timeline,* May–June, 2000, 17–25.

Lepper, Bradley T., and Richard W. Yerkes. "Hopewellian Occupations at the Northern Periphery of the Newark Earthworks: The Newark Expressway Sites Revisited." In *Ohio Hopewell Community Organization,* edited by William S. Dancey and Paul J. Pacheco, 175–205. Kent, OH: Kent State University Press, 1997.

Lepper, Bradley T., Richard W. Yerkes, and William H. Pickard. "Prehistoric Flint Procurement Strategies at Flint Ridge, Licking County, Ohio." *Midcontinental Journal of Archaeology* 26 (2001): 53–78.

Lewis, G. Malcolm. *Cartographic Encounters: Perspectives on Native American Mapmaking and Map Use.* Chicago: University of Chicago Press, 1998.

———, ed. "Maps, Mapmaking, and Map Use by Native North Americans." In *The History of Cartography,* vol. 2, book 3, *Cartography in the Traditional African, American, Arctic, Australian, and Pacific Societies,* edited by David Woodward and G. Malcolm Lewis, 51–182. Chicago: University of Chicago Press, 1998.

Lippert, Dorothy. "In Front of the Mirror: Native Americans and Academic Archaeology." In *Native Americans and Archaeologists: Stepping Stones to Common Ground,* edited by Nina Swidler, Kurt E. Dongoske, Roger Anyon, and Alan S. Downer, 120–27. Walnut Creek, CA: Altamira, 1997.

Llewellyn, Karl N., and E. Adamson Hoebel. *The Cheyenne Way: Conflict and Case Law in Primitive Jurisprudence.* Norman: University of Oklahoma Press, 1941.

Lockhart, James. *The Nahuas after the Conquest: A Social and Cultural History of the Indians of Central Mexico, Sixteenth through Eighteenth Centuries.* Stanford, CA: Stanford University Press, 1992.

Loubser, Johannes. "From Boulder to Mountain and Back Again: Self-Similarity between Landscape and Mindscape in Cherokee Thought, Speech and Action as Expressed by the Judaculla Rock Petroglyphs." *Time and Mind* 2, no. 3 (2009): 287–312.

Loveday, Roy. *Inscribed across the Landscape: The Cursus Enigma.* Stroud, Gloucestershire: Tempus, 2006.

Lucas, David M. "Our Grandmother of the Shawnee: Messages of a Female Deity." Paper presented at the Annual Meeting of the National Communication Association, Atlanta, GA, November 2001.

Lucas, Gavin. *The Archaeology of Time.* London: Routledge, 2005.

MacCannell, Dean. *The Tourist: A New Theory of the Leisure Class.* 3rd ed. Berkeley: University of California Press, 1999.
Mahoney, Nancy M. "Redefining the Scale of Chacoan Communities." In *Great House Communities across the Chacoan Landscape,* edited by John Kantner and Nancy M. Mahoney, 19–27. Tucson: University of Arizona Press, 2000.
Malo, David. *Hawaiian Antiquities.* Translated by Nathaniel B. Emerson. Honolulu: Bernice P. Bishop Museum, [1898] 1951.
Malville, J. McKim, ed. *Chimney Rock: The Ultimate Outlier.* Lanham, MD: Lexington, 2004.
Malville, J. McKim, and Nancy J. Malville. "Pilgrimage and Periodic Festivals as Processes of Social Integration in Chaco Canyon." *Kiva* 66, no. 3 (2001): 327–44.
Mann, Charles C. *1493: Uncovering the New World Columbus Created.* New York: Knopf, 2011.
Mason, Ronald J. *Inconstant Companions: Archaeology and North American Indian Oral Traditions.* Tuscaloosa: University of Alabama Press, 2006.
Mathers, Clay, Timothy Darvill, and Barbara J. Little. *Heritage of Value, Archaeology of Renown: Reshaping Archaeological Assessment and Significance.* Gainesville: University Press of Florida, 2005.
McCorriston, Joy. *Pilgrimage and Household in the Ancient Near East.* Cambridge: Cambridge University Press, 2011.
McCoy, Padraic. "The Land Must Hold the People: Native Modes of Territoriality and Contemporary Justifications for Placing Land into Trust through 25 C.F.R. Part 151." *American Indian Law Review* 27 (2003): 421–502.
Meller, Harald. "Nebra: vom logos zum Mythos—Biograhie eines Himmelsbildes." In *Der Griff nach den Sternen. Wie Europas Eliten zu Macht und Reichtum kamen. Internationales symposium in Halle (Saale) 16–21 Febuar 2005,* edited by Harald Meller and François Bertemes, 23–72. Halle [Saale]: Landesmuseum für Vorgeschichte, Landesamt für Denkmalpflege und Archäologie Sachsen-Anhalt, 2010.
Mendoza, Gerardo Gutiérrez. "Territorial Structure and Urbanism in Mesoamerica: The Huaxtec and Mixtec-Tlpanec-Nahua Cases." In *Urbanism in Mesoamerica,* vol. 1, edited by William T. Sanders, Mastache De Escobar Alba Guadalupe, and Robert H. Cobean. México, DF: Instituto Nacional De Antropología e Historia, 2003.
Merry, Sally. *Colonizing Hawai'i: The Cultural Power of Law.* Princeton, NJ: Princeton University Press, 2000.
"The Mesoamerican Origin of North American Stickball." http://peopleofonefire.com/?s=Mesoamerican+Origin+of+North+American+Stickball.
Mickelson, Michael E., and Bradley T. Lepper. "Archaeoastronomy at the Newark Earthworks." *Mediterranean Archaeology and Archaeometry* 6, no. 3 (2007): 175–80.
Middleton, Beth Rose. *Trust in the Land: New Directions in Tribal Conservation.* Tucson: University of Arizona Press, 2011.

Miller, William Ian. *Bloodtaking and Peacemaking: Feud, Law, and Society in Saga Iceland.* Chicago: University of Chicago Press, 1990.
Mills, Barbara J. "Recent Research on Chaco: Changing Views on Economy, Ritual and Society." *Journal of Archaeological Research* 10, no. 1 (2002): 65–117.
Mills, Lisa. "Mitochondrial DNA Analysis of the Ohio Hopewell of the Hopewell Mound Group." *West Virginia Archeologist* 53 (2001): 1–18.
———. "Mitochondrial DNA Analysis of the Ohio Hopewell of the Hopewell Mound Group." PhD diss., , Ohio State University, Columbus, 2003.
Mills, William C. *The Archaeological Atlas of Ohio.* Canal Winchester, OH: Ohio Archeological and Historical Society, 1914.
———. "Exploration of the Tremper Mound." *Ohio Archaeological and Historical Quarterly* 25 (1916): 263–398.
———. "Explorations of the Edwin Harness Mound." *Ohio Archaeological and Historical Quarterly* 16 (1907): 113–93.
———. "Explorations of the Seip Mound." *Ohio Archaeological and Historical Quarterly* 18 (1909): 269–321.
Milner, George R. *The Moundbuilders: Ancient Peoples of Eastern North America.* London: Thames and Hudson, 2004.
Mooney, James. "The Cherokee River Cult." *Journal of Cherokee Studies* 17 (1982): 30–36.
Moorehead, Warren K. *The American Indian in the United States, 1840–1914.* New York: Books for Libraries, 1914.
———. "Fort Ancient: The Great Prehistoric Earthwork of Warren County, Ohio." *Phillips Academy, Department of Archaeology, Bulletin* 4 (1908): 27–166.
Morgan, Lewis Henry. *Houses and House-Life of the American Aborigines.* Chicago: University of Chicago Press, 1965. Originally published as vol. 4 of *Contributions to North American Ethnology.* Washington, DC: Government Printing Office, 1881.
Morgan, R. G. "Ohio's Prehistoric 'Engineers.'" *Ohio State Engineer* 20, no. 6 (1937): 2–5.
Morinis, E. Alan. *Sacred Journeys: The Anthropology of Pilgrimage.* Westport, CT: Greenwood, 1992.
Mundy, Barbara E. "Mesoamerican Cartography," in *The History of Cartography,* vol. 2, book 3, *Cartography in the Traditional African, American, Arctic, Australian, and Pacific Societies,* ed. David Woodward and G. Malcolm Lewis. Chicago: University of Chicago Press, 1998.
Nabokov, Peter. *A Forest of Time: American Indian Ways of History.* Cambridge: Cambridge University Press, 2002.
Nabokov, Peter, and Robert Easton. *Native American Architecture.* New York: Oxford University Press, 1989.
Naone Hall, Dana, ed. *Mālama: Hawaiian Land and Water.* Honolulu: Bamboo Ridge, 1985.
———. "Sovereign Ground." In *The Value of Hawai'i: Knowing the Past, Shaping the*

Future, edited by Craig Howe and Jon Osorio, 195–201. Honolulu: University of Hawai'i Press, 2010.

National Museum of the American Indian. *Programs and Services Guide.* Washington, DC: NMAI–Smithsonian Institution, 2007.

Needham, Stuart, and Anne Woodward. "The Clandon Barrow Finery: A Synopsis of Success in an Early Bronze Age World." *Proceedings of the Prehistoric Society* 74 (2008): 1–52, 369, 372–73, 376–77.

Nelson, Ben A. "Aggregation, Warfare, and the Spread of the Mesoamerican Tradition." In *The Archaeology of Regional Interaction: Religion, Warfare, and Exchange across the American Southwest and Beyond,* edited by Michelle Hegmon, 317–37. Boulder: University Press of Colorado, 2000.

———. "Chronology and Stratigraphy at La Quemada, Zacatecas, Mexico." *Journal of Field Archaeology* 24, no. 1 (1997): 85–109.

———. "Complexity, Hierarchy, and Scale: A Controlled Comparison between Chaco Canyon, New Mexico and La Quemada, Zacatecas." *American Antiquity* 60 (1995): 597–618.

Nihipali, Kunani. "Stone by Stone, Bone by Bone: Rebuilding the Hawaiian Nation in the Illusion of Reality." *Arizona State Law Journal* 34, no. 1 (2002): 28–46.

Noodin, Margaret. "Bundling the Day and Unraveling the Night." *Studies in American Indian Literatures* 25, no. 2 (Summer 2013): 237–40.

Norberg-Schulz, Christian. *Genius Loci: Towards a Phenomenology of Architecture.* New York: Rizzoli, 1980.

Norris, Tina, Paula Vines, and Elizabeth Hoeffel. "The American Indian and Alaska Native Population: 2010." 2010 Census Briefs. http://www.census.gov/prod/cen2010/briefs/c2010br-10.pdf.

O'Donnell, James H. *Ohio's First Peoples.* Athens: Ohio University Press, 2004.

Ohio Historical Society. *Newark Earthworks Historic Site Management Plan.* http://ohsweb.ohiohistory.org/places/c08/hsmp2.shtml.

Ohio—Present-Day Tribes Associated with Indian Land Cessions 1784–1894. National NAGPRA Online Databases. http://www.nps.gov/history/nagpra/ONLINEDB/INDEX.HTM.

Pacheco, Paul J. "Ohio Middle Woodland Intracommunity Settlement Variability: A Case Study from the Licking Valley." In *Ohio Hopewell Community Organization,* edited by William S. Dancey and Paul J. Pacheco, 41–84. Kent, OH: Kent State University Press, 1997.

Park, Samuel. *Notes of the Early History of Union Township, Licking County, Ohio.* Terre Haute, IN: O. J. Smith, 1870.

Parker Pearson, Michael. "The Stonehenge Riverside Project: Excavations at the East Entrance of Durrington Walls." In *From Stonehenge to the Baltic: Living with Cultural Diversity in the Third Millennium BC,* edited by Mats Larsson and Michael Parker Pearson, 125–57. Oxford: Archaeopress, 2007.

Parker Pearson, Michael, Josh Pollard, Julian Thomas, and Kate Welham. "Newhenge." *British Archaeology* 110 (2010): 15–17.

Parker Pearson, Michael, and Ramilisonina. "Stonehenge for the Ancestors: The Stones Pass on the Message." *Antiquity* 72 (1998): 308–26.

Parker Pearson, Michael, and Colin Richards. "Ordering the World: Perceptions of Architecture, Space and Time." In *Architecture and Order: Approaches to Social Space,* edited by Michael Parker Pearson and Colin Richards, 1–37. London: Routledge, 1994.

Pásztor, Emilia, Judit P. Barna, and Curt Roslund. "The Orientation of *Rondels* of the Neolithic Lengyel Culture in Central Europe." *Antiquity* 82 (2008): 910–24.

Pasztory, Esther. "Andean Aesthetics." in *The Spirit of Ancient Peru: Treasures from the Museo Arqueológico Rafael Larco Herrera.* Edited by Kathleen Berrin. New York: Thames and Hudson, 1997.

Patton, Mark, and Warwick Rodwell, and Olga Finch. *La Hougue Bie, Jersey.* Saint Helier, Jersey: Société Jersiaise, 1999.

Pavúk, Juraj, and Vladimir Karlovský. "Astronomische Orientierung der spätneolithischen Kreisanlagen in Mitteleuropa." *Germania* 86, no. 2 (2008): 465–700.

Pearce, Margaret Wickens. "The Last Piece Is You." In "Cartography and Narratives," special issue, *Cartographic Journal* 51, no. 2 (May 2014): 12.

Pollard, Joshua, and Andrew Reynolds. *Avebury: The Biography of a Landscape.* Stroud, Gloucestershire: Tempus, 2002.

Pennefather-O'Brien, Elizabeth. "Biological Affinities among Middle Woodland Populations Associated with the Hopewell Horizon." PhD diss., Indiana University, 2006.

Perry, Richard Warren. "Remapping the Legal Landscapes of Native North America: Layered Identities in Comparative Perspective." *PoLAR: Political and Legal Anthropology Review* 25, no. 1 (2002): 129–50.

Petersen, Andrew. "The Archaeology of the Syrian and Iraqi Hajj Routes." *World Archaeology* 26, no. 1 (1994): 47–56.

Pickles, John. *A History of Spaces: Cartographic Reason, Mapping, and the Geo-coded World.* London: Routledge, 2004.

Piggott, Stuart. "The Sources of Geoffrey of Monmouth II: The Stonehenge Story." *Antiquity* 15 (1941): 305–19.

Pliny the Elder. *The Natural History.* Edited by John Bostock and Henry Thomas Riley. London: Taylor and Francis, 1855.

Plog, Stephen, and Carrie Heitman. "Hierarchy and Social Inequality in the American Southwest, AD 800–1200." *Proceedings of the National Academy of Sciences* 107, no. 46 (2010): 19,619–626.

Pollard, Joshua, and Andrew Reynolds. *Avebury: The Biography of a Landscape.* Stroud, Gloucestershire: Tempus, 2002.

Poolaw, Linda. "Moon." In CERHAS, *EarthWorks.*

Porter, Ventia, ed. *Hajj: Journey to the Heart of Islam.* Cambridge, MA: Harvard University Press, 2012.

Powell, Christopher. "The Shapes of Sacred Space: A Proposed System of Geometry

Used to Lay Out and Design Maya Art and Architecture and Some Implication Concerning Maya Cosmology." PhD diss., University of Texas, Austin, 2010.

Preston, James J. "Spiritual Magnetism: An Organizing Principle for the Study of Pilgrimage." In *Sacred Journeys: The Anthropology of Pilgrimage*, edited by Alan E. Morinis, 31–46. Westport, CT: Greenwood, 1992.

Price, Timothy. "Hopewell Road: GIS Solutions Towards Pathway Discovery." http://timothy-price.com/hopewell_thesis.htm.

Prufer, O. H. "The Hopewell Cult." In *New World Archaeology: Theoretical and Cultural Transformation: Readings from Scientific American*, edited by Ezra B. W. Zubrow, Margaret C. Fritz, and John M. Fritz, 222–30. San Francisco: W. H. Freeman, 1974.

Ragon, Michael. *The Space of Death: A Study of Funerary Architecture, Decoration, and Urbanism.* Translated by Alan Sheridan. Charlottesville: University Press of Virginia, 1983.

Renfrew, Colin. *The Emergence of Civilization: The Cyclades and the Aegean in the Third Millennium B.C.* London: Methuen, 1972.

———. *Prehistory: The Making of the Human Mind.* London: Weidenfeld and Nicolson, 2007.

———. "Production and Consumption in a Sacred Economy: The Material Correlates of High Devotional Expression at Chaco Canyon." *American Antiquity* 66, no. 1 (2001): 14–25.

Rice-Rollins, Julie. "The Cartographic Heritage of the Lakota Sioux." *Cartographic Perspectives* 48 (Spring 2004): 39–56.

Richards, Colin. "Monumental Choreography, Architecture and Spatial Representation in Later Neolithic Orkney." In *Interpretative Archaeology*, edited by Christopher Tilley. Providence: Berg, 1993.

———. "Monuments as Landscape: Creating the Center of the World in Late Neolithic Orkney." *World Archaeology* 28, no. 2 (1996): 190–208.

Richland, Justin B. "Hopi Tradition as Jurisdiction: On the Potentializing Limits of Hopi Sovereignty." *Law & Social Inquiry* 36 (2011): 201–34.

Richter, Daniel K. *Before the Revolution: America's Ancient Past.* Cambridge, MA: Harvard University Press, 2011.

Robb, David. "Indian Characteristics and Customs." In *Historical Collections of Ohio: An Encyclopedia of the State*, vol. 1, edited by Henry Howe, 299–300. Cincinnati: C. J. Krehbeil, 1904.

Romain, William F. "Hopewell Geometric Enclosures: Gatherings of the Fourfold." PhD diss., University of Leicester, 2004.

———. *Mysteries of the Hopewell: Astronomers, Geometers, and Magicians of the Eastern Woodlands.* Akron, OH: University of Akron Press, 2000.

———. "Summary Report on the Orientations and Alignments of the Ohio Hopewell Geometric Enclosures." Electronic appendix 3.1 to Carr and Case, *Gathering Hopewell.*

Romain, William F., and Jarrod Burks. "LiDAR Analyses of Prehistoric Earthworks in Ross County, Ohio." *Current Research in Ohio Archaeology,* Mar. 3, 2008. http://www.ohioarchaeology.org.

———. "LiDAR Assessment of the Newark Earthworks." *Current Research in Ohio Archaeology.* Feb. 4, 2008. http://www.ohioarchaeology.org.

———. "LiDAR Imaging of the Great Hopewell Road." *Current Research in Ohio Archaeology,* Feb. 7, 2008, http://www.ohioarchaeology.org.

Royal Commission on Historical Monuments (England). *Stonehenge and Its Environs: Monuments and Land Use.* Edinburgh: Edinburgh University Press, 1979.

Royce, Charles C. *Indian Land Cessions in the United States.* Washington, DC: Government Printing Office, 1899.

Ruggles, Clive L. N. "Astronomy and Stonehenge." In *Science and Stonehenge,* edited by Barry W. Cunliffe and Colin Renfrew, 203–24. Oxford: Oxford University Press, 1997.

———. *Astronomy in Prehistoric Britain and Ireland.* New Haven, CT: Yale University Press, 1999.

———. "Sun, Moon, Stars and Stonehenge." *Archaeoastronomy* 24 (1999): 83–88.

Rundstrom, Robert. "Mapping, Postmodernism, Indigenous People, and the Changing Direction of North American Cartography." *Cartographica* 28, no. 2 (1991): 1–12.

Russell, Miles. *Monuments of the British Neolithic: The Roots of Architecture.* Stroud, Gloucestershire: Tempus, 2002.

Salisbury, James H., and Charles B. Salisbury. "Accurate Surveys and Descriptions of the Ancient Earthworks at Newark, Ohio." MS on file, American Antiquarian Society, Worcester, MA, 1862.

Salomon, Frank. "The Introductory Essay: The Huarochirí Manuscript." In *The Huarochirí Manuscript: A Testament of Ancient and Colonial Andean Religion,* edited by Frank Salomon, Jorge Urioste, and Francisco De Avila. Austin: University of Texas Press, Austin, 1991.

Sayre, Gordon M. "The Mound Builders and the Imagination of American Antiquity in Jefferson, Bartram, and Chateaubriand." *Early American Literature* 33, no. 3 (1998): 225–49.

Scarre, Chris, ed. *Seventy Wonders of the Ancient World: The Great Monuments and How They Were Built.* London: Thames and Hudson, 1999.

Schelberg, John D. "Analogy, Complexity, and Regionally-Based Perspectives." In *Recent Research in Chaco Prehistory,* edited by W. James Judge and John D. Schelberg, Reports of the Chaco Center 8. Albuquerque: National Park Service, 1984.

Scott, James C. *The Art of Not Being Governed: An Anarchist History of Upland Southeast Asia.* New Haven, CT: Yale University Press, 2009.

Seidemann, Ryan M. "Altered Meanings: The Department of the Interior's Rewriting of the Native American Graves Protection and Repatriation Act to Regulate Culturally Unidentifiable Human Remains." *Temple Journal of Science, Technology & Environmental Law* 28, no. 1 (2009): 1–48.

Shetrone, Henry C. "Explorations of the Hopewell Group of Prehistoric Earthworks." *Ohio Archaeological and Historical Quarterly* 35 (1927): 1–227.

———. *The Mound Builders.* New York: American Museum of Natural History, 1930.

Shiels, Richard. "On the Pilgrim Road." *The Megalithic Portal,* Oct. 12, 2009. http://www.megalithic.co.uk/article.php?sid=2146413751.

Shook, Beth, Alison Schultz, and David Glenn Smith. "Using Ancient mtDNA to Reconstruct the Population History of Northeastern North America." *American Journal of Physical Anthropology* 137 (2008): 14–29.

Silva, Noenoe. *Aloha Betrayed: Native Hawaiian Resistance to American Colonialism.* Durham, NC: Duke University Press, 2004.

Silverberg, Robert. *The Mound Builders of Ancient America: The Archaeology of a Myth.* Abridged ed. Athens: Ohio University Press, 1986.

Silverman, Helaine. "The Archaeological Identification of an Ancient Peruvian Pilgrimage Center." *World Archaeology* 26, no. 1 (1994): 1–18.

———. "Beyond the Pampa: The Geoglyphs of the Valleys of Nazca." *National Geographic Research* 6, no. 4 (1990): 435–56.

———. *Cahuachi in the Ancient Nasca World.* Iowa City: University of Iowa Press, 1993.

———. "The Early Nasca Pilgrimage Center of Cahuachi and the Nazca Lines: Anthropological and Archaeological Perspectives." In *The Lines of Nazca,* edited by Anthony F. Aveni, 207–44. Philadelphia: American Philosophical Society, 1990.

———. "Nasca. Nazca. Continuities and Discontinuities on the South Coast of Peru." In *Global Perspectives on the Collapse of Complex Systems,* edited by Jim A. Railey and Richard Martin Reycraft, 83–100. Albuquerque: Maxwell Museum of Anthropology, 2008.

Sims, Lionel. "The 'Solarization' of the Moon: Manipulated Knowledge at Stonehenge." *Cambridge Archaeological Journal* 16, no. 2 (2006): 191–207.

———. "Which Way Forward for Archaeoastronomy?" *Journal of Cosmology* 9 (2010): 2160–71.

Smith, Linda Tuhiwai. *Decolonizing Methodologies: Research and Indigenous Peoples.* London: Zed Books, 1999.

Smith, Michael E. *Aztec City-State Capitals.* Gainesville: University Press of Florida, 2008.

———. "Aztec City States." In *A Comparative Study of Thirty City-State Cultures: An Investigation,* edited by Mogens Herman Hansen, 580–95. Copenhagen: Royal Danish Academy of Sciences and Letters, 2000.

Smith, Michael E., and Frances Berdan. *The Postclassic Mesoamerican World.* Salt Lake City: University of Utah, 2003.

Smith, Monica L. "Introduction: The Social Construction of Ancient Cities." In *The Social Construction of Ancient Cities,* edited by Monica L. Smith, 7–8. Washington, DC: Smithsonian Institution Press, 2003.

———. "What It Takes to Get Complex." In *The Comparative Archaeology of Complex Societies,* edited by Michael E. Smith. Cambridge: Cambridge University Press, 2012.

Sofaer, Anna. "The Primary Architecture of the Chacoan Culture: A Cosmological Expression." In *Anasazi Architecture and American Design,* edited by Baker H. Morrow and Vincent B. Price, 88–132. Albuquerque: University of New Mexico, 1997.

Spirn, Anne Whiston. *The Language of Landscape.* New Haven, CT: Yale University Press, 1998, 218–19.

Squier, Ephraim, and Edwin H. Davis. *Ancient Monuments of the Mississippi Valley: Comprising the Results of Extensive Original Surveys and Explorations.* New York: Bartlett and Welford, 1848.

Stannard, David E. *American Holocaust: The Conquest of the New World.* New York: Oxford University Press, 1993.

Stanzione, Vincent James. "My Walk with the Ancients." Unpublished document, 2010.

———. "Walking Is Knowing: Pilgrimage through the Pictorial History of the Cuauhtinchantlaca." In *Cave, City, and Eagle's Nest: An Interpretive Journey through the Mapa de Cuauhtinchan No. 2,* edited by David Carrasco and Scott Sessions, 317–33. Albuquerque: University of New Mexico Press, 2007.

Stein, John R. "The Chaco Roads—Clues to an Ancient Riddle?" *El Palacio* 94 (1989): 4–16.

Steinmetz, Paul B. "The Sacred Pipe in American Indian Religions." *American Indian Culture and Research Journal* 8, no. 3 (1984): 27–80.

Stranger, A. *A Six Days' Tour through the Isle of Man.* N.p.: Douglas, 1836.

Strong, William Duncan. *Paracas, Nazca, and Tiahuanacoid Cultural Relationships in South Coastal Peru.* Salt Lake City: Society for American Archaeology, 1957.

Sunderstrom, Linea. "Mirror of Heaven: Cross-Cultural Transference of the Sacred Geography of the Black Hills." *World Archaeology* 28, no. 2 (1996): 177–89.

Sutcliffe, Ron. *Moon Tracks: A Guide to Understanding Some of the Patterns We See with an Emphasis on Southwest Ancient Puebloan Cultures.* Pagosa Springs, CO: Moonspiral, 2006.

Swanton, John R. "The Creeks and Mound Builders." *American Anthropologist* 14, no. 2 (1912): 320–24.

Tacitus, Cornelius, and Harold Mattingly. *Tacitus on Britain and Germany: A New Translation of the "Agricola" and the "Germania."* Harmondsworth, Middlesex: Penguin, 1948.

Taves, Ann. *Religious Experience Reconsidered: A Building Block Approach to the Study of Religion and Other Special Things.* Princeton, NJ: Princeton University Press, 2009.

Taylor, William. "The Uncanny." In CERHAS, *EarthWorks.*

Tello, Julio C. "Un Modelo de Escenografía Plástica en el Arte Antiguo Peruano." *Wira Kocha* 1, no. 1 (1931): 87–112.

Tengan, Ty Kāwika. *Native Men Remade: Gender and Nation in Contemporary Hawai'i.* Durham, NC: Duke University Press, 2008.

Thom, A. S., J. M. D. Ker, and T. R. Burrows. "The Bush Barrow Gold Lozenge: Is It a Solar and Lunar Calendar for Stonehenge?" *Antiquity* 62 (1988): 492–502.

Thomas, Cyrus. *Report on the Mound Explorations of the Bureau of Ethnology: Extract from the Twelfth Annual Report of the Bureau of Ethnology.* Washington, DC: Government Printing Office, 1894.

Thomas, David Hurst. *Exploring Native North America.* Oxford: Oxford University Press, 2000.

———. *Skull Wars: Kennewick Man, Archaeology, and the Battle for Native American Identity.* New York: Basic Books, 2000.

Thomas, N. "The Thornborough Circles near Ripon, North Riding." *Yorkshire Archaeological Journal* 38, no. 4 (1955): 425–45.

Thompson, Victor D. "The Mississippian Production of Space through Earthen Pyramids and Public Buildings on the Georgia Coast, USA." *World Archaeology* 41, no. 3 (2009).

Thornton, Russell. *American Indian Holocaust and Survival: A Population History since 1492.* Norman: University of Oklahoma Press, 1987.

Toensing, Gale Courey. "Updated Federally Recognized Tribes List Published." Indian Country Today Media Network.com. http://indiancountrytodaymedia network.com/2014/02/10/updated-federally-recognized-tribes-list-published -153459.

Townsend, Richard F. *The Ancient Americas: Art from Sacred Landscapes.* Chicago: Art Institute of Chicago, 1992.

Trafzer, Clifford E. "Serra's Legacy: The Desecration of American Indian Burials at Mission San Diego." *American Indian Culture and Research Journal* 16, no. 2 (Spring 1992): 57–76.

Trombold, Charles D. "Causeways in the Context of Strategic Planning in the La Quemada Region, Zacatecas, Mexico." In *Ancient Road Networks and Settlement Hierarchies in the New World,* edited by Charles D. Trombold. Cambridge: Cambridge University Press, 1991.

Trope, Jack F., and Walter R. Echo-Hawk. "The Native American Graves Protection and Repatriation Act: Background and Legislative History." *Arizona State Law Journal* 24 (1992): 35.

Turner, Christopher S. "Ohio Hopewell Astraeoastronomy: A Meeting of Earth, Mind and Sky." *Time and Mind* 4, no. 3 (2011): 310–24.

———. "A Report on Archaeoastronomical Research at the Hopeton Earthworks, Ross County, Ohio." Unpublished paper presented to the Sixth Biennial History of Astronomy Workshop, University of Notre Dame, June 19–22, 2003.

Turner, Edith. "Pilgrimage." In *Encyclopedia of Religion,* 2nd ed., edited by Lindsay Jones, vol. 10, 7,145–48. Detroit: Macmillan Reference, 2005.

Turner, Victor. "Pilgrimages as Social Processes." In *Dramas, Fields, and Metaphors:*

Symbolic Action in Human Society, by Victor Turner, 166–230. Ithaca, NY: Cornell University Press, 1974.

———. *Process, Performance, and Pilgrimage: A Study in Comparative Symbology.* New Delhi: Concept, 1979.

Urton, Gary. "Andean Ritual Sweeping and the Nazca Lines." In *The Lines of Nazca,* edited by Anthony F. Aveni, 173–206. Philadelphia: American Philosophical Society, 1990.

Valdez, Lidio. "Alpacas en el Centro Ceremonial Nasca de Cahuachi." *Boletín del Museo de Arqueología y Antropología* 4, no. 3 (2001): 59–68.

Valeri, Valerio. *Kingship and Sacrifice: Ritual and Society in Ancient Hawaii.* Translated by Paula Wissig. Chicago: University of Chicago Press, 1985.

Van Dyke, Ruth M. *The Chaco Experience: Landscape and Ideology at the Center Place.* Santa Fe, NM: School for Advanced Research, 2007.

Veselay, Dalibor. *Architecture in the Age of Divided Representation: The Question of Creativity in the Shadow of Production.* Cambridge, MA: MIT Press, 2004.

Vickers, Daniel, ed. *A Companion to Colonial America.* Malden, MA: Blackwell, 2003.

Vitruvius. *The Ten Books on Architecture.* Translated by Morris Hicky Morgan. New York: Dover, 1960.

Vivian, Gordon R., and Tom W. Mathews. *Kin Kletso: A Pueblo III Community in Chaco Canyon, New Mexico.* Vol. 6, part I. Southwestern Monuments Association Technical Series. Tucson: Southwest Parks and Monuments Association, 1973.

Vivian, R. Gwinn. *The Chacoan Prehistory of the San Juan Basin.* San Diego: Academic, 1990.

Walker, Amelia Bell. "Tribal Towns, Stomp Grounds, and Land: Oklahoma Creeks after Removal." *Chicago Anthropology Exchange* 14 (1981): 50–69.

Ware, John A. "Chaco Social Organization: A Peripheral View." In *Chaco Society and Polity: Papers from the 1999 Conference,* edited by Linda S. Cordell, W. James Judge, and June-el Piper, 79–93. Special Publication 4. Albuquerque: New Mexico Archaeological Council, 2001.

———. "Descent Group and Sodality: Alternative Pueblo Social Histories." In *Traditions, Transitions, and Technologies: Themes in Southwestern Archaeology,* edited by Sarah H. Schlanger, 94–112. Boulder: University Press of Colorado, 2002.

Waselkov, Gregory A. "Indian Maps of the Colonial Southeast." In *Powhatan's Mantle: Indians in the Colonial Southeast,* edited by Gregory A. Waselkov, Peter H. Wood, and M. Thomas Hatley, 435–502. Lincoln: University of Nebraska Press, 2006.

———. "Indian Maps of the Colonial Southeast: Archaeological Implications and Prospects." In *Cartographic Encounters: Perspectives on Native American Mapmaking and Map Use,* edited by G. Malcolm Lewis, 205–21. Chicago: University of Chicago Press, 1998.

Waselkov, Gregory A., Peter H. Wood, and M. Thomas Hatley, eds. *Powhatan's*

Mantle: Indians in the Colonial Southeast. Lincoln: University of Nebraska Press, 2006.

Weslager, Clinton A. *The Delaware Indian Westward Migration.* Wallingford, CT: Middle Atlantic, 1978.

Wheatley, Paul. *The Pivot of the Four Quarters: A Preliminary Enquiry into the Origins and Character of the Ancient Chinese City.* Chicago: Aldine, 1971.

White, Nancy Marie, ed. *Gulf Coast Archaeology: The Southeastern United States and Mexico.* Gainesville: University Press of Florida, 2005.

Whittle, Alasdair W. R. *Europe in the Neolithic: The Creation of New Worlds.* Cambridge: Cambridge University Press, 1996.

———. "Historical and Archaeological Map of Ohio: Prepared at the Ohio Agricultural and Mechanical College, Columbus, Ohio, by Thomas Mathew." Unpublished map, VFM0620-2. On file, Ohio Historical Society, Columbus, 1876.

———. "People and the Diverse Past: Two Comments on 'Stonehenge for the Ancestors.'" *Antiquity* 72 (1998): 852–54.

Whittlesey, Charles. "Field book, July 1, 1868." Unpublished field notes, MSS 2872. On file, Western Reserve Historical Society, Cleveland, OH, 1868.

Wilcox, David R. "The Evolution of the Chacoan Polity." In *The Chimney Rock Archaeological Symposium,* edited by J. McKim Malville and Gary Matlock, 76–90. Fort Collins, CO: Rocky Mountain Forest and Range Experiment Station, US Department of Agriculture Forest Service, 1993.

Wilkinson, Charles. *The People Are Dancing Again: The History of the Siletz Tribe of Western Oregon.* Seattle: University of Washington Press, 2010.

Williams, Robert A. *The American Indian in Western Legal Thought: The Discourse of Conquest.* New York: Oxford University Press, 1990.

Williams, Stephen. *Fantastic Archaeology: The Wild Side of North American Prehistory.* Philadelphia: University of Pennsylvania Press, 1991.

Wills, W. H. "Political Leadership and the Construction of Chaco Great Houses." In *Alternative Leadership Strategies in the Prehispanic Southwest,* edited by Barbara J. Mills, 19–44. Tucson: University of Arizona Press, 2000.

Wilson, John N. "Mounds near Newark." Pioneer Papers, no. 25. In *Isaac Smucker Scrap Book,* vol. 1, 69–71. On file, Granville Public Library, Granville, Ohio, 1868.

Woodard, Stephanie. "A Pow Wow amongst the Largest Geometric Earthworks Complex in the World." *Indian Country Today,* June 9, 2012.

Woodward, David, and G. Malcolm Lewis, eds. *The History of Cartography.* Vol. 2, book 3, *Cartography in the Traditional African, American, Arctic, Australian, and Pacific Societies.* Chicago: University of Chicago Press, 1998.

Wymer, DeeAnne. "Where Do (Hopewell) Research Answers Come From?" In Byers and Wymer, *Hopewell Settlement Patterns, Subsistence, and Symbolic Landscapes,* 309–28.

Wyrick, David. "Ancient Works near Newark, Licking County, Ohio." In *Atlas of Licking County, Ohio from Actual Surveys by and under the Direction of F. W.*

Beers, Assisted by Beach Nichols and Others, edited by Frederick W. Beers and Beach Nichols. New York: Beers, Soule, 1866.

Yablon, Marcia. "Property Rights and Sacred Sites: Federal Regulatory Responses to American Indian Religious Claims on Public Land." *Yale Law Journal* 113 (2004): 1,623–62.

Yang, Y., and S. Lu. *Temple of Heaven.* Beijing: China National Photography Art, 2002.

Yarrow, H. C., and V. LaMonte Smith. *North American Indian Burial Customs.* Ogden, UT: Eagle's View, 1988.

Yoffee, Norman. "The Chaco 'Rituality' Revisited." In *Chaco Society and Polity: Papers from the 1999 Conference,* edited by Linda S. Cordell, W. James Judge, and June-el Piper, 63–78. Albuquerque: New Mexico Archaeological Council, 2001.

Yoffee, Norman, Suzanne K. Fish, and George R. Milner. "Communidades, Ritualities, Chiefdoms: Social Evolution in the American Southwest and Southeast." In *Great Towns and Regional Polities in the Prehistoric American Southwest and Southeast,* edited by Jill E. Neitzel, 261–71. Albuquerque: University of New Mexico Press, 1999.

Zolbrod, Paul G. *Diné bahane': The Navajo Creation Story.* Albuquerque: University of New Mexico Press, 1984.

Zuidema, R. Tom. *The Ceque System of Cuzco: The Social Organization of the Capital of the Inca.* Leiden: E. J. Brill, 1964.

CONTRIBUTORS

Thomas Barrie
Professor, School of Architecture, North Carolina State University

Thomas S. Bremer
Associate Professor, Department of Religious Studies, Rhodes College

Marti L. Chaatsmith
Associate Director, Newark Earthworks Center, The Ohio State University at Newark; Comanche Nation

Duane Champagne
Professor, Department of Sociology, American Indian Studies Program, and School of Law, University of California, Los Angeles; Turtle Mountain Band of Chippewa

Timothy Darvill
Professor of Archaeology, and Director, Center for Archaeology and Anthropology, Faculty of Science and Technology, Bournemouth University, Dorset, United Kingdom

Carole Goldberg
Jonathan D. Varat Distinguished Professor of Law, University of California, Los Angeles

John E. Hancock
Professor Emeritus, School of Architecture and Interior Design, and Director, Center for the Electronic Reconstruction of Historical and Archaeological Sites, University of Cincinnati

Ray Hively
Professor Emeritus, Department of Astronomy and Physics, Earlham College

Robert Horn
Professor Emeritus, Department of Philosophy, Earlham College

Greg Johnson
Associate Professor, Department of Religious Studies, University of Colorado, Boulder

Lindsay Jones
Professor, Department of Comparative Studies, The Ohio State University

Stephen H. Lekson
Curator of Archaeology, Museum of Natural History, and Professor, Department of Anthropology, University of Colorado, Boulder

Bradley T. Lepper
Curator of Archaeology,
Ohio History Connection

Mary N. MacDonald
Professor Emerita, Department of Religious Studies, Le Moyne College

Margaret Wickens Pearce
Associate Professor, Department of Geography, University of Kansas; Citizen Potawatomi

Richard D. Shiels
Associate Professor Emeritus, Department of History, and Director, Newark Earthworks Center, The Ohio State University at Newark

Helaine Silverman
Professor, Department of Anthropology, and Director, Collaborative for Cultural Heritage Management and Policy, University of Illinois

Winnifred Fallers Sullivan
Professor, Department of Religious Studies, Indiana University

Glenna Wallace
Chief, Eastern Shawnee Tribe of Oklahoma

INDEX

Adena culture, 219, 248
Adena pipe, 223
aesthetics, 199–200, 203, 207
alanui, 268
Algonquin, 253
Alligator Mound, 188
Alta Vista, 123
altepetl, 10, 112
American Antiquarian Society, 27
American Indian Religious Freedom Act, 245–47
American Indian Studies Program at The Ohio State University, 222
Ancient Ohio Trail, 11, 160, 218
Apollo Moon landing, 62
archaeoastronomy, 67, 91
Archaeological Atlas of Ohio (Mills), 25
archeology: "gulf coast archeology," 124; New Archaeology, 117–18, 119; post-processual approaches, 118
architecture: funerary, 167; sacred, 166; works, 156
astronomy: archaeoastronomy, 67, 91; astronomical alignments with ancient monuments, 142; lunar, 8, 25, 72, 77, 78, 80, 90, 172, 190; solar, 83
Atwater, Caleb, 47; *A Description of the Antiquities Discovered in Ohio and Other Western States*, 27
Avebury, 10, 131, 147
Aveni, Anthony, 103
Ayau, Halealoha, 270–73

Barrie, Thomas, 12, 290
Bear Butte, 237
Bell, Catherine, 202
Bender, Herman, 143
Bernardini, Wesley, 113
Black Hills, South Dakota, 136, 182, 246
Blair, Judge Park B., 32
Bones of Contention: Native American Archaeology (film), 236
Bradley, Richard, 139
Bremer, Thomas, 14, 195, 291
Brine, Lindesay, 200–201
Buffalo Bill's Wild West Show, 6, 28
Bureau of Indian Affairs, 284
Burks, Jarrod, 25–26
burial practices, 41–42, 51–52, 131. *See also* funerary practices
Burl, Aubrey, 169–70
Byers, Martin, 49, 54–55, 76, 100, 169

Cahuachi, 46, 49, 99–100, 107
capitalism, 208, 210–11
Captain of Pakana (Alabama chief), 185
Carr, Christopher, 165, 171, 174–75, 177
cartographies, indigenous and Western, 180–87
Case, D. Roy, 165, 171, 174–75, 177
Castleden, Rodney, 169
Catahöyük, Turkey, 11, 146
Center for the Electronic Reconstruction of Archaeological and Historical Sites (CERHAS), 11, 13, 153
ceramics, 102
ceremonial centers, 129, 130, 134, 231, 141; cosmological structuring of, 141; journeying to, 137; and links between life and death, 145; monuments, 129; seasonal communal gatherings at, 135–36
Chaatsmith, Marti, 15, 234, 265, 291

Chaco Canyon, 8–10, 111–28; as city, 116; great houses, 114–16; Great North Road, 122; Hohokam and Chaco, 121; and Hopewell, 112–13; interpretive models, 112–21; monuments, 115; rituality, 111–14; roads, 113–14, 122
Champagne, Duane, 16, 235–36, 291
charnel houses, 12, 170–72
Cherry Valley, 68, 79, 85, 88
Cherry Valley Ellipse, 23, 42, 47, 49–50, 54, 57, 64, 86, 131, 193, 216–17
Cherry Valley Mound Group, 50, 56–57; big house, 53; partially leveled, 51; "post holes," 52
Cheyenne, 252
Chickasaw map makers, 184–87
Chillicothe, Ohio, 24, 26
Chillicothe Conference, 44
Chumash Village, California, 258
Circleville, Ohio, 26–27
Civilian Conservation Corps (CCC), 43
Civilian Works Administration (CWA), 43
Coffman Knob, 85–86
Columbus Journal Dispatch, 33
commodification, 203, 207–8
Cover, Robert, 279
Crandall, Barbara, 282
Creek: migration stories, 184; Square Ground, 137, 176

Darvill, Timothy, 10, 290
DeBoer, Warren, 107
DeCerteau, Michel, 100–103
Delaware (Native American tribe), 249–50
Dille, Israel, 50
Dunbar, Robin, 136
Durrington Walls, 135, 139, 147

Eagle Mound, 41, 47, 72, 153, 172, 175
Earth Diver myth, 176
earthworks, 108, 153; Fort Ancient, 3, 55, 153, 219; High Bank Works, 64, 134, 146, 188, 190, 195; Hopewell Mound Group, 57; as monumental works of architecture, 153; Mound City Group, 26, 57, 170; Spiro Mound, 223; Tremper Mound, 51, 153; West Kennet Long Barrow, 169. *See also* geoglyphs; Newark Earthworks
Eastern Shawnee Tribe of Oklahoma, 14–15, 208–10, 226
Eliade, Mircea, 176–77
Ellipse, 23, 42, 47, 49–50, 54, 57, 64, 86, 131, 193, 216–17

Fairgrounds Circle. *See* Great Circle
Fatig, Richard, 34
Flint Ridge quarries, 44, 191
Forbidden City, 132
Fort Ancient, 3, 55, 153, 219
Fort Sherman, 28
Fowke, Gerard, 66, 73
Fowles, Severin, 117
Friends of the Mounds, 17, 35
funerary practices, 49; architecture, 167; rituals, 177. *See also* burial practices

Gadamer, Hans-Georg, 162, 173, 178
Galloway, Kay, 185
geoglyphs, 97, 103–4, 106–7; continuity and discontinuity in making of, 106–7; frame, scale, and perspective in, 104–5
geophysics, 91
Giedion, Sigfried, 159
Goldberg, Carole, 16, 235–36, 291
Goseck, Germany, 142
Grand Army of the Republic, 28
Great Circle (Fairgrounds Circle), 23–24, 26–29, 41, 47, 56, 64, 139, 175, 176, 191, 203, 217, 222, 231, 277; as Licking County Fairgrounds, 28, 47; water barrier, 54

Great Hopewell Road, 24, 55, 79, 122, 137, 190, 195, 252–53, 268
Great Miami River Valley, 43
Greber, N'omi, 170
Green Corn Ceremony, 138
Greenman, Emerson, 41, 47

Hall, Dana Naone, 265
Hancock, John, 11, 290
Harness Mound, 51, 57, 146
Harness Mound Big House, 53
Haudenosaunee, 15
Heidegger, Martin, 129, 157
heritage, intangible, 129
hermeneutics, 154–55; and archaeological sites, 155; and monumental works, 155; and places of continuing Native American meaning, 155
Hewett, Edgar, characterizes Chaco ruins as pueblos, 112–13
High Bank Works, 64, 134, 146, 188, 190, 195; High Bank Circle, 64; High Bank Pueblo, 112
Hill Earthwork, 86
Hitch, Neal, 30
Hively, Ray, 7, 25, 49, 113, 187, 232, 289
Hodder, Ian, 146
Holler, Jacob, 53
home, ontological significance of, 167
Hopewell culture, 7, 41, 43, 56, 58, 63, 68, 101, 219, 221, 248, 279–80; ceremonial centers, 3; interaction sphere, 43, 56
Hopewell Culture National Historical Park, 3, 26
Hopewell Mound Group, 57
Horn, Robert, 7, 25, 49, 113, 187, 232, 289
house tombs, 168
Howe, Le Anne, 223
Hull, Eli, 30

Iceland, 256, 280–82, 284
Idlewilde Park, 28–29
Indian Claims Commission Act, 256
Indian Country Conservancy, 259
indigenous peoples: cartography, 13, 192; "discourse of indigeneity," 234; legal studies, 284; rights, 15; stewardship, 258–59
Insoll, Timothy, 135
Intrusive Mound culture, 55
Iroquois, 249, 253

Johnson, Greg, 17, 236, 291
Jones, Clarence, 33

Kaho'olawe (Target Island), 263
Kaskaskia, 250
Kennewick Man, 236, 256
Kenton, Simon, 55
Kickapoo, 250
Kirshenblatt-Ginblett, Barbara, 209
Krupp, Tom, 205
Kū (Hawaiian diety), 268
Kula Ridge, 265

LaFarg, Oliver, 284
La Hougue Bie, 140
Lakota, 136, 145, 182
Large Hadron Collider, 41, 62
La Quemada, 123
Lekson, Stephen, 9, 286, 290
Lenape, 246
Lengyel culture, 142
Lepper, Bradley, 7, 289
Licking County: Agricultural Society, 27; Archaeological and Landmarks Society, 35; Board of Trade, 30, 32; Commissioners, 32; Conservation League, 33; Convention and Visitors Bureau, 26; Country Club, 30; Fairgrounds, 47; Historical Society, 33; Pioneer, Historical, and Antiquarian Society, 51
Licking River, 24, 131; south fork, 88

LiDAR (Light Detection and Ranging), 13, 24, 91
Lingafelter, James, 28
Little Miami River Valley, 43
Little Turtle, 250
Lono (Hawaiian diety), 268
Lucas, Gavin, 143
lunar astronomy, Newark Earthworks and, 8, 25, 72, 77, 78, 80, 90, 172, 190
Luxor, Egypt, 137

Macaulay, David, "Motel of the Mysteries," 124
MacDonald, Mary, 15, 291
Makena, Hawaiʻi, 269
mālama (care), 17–18, 262–67, 273–75
Marietta, Ohio, 50, 219
McPherson, Henry, 34
memory, 105, 108
Miami (Native American tribe), 250
Middleton, James, 66, 73; maps by, 45–46
Miller, William Ian, 280
Mills, William, 44, 56
Miloliʻi, Hawaiʻi, 270–72
Mingo Ouma (Chickasaw chief), 185–86
modernity, 199, 200–201
Moʻokini Heiau, 263
Moʻoloa (Long Lizard), 268–70
Moon. *See* lunar astronomy, Newark Earthworks and
Mooney, James, 130
Moorehead, Warren K., 44, 55–56
Morgan, Lewis Henry, 112
Morgan, R. G., 187
Mortuary Temple of King Zoser at Saqqara, 168, 175
Moundbuilders Country Club, 3, 6, 30, 217, 222, 232, 255, 256, 282–83
Moundbuilders Park, 29, 33–34, 203
Mound City Group, 26, 57, 170
Mount Rushmore, 206
Mundy, Barbara, 183
Muscogee (Creek) Nation, 227
Muskingum River Valley, 43

Nasca (ancient civilization), 98–100, 106, 108
Nassau (Nasaw) mapmakers, 184–87
National Aeronautics and Space Administration (NASA), 62
National Congress of American Indians, 225–26, 247
National Guard, 29–30
National Historic Landmarks, 26
National Historic Preservation Act, 245, 247, 255–56
National Museum of the American Indian, 254
National Parks Service, 34
Native American Graves Protection and Repatriation Act (NAGPRA), 16, 60, 118, 230, 235, 237, 245, 247, 254–59, 267, 283
Native American Indian Center of Central Ohio (NAICCO), 232
Native American Land Conservancy, 258
Native American Rights Fund, 225, 247
Nazca, Peru, 8–9, 97–109
Nazca Lines of Peru, 8, 100, 108; concepts of *kancha* and pampa, 104; as mnemonic representation, 105–6
Nebra Disk, 144
New Archaeology, 117–18, 119
Newark, Ohio, 32
Newark Advocate, 30, 32–33, 282
Newark Board of Trade, 30, 32
Newark Daily Advocate, 53
Newark Earthworks: alignments with lunar cycle, 8, 25, 72, 77, 78, 80, 90, 172, 190; Alligator Mound, 188; and American Indian perspectives, 216–18; artifacts from distant places found at, 191; as ceremonial center, 42, 49,

176; challenges for modern audience, 153–55; Cherry Valley Ellipse, 23, 42, 47, 49–50, 54, 57, 64, 86, 131, 193, 216–17; contemporary stewardship of, 17; as cosmogram, 175; creating public engagement with, 157, 159; Eagle Mound, 41, 47, 72, 153, 172, 175; as fairgrounds, 6, 27, 28, 47; Great Hopewell Road, 24, 55, 79, 122, 137, 190, 195, 252–53, 268; Hill Earthwork, 86; and hilltop observing stations, 68, 77, 79, 80; interpretation of, 215; as liminal place, 12, 164; as "machine for world renewal," 49–50, 54–55, 176; as mnemonic representation, 105–6; Moundbuilders Country Club, 3, 6, 30, 217, 222, 232, 255, 256, 282–83; Moundbuilders Park, 29, 33–34, 203; Observatory Circle, 64, 80, 188; Observatory Mound, 48; Octagon, 23–26, 29, 56, 64, 75, 80, 139, 142, 188, 190, 195, 216–17, 222, 231–32, 280; Octagon State Park, 34; Ohio's official prehistoric monument, 43; oriented to cycles of sun and moon, 68; as pilgrimage destination, 9, 16–17; Salisbury Square, 64, 86, 186, 193; Wright Square, 23, 47–48, 64, 81, 86, 139, 193, 217, 231. *See also* Cherry Valley Mound Group; Great Circle

"Newark Earthworks and World Heritage, The: One Site, Many Contexts" (symposium), 3

Newark Earthworks Center of The Ohio State University, 2, 6, 16, 35, 204, 222, 237

Newark Earthworks Day Symposium, 208

Nienast, Bob, 205

Northwest Ordinance, 283

Observatory Circle, 64, 80, 188

Observatory Circle Diameter (OCD), 64

Observatory Mound, 48

Octagon, 23–26, 29, 56, 64, 75, 139, 142, 188, 190, 195, 216–17, 222, 231–32, 280

Octagon State Park, 34

Ohio and Erie Canal, 27

Ohio History Connection (formerly Ohio Historical Society, originally Ohio Archaeological and Historical Society), 26, 34–35, 222, 237, 255, 283

Ohio State Center for the Study of Religion, 2

Ohio State Museum, 4

Ohio World Heritage Committee, 224

Olmec, 124–25

Ottawa (Native American tribe), 250, 255

pampa, 98–100, 103–4

Papahānaumokuakea, 263

Parker Pearson, Mike, 135

Pearce, Margaret Wickens, 13, 290

Piankeshaw, 250

pilgrimage, 9, 16–17, 100, 103, 111, 252; ritual paths, 102

Pi'ilani Heiau, 263

Pliny the Elder, 72–73

Pond, Bob, 205–6

Poolaw, Linda, 154

Potawatomi, 250

Poverty Point, 124–25

Powhatan's mantle (map), 184

Prufer, Olaf, 174

pueblos, 119; "unit pueblos," 115

Pu'u Kohalā, 263

Raccoon Creek, 24, 88

Raccoon Creek Valley, 86

Ragon, Michel, 168

Ramilisonina, 135

Ramp Creek, 24

Read, Kay, 278

Removal Act, 255

Renfrew, Colin, 129, 136
Richards, Colin, 146
Richland, Justin, 284–85
Richter, Daniel, 117
ritual, 18, 111–28, 202–3
roads, 123, 137
Robb, David, 250–51
Ruggles, Clive, 142
Ruhl, Katharine, 170
Running, Gilly, 204, 207
Russell, Miles, 140

sacred landscapes, 98, 99, 105. *See also* ceremonial centers
Sagan, Carl, 49
Salisbury, James and Charles, map by, 22, 27
Salisbury Square, 64, 86, 186, 193
Scioto River Valley, 43
Serpent Bridge, 177
Shaman of Newark, 53
shamans, 102, 172, 277–78, 280
Sharon Valley, 85
Shaw, Julie, 282
Shawnee, 250, 255
Shetrone, Henry, *The Mound Builders,* 41, 44
Shiels, Richard, 2, 6, 206–7, 232, 289
Silverman, Helaine, 8, 289
Simms, Lionel, 142
Sinkoye Intertribal Wilderness Area Native American Land Conservancy, 258
Sioux, 253
Smith, John, 184
Smithsonian Institution, 27, 66; surveys, 67; *Twelfth Annual Report of the Smithsonian Bureau of Ethnology,* 66, 73
Society of American Indians Centennial Symposium, 222, 227
spirituality, 211n25
Spiro Mound, 223
Squire, Ephraim, and Edwin Davis, *The Ancient Monuments of the Mississippi Valley,* 27, 50, 73, 220
Stadden, Isaac and Catherine, 27
standstills, lunar, 72, 78, 80; major and minor, 69
Stanzione, Vincent, 204–7
Stasel, Albert A., 32
Stonehenge, 8, 10–11, 131, 139, 142–43, 145; bluestones, 135, 140; Bush Barrow, 144–45
Stoufer, Joan, 205
Sullivan, Winnifred Fallers, 14, 292
Sunderstrom, Linea, 136

Tacitus, 136
Target Island (Kaho'olawe), 263
Taves, Ann, 199
Taylor, William, 154, 160
Tecumseh, 209
Temple of Heaven (Tian Tan), 10–11, 132–34, 136, 143, 145; Haiman Road, 132
Teotihuacan, 122–23
Thing (Icelandic national assembly), 280–81; lawspeaker at, 280
Thom, Alexander, 145
Thomas, Cyrus, 45–46
Thompson, Victor, 140
"To Bridge a Gap" (conference), 226
tourism, 14, 56; touristic engagement, 202
Treaty of Greenville, 250, 256, 283
Tremper Mound, 51, 153
tribal outreach and stewardship, 15, 225
Tyn Wald Hill, Isle of Man, 140–41

UNESCO Convention on the Means of Prohibiting and Preventing the Illicit Import, Export, and Transfer of Ownership of Cultural Property, 235
UNESCO World Heritage list, 17, 26, 43,

35, 253; designation for, 3, 11, 35, 156, 226, 233; "outstanding universal value" criterion, 1, 6, 13, 16
United Nations Declaration on the Rights of Indigenous Peoples (UNDRIP), 16, 230, 235, 237–39, 267, 283
United Nations Educational, Scientific, and Cultural Organization (UNESCO), 23, 153
United Nations Economic and Social council, Working Group on Indigenous Populations, 238
United Nations Permanent Forum on Indigenous Issues, 238

Vitruvius, 167
Vivain, Gordon, characterizes Chaco ruins as pueblos, 113
Volcanoes National Park, 263

"Walk with the Ancients" pilgraimage, 14, 195, 203–8
Wallace, Chief Glenna, 204, 208–10, 226
Wapatomica, 209–10, 212n27
Waselkov, Gregory, 185
Wea, 250
Webster, Daniel, 43
West Kennet Long Barrow, 169
Western Indian Confederacy, 250, 255
Wheatley, Paul, 130
Whittlesey, Charles, map by, 44, 51, 73
Wilson, John, 50–52
Wishtoyo, 258
World Convention on the Protection of the World Cultural and Natural Heritage, 283
Woodard, Stephanie, 232
Woolson, Frank A., 33
World Heritage site. *See* UNESCO World Heritage list
Wray Figurine, 53, 145, 216, 221
Wright, James C., 53
Wright Square, 23, 47–48, 64, 81, 86, 139, 193, 217, 231
Wyandotte, 219, 250, 255
Wyrick, David, map by, 27

Yoffee, Norman, on Chaco and rituality, 111

RECENT BOOKS IN THE STUDIES IN RELIGION AND CULTURE SERIES

Michael L. Raposa
Meditation and the Martial Arts

John D. Barbour
The Value of Solitude:
The Ethics and Spirituality of
Aloneness in Autobiography

David M. Craig
John Ruskin and the Ethics
of Consumption

Clayton Crockett, editor
Religion and Violence in a Secular World:
Toward a New Political Theology

Greg Johnson
Sacred Claims:
Repatriation and Living Tradition

Frederick J. Ruf
Bewildered Travel:
The Sacred Quest for Confusion

Jonathan Rothchild, Matthew Myer Boulton, and Kevin Jung, editors
Doing Justice to Mercy:
Religion, Law, and Criminal Justice

Gabriel Vahanian
Praise of the Secular

William B. Parsons, Diane Jonte-Pace, and Susan E. Henking, editors
Mourning Religion

Victor E. Taylor
Religion after Postmodernism:
Retheorizing Myth and Literature

J. Heath Atchley
Encountering the Secular:
Philosophical Endeavors in
Religion and Culture

Jeffrey S. Bennett
When the Sun Danced:
Myth, Miracles, and Modernity in
Early Twentieth-Century Portugal

Wesley A. Kort
Textual Intimacy:
Autobiography and Religious Identities

Carl Raschke
Postmodernism and the
Revolution in Religious Theory:
Toward a Semiotics of the Event

Daniel Boscaljon
Vigilant Faith:
Passionate Agnosticism
in a Secular World

William B. Parsons
Freud and Augustine in Dialogue:
Psychoanalysis, Mysticism, and the
Culture of Modern Spirituality

Zhange Ni
The Pagan Writes Back:
When World Religion Meets
World Literature

Giles Gunn
Ideas to Live For:
Toward a Global Ethics

Lindsay Jones and Richard D. Shiels, editors
The Newark Earthworks:
Enduring Monuments,
Contested Meanings